Topological methods in
Euclidean spaces

Topological methods in Euclidean spaces

Gregory L. Naber
Department of Mathematics
Pennsylvania State University

Cambridge University Press

Cambridge
London   New York   New Rochelle
Melbourne   Sydney

Published by the Press Syndicate of the University of Cambridge
The Pitt Building, Trumpington Street, Cambridge CB2 1RP
32 East 57th Street, New York, NY 10022, USA
296 Beaconsfield Parade, Middle Park, Melbourne 3206, Australia

First published 1980

Printed in the United States of America
Typeset by Progressive Typographers, Inc., Emigsville, Pennsylvania
Printed and bound by Vail-Ballou Press, Inc., Binghamton, New York

Library of Congress Cataloging in Publication Data
Naber, Gregory, 1948–
Topological methods in Euclidean spaces.
Bibliography: p.
Includes index.
1. Topology I. Title.
QA611.N22   514   79-9225
ISBN 0 521 22746 1 hard covers
ISBN 0 521 29632 3 paperback

For Liz

Contents

Preface

Topology evolved as an independent discipline in response to certain
rather specific problems in classical analysis. Of course, it is characteristic of any fruitful branch of mathematics that the subject develop and
take on a significance independent of those problems from which it arose.
In the case of topology, however, this development has been so extensive
and so rapid that, unfortunately, its origins and relations to other areas of
mathematics are often lost sight of entirely, and, even less desirable, the
essential unity of the subject itself is sacrificed to the demands of speciali-
zation.

It is the intention of this introduction to the methods of topology in
Euclidean spaces to persuade students of mathematics, at the earliest possible point in their studies, that the evolution of topology from analysis
and geometry was natural and, indeed, inevitable; that the most fruitful
concepts and most interesting problems in the subject are still drawn from
independent branches of mathematics; and that, underlying its sometimes
overwhelming diversity of ideas and techniques, there is a fundamental
unity of purpose. To this end an ambitious agenda of topics from point-set,
algebraic, and differential topology has been included, although much of
the material familiar from standard introductions to topology is omitted al-
together. Indeed, metric space and topological space are never defined.
Rather, we restrict attention exclusively to subspaces of Euclidean spaces
where geometrical intuition remains strong so that we can avoid the tire-
some technicalities inherent in axiomatic treatments. In this way it is possible to go rather far in the development of those techniques that are central to topology itself as well as its applications in other areas of mathematics and the sciences.

A very considerable emphasis has been placed on motivation, which we
draw primarily from the student's background in differential equations,
linear algebra, modern algebra, and advanced calculus. We assume this
background to be rather strong and, in addition, that our readers are possessed of a healthy supply of that elusive quality known as "mathematical
maturity." A great many arguments are left to the reader in the form of
exercises embedded in the body of the text and no asterisk appears to

designate those that are used in the sequel – they are all used and must be worked conscientiously. Of the 214 exercises in the text, 162 are of this variety, while 52 are included in Supplementary Exercises at the ends of chapters; the latter, although no less important, are not specifically called upon in the development. A Guide to Further Study has been included at the end of the book to suggest several directions in which to proceed to obtain a deeper understanding of various aspects of the subject.

Gregory L. Naber

October 1979

Chapter 1
Point-set topology of Euclidean spaces

1-1   Introduction

Geometry, in the broadest possible sense, emerged before the written and
perhaps even the spoken word as a gradual accumulation of subconscious
notions about physical space based on the ability of our species to recog-
nize "forms" and compare shapes and sizes. Until approximately 600 B.C.
the study of geometrical figures proceeded in the manner of an experi-
mental science in which induction and empirical procedure were the tools
of discovery. General properties and relationships were extracted from
observations necessitated by the demands of daily life, the result being a
rather formidable collection of "laboratory" results on areas, volumes,
and relations between various figures. It was left to the Greeks to trans-
form this vast array of empirical data into the very beautiful intellectual
discipline we now know as Euclidean geometry. The transformation required approximately three centuries to complete and culminated, around
300 B.C., with the appearance of Euclid's *Elements*. It is difficult indeed
to exaggerate the importance of this event for the development of mathematics. So decisive was the influence of Euclid that it was not until the
seventeenth century that mathematicians found themselves capable of
adopting essentially new attitudes toward their subject. Slowly, at times it
seems unwillingly, mathematics began to free itself from the constraints
imposed by the strict axiomatic method of the *Elements*. New and remarkably powerful concepts and techniques evolved that eventually led
to an expanded and more lucid view of mathematics in general and geometry in particular. The object in this introductory section is to indicate
how the subject of interest to us here (topology) arose as a branch of
geometry in this expanded sense.

Perhaps the most fundamental concept of the earlier books of Euclid's
Elements is that of congruence. Intuitively, two plane geometric figures
(arbitrary subsets of the plane from our point of view) are congruent if
they differ only in the position they occupy in the plane, that is, if they
can be made to coincide by the application of some rigid motion in the
plane. Somewhat more precisely, two figures F_1 and F_2 are said to be

congruent if there is a mapping f of the plane onto itself that leaves invariant the distance between each pair of points (i.e., $d(f(p), f(q)) = d(p, q)$ for all p and q) and carries F_1 onto F_2 (i.e., $f(F_1) = F_2$). A map that preserves the distance between any pair of points is called an *isometry* and is the mathematical analog of a *rigid motion;* the study of congruent figures in the plane is, for this reason, often referred to as *plane Euclidean metric geometry*. If we construct an orthogonal Cartesian coordinate system in the plane, we can show that the isometries of the plane are precisely the maps $(x, y) \rightarrow (x', y')$, where

(1) $\begin{aligned} x' &= Ax + By + C \\ y' &= \pm(-Bx + Ay) + D, \end{aligned}$

A, B, C, and D being real constants with $A^2 + B^2 = 1$ (see Gans, p. 65). Observe that the composition of any two isometries is again an isometry and that each isometry has an inverse that is again an isometry. Now, any collection of invertible mappings of a set S onto itself that is closed under the formation of compositions and inverses is called a *group of transformations* on S; the collection of all maps of the form (1) is therefore referred to as the *group of planar isometries*. From the point of view of plane Euclidean metric geometry the only properties of a geometric figure F that are of interest are those that are possessed by all figures congruent to F, that is, those properties that are invariant under the group of planar isometries. Since any map of the form (1) carries straight lines onto straight lines, the property of being a straight line is one such property. Similarly, the property of being a square or, more generally, a polygon of a particular type is invariant under the group of planar isometries, as is the property of being a conic of a particular type. The length of a line segment, area of a polygon, and eccentricity of a conic are likewise all invariants and are thus legitimate objects of study in plane Euclidean metric geometry.

Of course, the point of view of plane Euclidean metric geometry is not the only point of view. Indeed, in Book VI of the *Elements* itself, emphasis shifts from congruent to similar figures. Roughly speaking, two geometric figures are similar if they have the same shape, but not necessarily the same size. In order to formulate a more precise definition, let us refer to a map f of the plane onto itself under which each distance is multiplied by the same positive constant k (i.e., $d(f(p), f(q)) = k\,d(p, q)$ for all p and q) as a *similarity transformation* with *similarity ratio* k. It can be shown that, relative to an orthogonal Cartesian coordinate system, each such map has the form

(2) $\begin{aligned} x' &= ax + by + m \\ y' &= \pm(-bx + ay) + n, \end{aligned}$

where $(a^2 + b^2)^{1/2} = k$ (see Gans, p. 77). Two plane geometric figures F_1 and F_2 are then said to be similar if there exists a similarity transformation of the plane onto itself that carries F_1 onto F_2. Again, the set of all similarity transformations is easily seen to be a transformation group, and we might reasonably define *plane Euclidean similarity geometry* as the study of those properties of geometric figures that are invariant under this group, that is, those properties that, if possessed by some figure, are necessarily possessed by all similar figures. Since any isometry is also a similarity transformation (with $k = 1$), any such property is necessarily an invariant of the group of planar isometries; but the converse is false since, for example, the length of a line segment and area of a polygon are not preserved by all similarity transformations.

At this point it is important to observe that, in each of the two geometries discussed thus far, certain properties of geometric figures were of interest while others were not. In plane Euclidean metric geometry we are interested in the shape and size of a given figure, but not in its position or orientation in the plane, while similarity geometry concerns itself only with the shape of the figure. Those properties that we deem important depend entirely on the particular sort of investigation we choose to carry out. Similarity transformations are, of course, capable of "distorting" geometric figures more than isometries, but this additional distortion causes no concern as long as we are interested only in properties that are not effected by such distortions. In other sorts of studies the permissible degree of distortion may be even greater. For example, in the mathematical analysis of perspective it was found that the "interesting" properties of a geometric figure are those that are invariant under a class of maps called *plane projective transformations,* each of which can be represented, relative to an orthogonal Cartesian coordinate system, in the following form (see Gans, p. 174):

$$
(3) \qquad \begin{aligned} x' &= \frac{a_1 x + a_2 y + a_3}{c_1 x + c_2 y + c_3} \\[2mm] y' &= \frac{b_1 x + b_2 y + b_3}{c_1 x + c_2 y + c_3} \end{aligned} \qquad \text{where } \begin{vmatrix} a_1 & a_2 & a_3 \\ b_1 & b_2 & b_3 \\ c_1 & c_2 & c_3 \end{vmatrix} \neq 0.
$$

The collection of all such maps can be shown to form a transformation group, and we define *plane projective geometry* as the study of those properties of geometric figures that are invariant under this group. Two figures are said to be "projectively equivalent" if there is a projective transformation that carries one onto the other. Since any similarity transformation is also a projective transformation, any invariant of the projective group is also an invariant of the similarity group. The converse, however, is false since projective maps are capable of greater distortions of geometric figures than are similarities. For example, two conics are always projectively equivalent, but they are similar only if they have the same eccentricity.

Needless to say, the approach we have taken here to these various geometrical studies is of relatively recent vintage. Indeed, it was Felix Klein, in his famous Erlanger Program of 1872, who first proposed that a "geometry" be defined quite generally as the study of those properties of a set S that are invariant under some specified group of transformations of S. Plane Euclidean metric, similarity, and projective geometries and their obvious generalizations to three and higher dimensional spaces all fit quite nicely into Klein's scheme, as did the various other offshoots of classical Euclidean geometry known at the time. Despite the fact that, during this century, our conception of geometry has expanded still further and now includes studies that cannot properly be considered "geometries" in the Kleinian sense, the influence of the ideas expounded in the Erlanger Program has been great indeed. Even in theoretical physics Klein's emphasis on the study of invariants of transformation groups has had a profound impact. The special theory of relativity, for example, is perhaps best regarded as the invariant theory of the so-called Lorentz group of transformations on Minkowski space.

Based on his appreciation of the importance of Riemann's work in complex function theory, Klein was also able to anticipate the rise of a new branch of geometry that would concern itself with those properties of a geometric figure that remain invariant when the figure is bent, stretched, shrunk or deformed in any way that does not create new points or fuse existing points. Such a deformation is accomplished by any bijective map that, roughly speaking, "sends nearby points to nearby points," that is, a continuous one. In dimension two, then, the relevant group of transformations is the collection of all one-to-one maps of the plane onto itself that are continuous and have continuous inverse; such maps are called *homeomorphisms* or *topological maps* of the plane. Consider, for example, the map f of the plane onto itself, which is given by $f(x, y) = (x, y^3)$. Now, f is continuous and has inverse $f^{-1}(x, y) = (x, y^{1/3})$ that is also continuous, so f is indeed a homeomorphism of the plane. What sort of properties of a plane geometric figure are preserved by f? Certainly, the property of being a straight line is not since, for example, the line given by the equation $y = x$ is mapped by f onto the curve $y = x^3$ (see Figure 1-1 (a)). Similarly, the property of being a conic is not invariant since the circle $x^2 + y^2 = 1$ is carried by f onto the locus of $x^2 + y^{2/3} = 1$, which is shown in Figure 1-1 (b).

Topological transformations are clearly capable of a very great deal of distortion. Indeed, virtually all of the properties the reader is accustomed to associating with plane geometric figures are destroyed by even the rela-

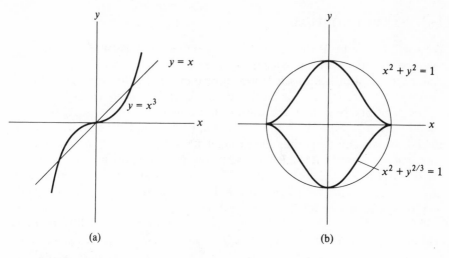

Figure 1-1

tively simple map f. Nevertheless, f does preserve a number of very important, albeit less obvious properties. For example, although a straight line need not be mapped by f onto another straight line, its image must also be "one-dimensional" and consist of one "connected" piece. The image of the circle $x^2 + y^2 = 1$, although not a conic, shares with the circle the property of being a "simple closed curve." Properties of plane geometric figures such as these that are invariant under the group of topological transformations of the plane are called *extrinsic topological properties*.

During the past one hundred years topology has outgrown its geometrical origins and today stands alongside analysis and algebra as one of the most fundamental branches of mathematics. Roughly speaking, topology might now be defined simply as the study of continuity. The approach we take here to this subject, while less general than it might be, is somewhat more general than that just outlined. We observe that the ambient space in which our geometrical figures are thought of as existing is, to a large extent, arbitrary (e.g., any plane figure can also be regarded as a subset of 3-space) and that, by insisting that the topological transformations be defined on this entire space, we have imposed rather unnatural restrictions on our study. We therefore choose to take a broader view of topological maps, allowing them to be defined on the given geometric figure itself without reference to the space in which it happens to be embedded, thus turning our attention from "extrinsic" to "intrinsic" topological properties, that is, properties of the figure itself that do not depend on the particular space in which it happens to reside.

1-2   Preliminaries

We shall denote by $\mathbf{R}$ the set of all real numbers and assume that the reader is familiar with the basic properties of this set (specifically, that under the usual operations $\mathbf{R}$ is a complete ordered field; see Apostol, Sections 1-1 through 1-9, or Buck, Appendix I). Recall that if $A_1, \ldots,$ A_n are arbitrary sets, then the Cartesian product $A_1 \times \cdots \times A_n$ is defined by $A_1 \times \cdots \times A_n = \{(a_1, \ldots, a_n) : a_i \in A_i \text{ for } i = 1, \ldots, n\}$. Euclidean n-space $\mathbf{R}^n$ is thus the product $\mathbf{R}^n = \mathbf{R} \times \cdots \times \mathbf{R}$ (n factors). As usual we identify $\mathbf{R}^n \times \mathbf{R}^m$ and $\mathbf{R}^{n+m}$ by not distinguishing between the ordered pair $((a_1, \ldots, a_n), (b_1, \ldots, b_m))$ and the $(n + m)$-tuple $(a_1, \ldots, a_n, b_1, \ldots, b_m)$. Thus, if $A \subseteq \mathbf{R}^n$ and $B \subseteq \mathbf{R}^m$, we may regard $A \times B$ as a subset of $\mathbf{R}^{n+m}$.

If $x = (x_1, \ldots, x_n)$ and $y = (y_1, \ldots, y_n)$ are points of $\mathbf{R}^n$ and a is a real number, we define $x + y = (x_1, \ldots, x_n) + (y_1, \ldots, y_n) = (x_1 + y_1, \ldots, x_n + y_n)$ and $ax = a(x_1, \ldots, x_n) = (ax_1, \ldots, ax_n)$ and thus endow $\mathbf{R}^n$ with the structure of a real vector space of algebraic dimension n. (We assume the reader to be acquainted with basic linear algebra.) We denote by 0 the additive identity $(0, 0, \ldots, 0, 0)$ in $\mathbf{R}^n$ and let $e_1 = (1, 0, \ldots, 0, 0)$, $e_2 = (0, 1, \ldots, 0, 0)$, $\ldots$, $e_n = (0, 0, \ldots, 0, 1)$ be the standard basis vectors for $\mathbf{R}^n$. A map $S : \mathbf{R}^n \to \mathbf{R}^m$ is said to be *affine* if there is a $y_0 \in \mathbf{R}^m$ and a linear map $T : \mathbf{R}^n \to \mathbf{R}^m$ such that $S(x) = y_0 + T(x)$ for each x in $\mathbf{R}^n$. Since the range of a linear map is a linear subspace, the range of an affine map must be of the form $y_0 + V = \{y_0 + v : v \in V\}$ for some linear subspace V of $\mathbf{R}^m$; such a "translation" of a linear subspace of $\mathbf{R}^m$ is called an *affine subspace* or *hyperplane* in $\mathbf{R}^m$ (see Section 2-2 for more details).

If $x = (x_1, \ldots, x_n)$ and $y = (y_1, \ldots, y_n)$ are arbitrary points of $\mathbf{R}^n$, then their inner product (or dot product) is defined, as usual, by $x \cdot y = x_1 y_1 + \cdots + x_n y_n$. The norm of x, denoted $\|x\|$, is then given by $\|x\| = (x \cdot x)^{1/2}$. Finally, the distance $d(x, y)$ between x and y is defined by $d(x, y) = \|y - x\|$. Standard properties of the inner product and norm (Apostol, Section 3-6, and Buck, Section 1.3) translate immediately to the following result on the "metric function" d.

Theorem 1-1. Let x, y, and z be arbitrary points in $\mathbf{R}^n$. Then
(a) $d(x, y) \geq 0$ and $d(x, y) = 0$ iff $x = y$.
(b) $d(x, y) = d(y, x)$.
(c) $d(x, y) \leq d(x, z) + d(z, y)$.

Now let x_0 be a point in $\mathbf{R}^n$ and $r > 0$ a real number. The *open ball* of radius r about x_0 is defined by $U_r(x_0) = \{x \in \mathbf{R}^n : d(x_0, x) < r\}$; the *closed ball* of radius r about x_0 is $B_r(x_0) = \{x \in \mathbf{R}^n : d(x_0, x) \leq r\}$. The ball $B_1(0) = \{x \in \mathbf{R}^n : \|x\| \leq 1\}$ is called the *closed n-ball* and denoted B^n, while the subset $S^{n-1} = \{x \in \mathbf{R}^n : \|x\| = 1\}$ is called the $(n - 1)$-sphere. If A is an arbitrary subset of $\mathbf{R}^n$, the *diameter* of A is defined by diam $A = \sup\{d(x, y) : x, y \in A\}$ if $A \neq \varnothing$ and diam $\varnothing = 0$; A is said to be *bounded* if diam A is finite (this is the case iff $A \subseteq B_r(0)$ for some $r > 0$). If B is another subset of $\mathbf{R}^n$, then the *distance* between A and B is defined by dist$(A, B) = 0$ if $A = \varnothing$ or $B = \varnothing$ and dist$(A, B) = \inf\{d(x, y) : x \in A, y \in B\}$ if $A \neq \varnothing$ and $B \neq \varnothing$.

If x and y are any two points in $\mathbf{R}^n$, then the *open line segment* joining x and y is denoted (x, y) and defined by $(x, y) = \{tx + (1 - t)y : 0 < t < 1\}$; the *closed line segment* joining x and y is $[x, y] = \{tx + (1 - t)y : 0 \leqslant t \leqslant 1\}$. A subset A of $\mathbf{R}^n$ is *convex* if $[x, y] \subseteq A$ whenever x and y are in A.

Exercise 1-1. Let x_0 be a point in $\mathbf{R}^n$ and $r > 0$ a real number. Show that $U_r(x_0)$ and $B_r(x_0)$ are both convex.

Observe that any intersection of convex sets is also convex and that any subset A of $\mathbf{R}^n$ is contained in a convex set (e.g., $\mathbf{R}^n$ itself). We may therefore define the *convex hull* $H(A)$ of A as the intersection of all convex subsets of $\mathbf{R}^n$ containing A and be assured that $H(A)$ is convex for every A.

The final preliminary matter we must consider is the distinction, no doubt already familiar to the reader (see Apostol, Section 2-11, or Buck, p. 30), between countable and uncountable sets. Let us say that two non-empty sets S_1 and S_2 are *numerically equivalent*, or of the same *cardinality*, if there is a one-to-one mapping of S_1 onto S_2. A set is *finite* if it is either empty or numerically equivalent to $\{1, \ldots, n\}$ for some positive integer n. A set is *countably infinite* if it is numerically equivalent to the set $\mathbf{N} = \{1, 2, \ldots, n, \ldots\}$ of all positive integers. If a set is either finite or countably infinite we say that it is *countable*. Intuitively, a set is countable if it is either empty or if its elements can be listed in a (perhaps terminating) sequence. Finally, a set that is not countable is *uncountable*.

Lemma 1-2. Every subset A of a countable set S is countable.

Proof: Since every subset of a finite set is finite (and therefore countable), we may assume without loss of generality that S is countably infinite. Let $f : \mathbf{N} \to S$ be a bijection, where $\mathbf{N} = \{1, 2, \ldots, n, \ldots\}$, and define $g : \mathbf{N} \to \mathbf{N}$ inductively as follows: Let $g(1)$ be the least positive integer for which $f(g(1))$ is in A and assume that $g(1), \ldots, g(n - 1)$ have been defined. Let $g(n)$ be the least positive integer greater than $g(n - 1)$ such that $f(g(n))$ is in A. The composition $f \circ g : \mathbf{N} \to A$ is a bijection, so A is countable. Q.E.D.

Lemma 1-3. The union of countably many countable sets is countable.

Proof: By Lemma 1-2 it will suffice to show that the union of a countably infinite collection $\{A_1, A_2, \ldots, A_n, \ldots\}$ of countably infinite sets is countable. Define $B_1 = A_1$ and, for $n > 1$, let $B_n = A_n - \cup_{k=1}^{n-1}A_k$. Then each B_n is countable by Lemma 1-2, $B_i \cap B_j = \varnothing$ if $i \neq j$ and $\cup_{n=1}^{\infty} B_n = \cup_{n=1}^{\infty}A_n$. Again by Lemma 1-2, we need only consider the case in which each B_n is countably infinite. Thus, we may enumerate the elements of each B_n as indicated:

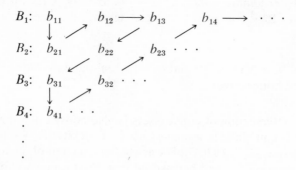

Now define $f : \mathbf{N} \to \cup_{n=1}^{\infty} B_n$ by $f(1) = b_{11}, f(2) = b_{21}, f(3) = b_{12}, f(4) = b_{13}, f(5) = b_{22}, \ldots$, and so on, following the scheme indicated by the arrows. Then f is surjective. Moreover, since the B_n are disjoint, f is one-to-one and the result follows.   Q.E.D.

Lemma 1-4. Let $S_1, \ldots, S_k$ be countable sets. Then $S_1 \times \cdots \times S_k$ is countable.

Exercise 1-2. Prove Lemma 1-4. Hint: Use Lemma 1-3 and induction.

Example 1-1. Countable and Uncountable Subsets of R. (a) The set Z of integers is countable. This follows immediately from the enumeration indicated:

$$
\begin{array}{ccccccc}
1 & 2 & 3 & 4 & 5 & 6 & 7 \cdots \\
\downarrow & \downarrow & \downarrow & \downarrow & \downarrow & \downarrow & \downarrow \\
0 & 1 & -1 & 2 & -2 & 3 & -3 \cdots
\end{array}
$$

(b) The set $\mathbf{Q}$ of rational numbers is countable. To see this, write each element of $\mathbf{Q}$ as m/n, where m and n are integers with no common factors and n is positive. The map that carries m/n to the ordered pair (m, n) thus maps $\mathbf{Q}$ bijectively onto a subset of $Z \times \mathbf{N}$. But $Z \times \mathbf{N}$ is countable by (a) and Lemma 1-4, so each of its subsets is countable by Lemma 1-2. It follows that $\mathbf{Q}$ is countable.

(c) The closed unit interval $I = [0, 1]$ is uncountable. To see this, let $f : \mathbf{N} \to [0, 1]$ be any one-to-one map. We show that f is not surjective. For each n in $\mathbf{N}$ let $0.a_{n1}a_{n2}a_{n3} \ldots$ be a decimal expansion for $f(n)$. Define a number $0.b_1b_2b_3 \ldots$ in $[0, 1]$ as follows: $b_k = 5$ if $a_{kk} \neq 5$ and $b_k = 7$ if $a_{kk} = 5$. Then $0.b_1b_2b_3 \ldots$ is not in the image of f since it has a unique decimal expansion that differs from $f(n)$ in the nth place for each n in $\mathbf{N}$.

(d) If a and b are real numbers with $a < b$, then the interval $[a, b]$ is uncountable. Since the map $f : [a, b] \to [0, 1]$ defined by $f(x) = (x - a)/(b - a)$ is bijective, this follows immediately from (c).

(e) From (d) and Lemma 1-2 it follows that any subset of $\mathbf{R}$ that contains an interval $[a, b]$, where $a < b$, is uncountable. In particular, $\mathbf{R}$ itself is uncountable. However, an uncountable subset of $\mathbf{R}$ need not contain an interval, for example, the set $\mathbf{P}$ of irrational numbers is uncountable since $\mathbf{Q}$ is countable and $\mathbf{R} = \mathbf{Q} \cup \mathbf{P}$. Another example is constructed in (f).

(f) Recall that for each x in $[0, 1]$ there exists a sequence $s_1, s_2, s_3, \ldots$ with $s_i \in \{0, 1, 2\}$ for each i such that $x = \Sigma_{i=1}^{\infty} s_i / 3^i$. (The procedure for determining the s_i will become clear shortly.) We shall write $x = :s_1s_2s_3 \ldots$ and call $:s_1s_2s_3 \ldots$ the *triadic expansion* of x. Some numbers have two such expansions. For example, $:2000 \ldots$ and $:1222 \ldots$ both represent the number $2/3$ since $2/3 + 0/3^2 + 0/3^3 + \cdots = 2/3$ and $1/3 + 2/3^2 + 2/3^3 + \cdots = 1/3 + 2 \Sigma_{i=2}^{\infty} (1/3)^i = 1/3 + 2 [\Sigma_{i=0}^{\infty} (1/3)^i - 1 - (1/3)] = 1/3 + 2 [(3/2) - 1 - (1/3)] = 2/3$. This situation will occur only when one of the expansions repeats 0's and the other repeats 2's from some point on. We define the *Cantor set C* to be the set of all those x's in $[0, 1]$ that have a triadic expansion in which the digit 1 does not occur. This set has a simple geometrical interpretation that we obtain as follows: Let F_1 denote the closed interval $[0, 1]$. Delete the open interval $(\frac{1}{3}, \frac{2}{3})$ from F_1 to obtain the set $F_2 = [0, \frac{1}{3}] \cup [\frac{2}{3}, 1]$ (see Figure 1-2). Note that the "middle third" $(\frac{1}{3}, \frac{2}{3})$ of $[0, 1]$ consists precisely of those x's in $[0, 1]$ whose triadic expansions must have a 1 in the first digit. Thus, F_2 consists of those x's in $[0, 1]$ that have a triadic expansion with $s_1 \neq 1$. Now delete from F_2 the middle thirds $(\frac{1}{9}, \frac{2}{9})$ and $(\frac{7}{9}, \frac{8}{9})$ of each of the two closed intervals $[0, \frac{1}{3}]$ and $[\frac{2}{3}, 1]$ to obtain the set $F_3 = [0, \frac{1}{9}] \cup [\frac{2}{9}, \frac{1}{3}] \cup [\frac{2}{3}, \frac{7}{9}] \cup [\frac{8}{9}, 1]$ (see Figure 1-2).

Observe that $(\frac{1}{9}, \frac{2}{9})$ and $(\frac{7}{9}, \frac{8}{9})$ consist precisely of those x's in $[0, 1]$ whose triadic expansions must have a 1 in the second digit, but not in the first. Thus, F_3 consists of those x's in $[0, 1]$ that have a triadic expansion $:s_1s_2s_3 \ldots$ with $s_1 \neq 1$ and $s_2 \neq 1$. We now continue this process inductively, at each stage deleting the open middle third of each closed interval

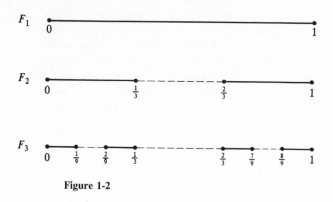

Figure 1-2

remaining from the previous stage. We therefore obtain a descending sequence $F_1 \supseteq F_2 \supseteq F_3 \supseteq \cdots$ of subsets of $[0, 1]$, each of which is a finite union of disjoint closed intervals (e.g., F_{25} consists of 16,777,216 such intervals). The Cantor set C is then $\cap_{n=1}^{\infty} F_n$.

Remark: The sum of the lengths of all the open intervals removed from $[0, 1]$ to form C is 1 since $\frac{1}{3} + \frac{2}{9} + \frac{4}{27} + \cdots = (\frac{1}{3}) \sum_{n=0}^{\infty} (\frac{2}{3})^n = (\frac{1}{3})(3) = 1$. It follows that C cannot contain an interval.

Finally, we show that C is uncountable by exhibiting a bijective map of $[0, 1)$ onto C. For each x in $[0, 1)$ let $x = :b_1 b_2 b_3 \ldots$ be a binary expansion for x. Thus, each b_i is either 0 or 1 and $x = \sum_{i=1}^{\infty} b_i / 2^i$. Let $s_i = 2b_i$ for each i, and let $f(x)$ be the point in $[0, 1]$ whose triadic expansion is $:s_1 s_2 s_3 \ldots$. Then f is one-to-one, $f(x)$ is in C for each x in $[0, 1)$, and, moreover, every element of C is the image under f of some x in $[0, 1)$ so f is surjective. It follows from (e) that C is uncountable.

1-3   Open sets, closed sets, and continuity

You will recall (Apostol, Definition 3-24, or Buck, Section 1.5) that a subset U of $\mathbf{R}^n$ is said to be *open in* $\mathbf{R}^n$ if, for each $x_0 \in U$, there is an $r > 0$ such that the open ball $U_r(x_0)$ is contained entirely in U.

Theorem 1-5. (a) $\varnothing$ and $\mathbf{R}^n$ are open in $\mathbf{R}^n$.
(b)   Any union of open subsets of $\mathbf{R}^n$ is open in $\mathbf{R}^n$.
(c)   Any finite intersection of open subsets of $\mathbf{R}^n$ is open in $\mathbf{R}^n$.

Exercise 1-3. Prove Theorem 1-5.   Q.E.D.

A set C in $\mathbf{R}^n$ is *closed in* $\mathbf{R}^n$ if its complement $\mathbf{R}^n - C$ is open in $\mathbf{R}^n$ (see Apostol, Theorem 3-31, or Buck, Section 1.5).

Theorem 1-6. (a) $\emptyset$ and $\mathbf{R}^n$ are closed in $\mathbf{R}^n$.

(b)   Any intersection of closed subsets of $\mathbf{R}^n$ is closed in $\mathbf{R}^n$.

(c)   Any finite union of closed subset of $\mathbf{R}^n$ is closed in $\mathbf{R}^n$.

Exercise 1-4. Prove Theorem 1-6.   Q.E.D.

An open ball in $\mathbf{R}^n$ is certainly open in $\mathbf{R}^n$. To see that a closed ball $B_r(x_0)$ is closed in $\mathbf{R}^n$, consider a point $x \in \mathbf{R}^n - B_r(x_0)$. Then $d(x_0, x) > r$, so $\epsilon = d(x_0, x) - r$ is a positive real number. We claim that the open ball $U_\epsilon(x)$ is contained in $\mathbf{R}^n - B_r(x_0)$. To see this, let y be any point in $U_\epsilon(x)$. Then $d(y, x) < \epsilon$ so that, since $d(x_0, x) \le d(x_0, y) + d(y, x)$, we have $d(y, x_0) \ge d(x_0, x) - d(y, x) > d(x_0, x) - \epsilon = d(x_0, x) - (d(x_0, x) - r)$ $= r$. Thus, $y \in \mathbf{R}^n - B_r(x_0)$ and the proof is complete.

If $a_i < b_i$ for each $i = 1, \ldots, n$, then the subset $(a_1, b_1) \times (a_2, b_2) \times \cdots \times (a_n, b_n)$ of $\mathbf{R}^n$ is called an *open rectangle* in $\mathbf{R}^n$, while $[a_1, b_1] \times [a_2, b_2] \times \cdots \times [a_n, b_n]$ is a *closed rectangle* in $\mathbf{R}^n$.

Exercise 1-5. Show that an open rectangle is an open subset and a closed rectangle is a closed subset of $\mathbf{R}^n$.

Exercise 1-6. Show that a subset U of $\mathbf{R}^n$ is open in $\mathbf{R}^n$ iff, for each $x_0 \in U$, there is an open rectangle in $\mathbf{R}^n$ containing x_0 and contained in U.

Observe that the subset $\{1/n\}_{n=1}^{\infty} = \{1, \frac{1}{2}, \frac{1}{3}, \ldots \}$ of $\mathbf{R}$ is neither open nor closed.

Next let X be an arbitrary subset of $\mathbf{R}^n$, Y an arbitrary subset of $\mathbf{R}^m$, $x_0 \in X$ and $f : X \to Y$ a map. Recall (Apostol, Section 4-5, or Buck, Section 2.2) that f is said to be *continuous at* x_0 if for each $\epsilon > 0$ there exists a $\delta > 0$ such that $d(f(x_0), f(x)) < \epsilon$ for all x in X with $d(x_0, x) < \delta$; f is *continuous* on X if it is continuous at each x_0 in X. In particular, $f : \mathbf{R}^n \to \mathbf{R}^m$ is continuous at $x_0 \in \mathbf{R}^n$ iff for each $\epsilon > 0$ there is a $\delta > 0$ such that $f(U_\delta(x_0)) \subseteq U_\epsilon(f(x_0))$.

Theorem 1-7. A map $f : \mathbf{R}^n \to \mathbf{R}^m$ is continuous at $x_0 \in \mathbf{R}^n$ iff for each open subset V of $\mathbf{R}^m$ containing $f(x_0)$ there is an open subset U of $\mathbf{R}^n$ containing x_0 such that $f(U) \subseteq V$.

Exercise 1-7. Prove Theorem 1-7.   Q.E.D.

Theorem 1-8. Let $f : \mathbf{R}^n \to \mathbf{R}^m$ be a map. Then the following are equivalent:

(a) f is continuous on $\mathbf{R}^n$.

(b)   For each open subset V of $\mathbf{R}^m$, $f^{-1}(V)$ is open in $\mathbf{R}^n$.

(c)   For each closed subset C of $\mathbf{R}^m$, $f^{-1}(C)$ is closed in $\mathbf{R}^n$.

Exercise 1-8. Prove Theorem 1-8. Hint: Prove the equivalence of (a) and (b) and note that (b) ⇔ (c) follows from $f^{-1}(\mathbf{R}^m - C) = \mathbf{R}^n - f^{-1}(C)$.
    Q.E.D.

Since our concern is primarily with maps between subsets of Euclidean spaces, it would behoove us to reformulate these last few results in this more general context. To this end we consider a map $f : X \to Y$, where $X \subseteq \mathbf{R}^n$ and $Y \subseteq \mathbf{R}^m$. By definition f is continuous at x_0 in X if for each $\epsilon > 0$ there is a $\delta > 0$ such that $f(U_\delta(x_0) \cap X) \subseteq U_\epsilon(f(x_0)) \cap Y$. This definition assumes a simpler form if we introduce some new terminology. Let Z be an arbitrary subset of some Euclidean space $\mathbf{R}^p$ and $z_0 \in Z$. A subset of Z of the form $U_r(z_0) \cap Z = \{z \in Z : d(z_0, z) < r\}$ is called an *open ball in Z* of radius r about z_0. A subset U of Z is said to be *open in Z* if for each $z_0 \in U$ there is an open ball in Z about z_0 that is contained in U.

Exercise 1-9. Show that $U \subseteq Z$ is open in Z iff there is an open subset U' of $\mathbf{R}^p$ with $U = U' \cap Z$.

A subset C of Z is *closed in Z* if its complement $Z - C$ is open in Z. Thus, C is closed in Z iff there is a closed subset C' of $\mathbf{R}^p$ such that $C = C' \cap Z$.

Theorem 1-9. Let Z be an arbitrary subset of $\mathbf{R}^p$.
(a) $\varnothing$ and Z are open in Z.
(b)   Any union of open subsets of Z is open in Z.
(c)   Any finite intersection of open subsets of Z is open in Z.

Theorem 1-10. Let Z be an arbitrary subset of $\mathbf{R}^p$.
(a) $\varnothing$ and Z are closed in Z.
(b)   Any intersection of closed subsets of Z is closed in Z.
(c)   Any finite union of closed subsets of Z is closed in Z.

The proofs of Theorems 1-9 and 1-10 are identical to those of Theorems 1-5 and 1-6, respectively. Similarly, the next two results follow in the same way as Theorems 1-7 and 1-8.

Theorem 1-11. Let $f : X \to Y$ be a map, where $X \subseteq \mathbf{R}^n$ and $Y \subseteq \mathbf{R}^m$. Then f is continuous at $x_0 \in X$ iff for each open set V in Y containing $f(x_0)$ there is an open set U in X containing x_0 such that $f(U) \subseteq V$.

Theorem 1-12. Let $f : X \to Y$ be a map, where $X \subseteq \mathbf{R}^n$ and $Y \subseteq \mathbf{R}^m$. Then the following are equivalent:
(a) f is continuous.

(b)   For each open subset V of Y, $f^{-1}(V)$ is open in X.

(c)   For each closed subset C of Y, $f^{-1}(C)$ is closed in X.

We denote by $C(X, Y)$ the set of all continuous maps of X into Y. Note that if $F : \mathbf{R}^n \to Y$ is continuous, then the restriction $f = F|X$ is in $C(X, Y)$. However, it need not be the case that every element of $C(X, Y)$ is the restriction of some continuous map of $\mathbf{R}^n$ into Y. For example, if $X = \mathbf{R} - \{0\}$ and $f : X \to \mathbf{R}$ is defined by $f(x) = 1/x$ for each $x \in X$, then f is continuous, but is not the restriction of any continuous map of $\mathbf{R}$ into $\mathbf{R}$. For this reason and because all of the concepts relevant to our study of continuity can be defined without referrence to the ambient Euclidean spaces, we shall henceforth adopt the attitude that the sets X and Y are objects of interest in themselves regardless of the particular Euclidean spaces in which they happen to reside. In this spirit we shall use the word *space* to refer to any subset of some Euclidean space whose intrinsic topological properties are of interest to us. If X is a space and Z is a subset of X, then Z is called a *subspace* of X. For us then the term "space" means "subspace of $\mathbf{R}^n$" for some Euclidean space $\mathbf{R}^n$. (The reader is warned that this usage is quite unorthodox and should not be carried over to any other texts.)

Let X be an arbitrary space and A a subset of X. By a *neighborhood* (nbd) of A in X we mean any subset of X that contains an open subset of X containing A. (Note that a neighborhood itself need not be open.) In particular, a nbd of a point x_0 in X is a subset of X that contains some open ball in X about x_0. A point x_0 in X is said to be *isolated* in X if $\{x_0\}$ is a nbd of x_0 in X, that is, if there is some $r > 0$ such that the open ball in X of radius r about x_0 contains no point of X other than x_0. If every point of X is isolated in X, then X is called a *discrete* space. For example, the subspace $\{1/n\}_{n=1}^{\infty}$ of $\mathbf{R}$ is discrete. However, the space $\{1/n\}_{n=1}^{\infty} \cup \{0\}$ consists of infinitely many isolated points and one nonisolated point and is therefore not discrete.

Exercise 1-10. Let X be a discrete space and Y an arbitrary space. Show that every map $f : X \to Y$ is continuous.

A point $x_0 \in A \subseteq X$ is said to be an *interior point* of A if there is a nbd of x_0 in X that is entirely contained in A; the set of all such points is called the *interior* of A in X and is denoted $\text{int}_X A$ or simply $\text{int } A$. A point $x_0 \in X$ is a *boundary point* of A if every nbd of x_0 in X intersects both A and $X - A$; the set of all such points is called the *boundary* of A in X and denoted $\text{bdy}_X A$ or $\text{bdy } A$. A point $x_0 \in X$ is called an *accumulation point* of A if every nbd of x_0 in X contains at least one point of A different from x_0. A sequence $\{x_n\}_{n=1}^{\infty}$ of points in X is said to *converge* to a point x_0 in X

if for each nbd U of x_0 in X there is a positive integer N such that $x_n \in U$ for all $n \geqslant N$. One easily verifies that $\{x_n\}_{n=1}^\infty$ converges to x_0 iff the sequence $\{d(x_0, x_n)\}_{n=1}^\infty$ of real numbers converges to zero.

Exercise 1-11. Show that a point x_0 in X is an accumulation point of the subset A of X iff there is a sequence $\{x_n\}_{n=1}^\infty$ of distinct points in A that converges to x_0 in X.

Theorem 1-13. Let X be an arbitrary space and A a subset of X. Then A is closed in X iff it contains all of its accumulation points in X.

Proof: First assume that A is closed in X. Then $X - A$ is open in X so each point of $X - A$ has a nbd in X which is entirely contained in $X - A$. Consequently, no point of $X - A$ can be an accumulation point of A so A contains all of its accumulation points.

Exercise 1-12. Complete the proof by showing that if A contains all of its accumulation points in X, then $X - A$ is open in X.   Q.E.D.

The *closure* of A in X, denoted $\mathrm{cl}_X A$ or $\mathrm{cl}\, A$ or $\overline{A}$, is the union of A and all of its accumulation points in X. By Theorem 1-13 $\overline{A}$ is closed in X. Note that a subset of X is closed in X iff it equals its closure. More generally, for any $A \subseteq X$, $\mathrm{cl}_X A$ is the intersection of all closed subsets of X that contain A. It follows that $\overline{\varnothing} = \varnothing$, $\overline{X} = X$, $\overline{\overline{A}} = \overline{A}$ and $\overline{A \cup B} = \overline{A} \cup \overline{B}$.

Proposition 1-14. Let A be a convex subset of $\mathbf{R}^n$. Then the closure of A in $\mathbf{R}^n$ is also convex.

Proof: Let $x, y \in \overline{A}$ and $t \in [0, 1]$. Choose sequences $\{x_n\}_{n=1}^\infty$ and $\{y_n\}_{n=1}^\infty$ of points in A with $x_n \to x$ and $y_n \to y$. Now consider the sequence $\{tx_n + (1 - t)y_n\}_{n=1}^\infty$. Since A is convex, each $tx_n + (1 - t)y_n$ is in A. Moreover, for each $n = 1, 2, \ldots$, we have $d(tx_n + (1 - t)y_n, tx + (1 - t)y) = \|tx + (1 - t)y - tx_n - (1 - t)y_n\| = \|t(x - x_n) + (1 - t)(y - y_n)\| \leqslant t\|x - x_n\| + (1 - t)\|y - y_n\| = td(x, x_n) + (1 - t)d(y, y_n)$, which goes to zero as $n \to \infty$. Thus, $tx_n + (1 - t)y_n \to tx + (1 - t)y$ so $tx + (1 - t)y \in \overline{A}$ and $\overline{A}$ is convex.   Q.E.D.

If A is an arbitrary subset of $\mathbf{R}^n$, then the closure of the convex hull of A is convex by Proposition 1-14 and is called the *closed convex hull* of A and is the intersection of all closed convex subsets of $\mathbf{R}^n$ that contain A.

Proposition 1-15. Let A be an arbitrary subset of $\mathbf{R}^n$ and $A = \text{cl}_{\mathbf{R}^n}\, A$. Then diam $\overline{A}$ = diam A.

Proof: The result is trivial if $A = \varnothing$, so we may assume that A is non-empty. Thus, diam $A = \sup\{d(x, y) : x, y \in A\}$ and diam $\overline{A} = \sup\{d(x, y) : x, y \in \overline{A}\}$. It follows immediately that diam $A \leqslant$ diam $\overline{A}$. Moreover, if either diam A or diam $\overline{A}$ is infinite the equality is trivial, so we may assume that diam $\overline{A} < \infty$. To show that diam $\overline{A} \leqslant$ diam A, it will suffice to show that $d(x, y) \leqslant$ diam A for all x and y in $\overline{A}$. Choose sequences $\{x_n\}_{n=1}^{\infty}$ and $\{y_n\}_{n=1}^{\infty}$ of points in A with $x_n \rightarrow x$ and $y_n \rightarrow y$.

Exercise 1-13. Show that $\lim_{n \rightarrow \infty} d(x_n, y_n) = d(x, y)$.

But $\lim_{n \rightarrow \infty} d(x_n, y_n)$ is less than or equal to diam A, so $d(x, y) \leqslant$ diam $\overline{A}$ as required. Thus, diam $\overline{A} \leqslant$ diam A, and we have diam $\overline{A}$ = diam A.
Q.E.D.

Theorem 1-16. Let X and Y be arbitrary spaces and $f : X \rightarrow Y$ a map. Then the following are equivalent:
(a) f is continuous.
(b)   If $\{x_n\}_{n=1}^{\infty}$ is a sequence of points in X that converges to x_0 in X, then the sequence $\{f(x_n)\}_{n=1}^{\infty}$ converges to $f(x_0)$ in Y.
(c) $f(\text{cl}_X\, A) \subseteq \text{cl}_Y\, f(A)$ for each subset A of X.

Proof: We first show that (a) implies (b). Thus, we assume that f is continuous and let $\{x_n\}_{n=1}^{\infty}$ be a sequence converging to x_0 in X. Let $\epsilon > 0$ be given and consider the open ball $U_\epsilon(f(x_0)) \cap Y$ of radius ϵ about $f(x_0)$ in Y. By continuity there exists a $\delta > 0$ such that $f(U_\delta(x_0) \cap X) \subseteq U_\epsilon(f(x_0)) \cap Y$. Since $x_n \rightarrow x_0$ in X, there is an $N > 0$ such that $x_n \in U_\delta(x_0) \cap X$ for all $n \geqslant N$. Consequently, $f(x_n) \in f(U_\delta(x_0) \cap X) \subseteq U_\epsilon(f(x_0)) \cap Y$ for all $n \geqslant N$, so $f(x_n) \rightarrow f(x_0)$ in Y as required.

To see that (b) implies (c), we let A be an arbitrary subset of X and consider a point y in $f(\text{cl}_X\, A)$. Then $y = f(x_0)$ for some x_0 in $\text{cl}_X\, A$. Let $\{x_n\}_{n=1}^{\infty}$ be a sequence of points in A that converges to x_0 in X. Then $f(x_n) \in f(A)$ for each n and, by (b), $f(x_n) \rightarrow f(x_0)$, so $f(x_0) = y$ is in $\text{cl}_Y\, f(A)$. Thus, $f(\text{cl}_X\, A) \subseteq \text{cl}_Y\, f(A)$.

Finally, we show that (c) implies (a) as follows: Let x_0 be an arbitrary point of X and V an open nbd of $f(x_0)$ in Y. Set $A = X - f^{-1}(V)$ and $U = X - \text{cl}_X\, A$. Then U is open in X.

Exercise 1-14. Complete the proof by showing that by part (c) $x_0 \in U$ and $f(U) \subseteq V$.   Q.E.D.

A subset D of a space X is said to be *dense* in X if $cl_X D = X$. Consider, for example, the set $\mathbf{Q}$ of rational numbers in $\mathbf{R}$. To show that $\mathbf{Q}$ is dense in $\mathbf{R}$, it suffices to show that for any real number r there exists a sequence of rational numbers converging to r in $\mathbf{R}$. Let $a_0.a_1a_2a_3 \ldots$ be a decimal expansion for r. Then, by definition

$$r = \sum_{n=0}^{\infty} a_n/10^n = \lim_{m \to \infty} \sum_{n=0}^{m} a_n/10^n.$$

But each $\sum_{n=0}^{m} a_n/10^n$ is rational so the result follows. A simple consequence of this and Exercise 1-6 is that the set $\mathbf{Q}^n$ of points in $\mathbf{R}^n$ with rational coordinates is dense in $\mathbf{R}^n$.

Exercise 1-15. Show that the set $\mathbf{P}$ of irrational numbers is dense in $\mathbf{R}$.

Exercise 1-16. Let X and Y be arbitrary spaces, D a dense subset of X and $f, g : X \to Y$ continuous maps. Show that if $f|D = g|D$, then $f = g$.

A set is dense in X if its closure contains every open subset of X. At the other extreme a subset A of X whose closure contains no open subset of X is said to be *nowhere dense* in X. Thus, A is nowhere dense in X if $int_X (cl_X A) = \varnothing$. As a simple example we consider the Cantor set C in $[0, 1]$. Being an intersection of closed subsets of $[0, 1]$, C is itself closed in $[0, 1]$, so $cl\ C = C$ and we need only show that C has an empty interior in $[0, 1]$. But this is clear since if C had an interior point in $[0, 1]$ it would necessarily contain an interval, and we have already seen that this is not the case (see the Remark in Example 1-1 (f)).

A collection $\mathscr{B}$ of open subsets of a space X is called a *base for the open sets in X* (or an *open base* for X) if every open set in X can be written as a union of elements of $\mathscr{B}$, that is, iff whenever U is open in X and $x_0 \in U$, there exists a $B \in \mathscr{B}$ with $x_0 \in B \subseteq U$. For example, the collection of all open balls $U_r(x_0)$ for $x_0 \in \mathbf{R}^n$ and $r > 0$ is an open base for $\mathbf{R}^n$, as is the collection of all open rectangles in $\mathbf{R}^n$ by Exercise 1-6. Each of these bases, however, has the disadvantage of being rather "large" (i.e., uncountable). We show next that $\mathbf{R}^n$ has a countable base for its open sets.

Lemma 1-17. Let $\mathscr{B}$ denote the collection of all open balls $U_r(x_0)$ in $\mathbf{R}^n$, where $x_0 \in \mathbf{Q}^n$ and r is a positive rational number. Then $\mathscr{B}$ is a countable base for the open sets in $\mathbf{R}^n$.

Proof: Since $\mathbf{Q}$ is countable by Example 1-1 (b) and $\mathbf{Q}^n$ is countable by Lemma 1-4, it follows from Lemma 1-3 that $\mathscr{B}$ is countable. To see that $\mathscr{B}$

is an open base for $\mathbf{R}^n$, we let x_0 be an arbitrary point of $\mathbf{R}^n$ and U an open subset of $\mathbf{R}^n$ containing x_0. Then there is an $\epsilon > 0$ such that $x_0 \in U_\epsilon(x_0) \subseteq U$. Since $\mathbf{Q}^n$ is dense in $\mathbf{R}^n$, we may select a point y with rational coordinates in $U_{\epsilon/4}(x_0)$. Since $\mathbf{Q}$ is dense in $\mathbf{R}$, we may choose a rational number r with $\epsilon/4 < r < \epsilon/2$. Then $x_0 \in U_r(y)$ since $d(x_0, y) < \epsilon/4$ and $r > \epsilon/4$, and $U_r(y) \subseteq U_\epsilon(x_0)$ since $r < \epsilon/2$. But $U_r(y) \in \mathcal{B}$ so the proof is complete.   Q.E.D.

Now, if X is an arbitrary subspace of $\mathbf{R}^n$, we may intersect each element of $\mathcal{B}$ with X to obtain a countable collection of open sets in X which by Exercise 1-9 is an open base for X. Thus, we have proved the next theorem:

Theorem 1-18. Let X be an arbitrary subspace of $\mathbf{R}^n$. Then X has a countable open base.

Exercise 1-17. Let X and Y be arbitrary spaces, $f: X \to Y$ a map, and $\mathcal{B}$ a base for the open sets in Y. Show that f is continuous iff $f^{-1}(B)$ is open in X for every $B \in \mathcal{B}$.

A collection $\mathcal{S}$ of open subsets of a space X is called a *subbase* for the open sets in X if the collection of all finite intersections of elements of $\mathcal{S}$ forms an open base for X. For example, the collection of all rays of the form (a, ∞) and $(-\infty, b)$ form a subbase for the open sets in $\mathbf{R}$.

Exercise 1-18. Let X and Y be arbitrary spaces, $f: X \to Y$ a map, and $\mathcal{S}$ a subbase for the open sets in Y. Show that f is continuous iff $f^{-1}(S)$ is open in X for each $S \in \mathcal{S}$.

Finally, if x_0 is a point in X, then a collection $\mathcal{U}_{x_0}$ of nbds (not necessarily open nbds) of x_0 in X is a *neighborhood base* (nbd base) or *local base of nbds* at x_0 if every nbd of x_0 in X contains an element of $\mathcal{U}_{x_0}$. For example, the collection of all open (or closed) balls of rational radius about x_0 is a local base at x_0 in $\mathbf{R}^n$.

Next suppose that X_i is a subspace of $\mathbf{R}^{n_i}$ for each $i = 1, \ldots, k$. Then the product $X_1 \times \cdots \times X_k$ is a subspace of $\mathbf{R}^{(n_1 + \cdots + n_k)}$. Let U_i be an open subset of X_i for each i. We claim that $U_1 \times \cdots \times U_k$ is open in $X_1 \times \cdots \times X_k$. To see this, consider an arbitrary point $x = (x_1, \ldots, x_k)$ in $U_1 \times \cdots \times U_k$. For each $i = 1, \ldots, k$ we may select by Exercise 1-6 an open rectangle R_i in $\mathbf{R}^{n_i}$ that contains x_i and whose intersection with X_i is contained in U_i. Then $R_1 \times \cdots \times R_k$ is an open rectangle in $\mathbf{R}^{(n_1 + \cdots + n_k)}$ that contains x and whose intersection with $X_1 \times \cdots \times X_k$ is contained in $U_1 \times \cdots \times U_k$. Thus, $U_1 \times \cdots \times U_k$ is open in $X_1 \times \cdots \times X_k$. It follows that $C_1 \times \cdots \times C_k$ is closed in

$X_1 \times \cdots \times X_k$ if C_i is closed in X_i for each $i = 1, \ldots, k$. Now, if $\mathscr{B}_i$ is an open base for X_i for each i, then the collection of all sets of the form $B_1 \times \cdots \times B_k$, where $B_i \in \mathscr{B}_i$, is a base for the open sets in $X_1 \times \cdots \times X_k$. Note that the collection of all sets of the form $U_1 \times \cdots \times U_k$, where each U_i is open in X_i and $U_i = X_i$ for all but one i, is a subbase for the open sets in $X_1 \times \cdots \times X_k$. We define the *projection* of $X_1 \times \cdots \times X_k$ onto the factor space X_i, denoted π_{X_i} or π_i, by $\pi_{X_i}(x_1, \ldots, x_k) = \pi_i(x_1, \ldots, x_k) = x_i$. If U_i is an open subset of X_i, then $\pi_i^{-1}(U_i) = X_1 \times \cdots \times U_i \times \cdots \times X_k$, which is open in $X_1 \times \cdots \times X_k$ so we find that π_i is continuous. More generally, suppose $f: X \to X_1 \times \cdots \times X_k$ is a map. Define the ith *coordinate function* f_i of f to be the composition $\pi_i \circ f: X \to X_i$. Then we have the next proposition.

Proposition 1-19. A map $f: X \to X_1 \times \cdots \times X_k$ is continuous iff each coordinate function f_i is continuous.

Proof. If f is continuous, then $\pi_i \circ f$ is the composition of two continuous functions and is therefore continuous. (This is a trivial consequence of Theorem 1-12.) Conversely, suppose $\pi_i \circ f$ is continuous for each $i = 1, \ldots, k$. The sets of the form $\pi_i^{-1}(U_i)$, $i = 1, \ldots, k$, with U_i open in X_i form a subbase for the open sets in $X_1 \times \cdots \times X_k$. But $f^{-1}(\pi_i^{-1}(U_i)) = (\pi_i \circ f)^{-1}(U_i)$. Thus, the inverse images under f of these subbasic open sets are open in X by continuity of $\pi_i \circ f$. The continuity of f therefore follows from Exercise 1-18.   Q.E.D.

A map $f: X \to Y$ is said to be *open* (respectively, *closed*) if, for each open (respectively, closed) set A in X, $f(A)$ is open (respectively, closed) in Y.

Exercise 1-19. Show that the projection map $\pi_i: X_1 \times \cdots \times X_k \to X_i$ is an open map, but need not be a closed map. Hint: The projection of $\mathbf{R}^2$ onto $\mathbf{R}$ is not closed.

Similarly, a closed map need not be open since any constant map is closed. For maps that are one-to-one and onto, however, the concepts of "open" and "closed" are equivalent.

Lemma 1-20. Let $f: X \to Y$ be a bijective map. Then f is an open map iff it is a closed map.

Proof: Suppose first that f is open and let $A \subseteq X$ be a closed set. Then $X - A$ is open in X, so by assumption $f(X - A)$ is open in Y. But, because f is bijective, $f(X - A) = f(X) - f(A) = Y - f(A)$, so $f(A)$ is closed in Y. The proof that "f closed" implies "f open" is similar.   Q.E.D.

A continuous map need not be either open or closed. For example, if A is a subset of X that is neither open nor closed, then the inclusion map $i : A \hookrightarrow X$ defined by $i(x) = x$ for each $x \in A$ is neither an open map nor a closed map, but is continuous, being the restriction to A of the identity map. A map $h : X \to Y$ that is continuous, bijective, and has continuous inverse $h^{-1} : Y \to X$ is called a *homeomorphism* or *topological map;* if such a map exists, we say that X and Y are *homeomorphic* or *topologically equivalent* and write $X \cong Y$. If $h : X \to Y$ is a homeomorphism, then so is $h^{-1} : Y \to X$ and, moreover, if $g : Y \to Z$ is another homeomorphism, then the composition $g \circ h$ is a homeomorphism of X onto Z. Any property that is invariant under homeomorphism, one that if possessed by one space is necessarily possessed by all homeomorphic spaces, is called an (intrinsic) *topological property*. We begin the investigation of topological properties in earnest in the next section, but first we record one simple theorem and a number of examples.

Theorem 1-21. Let X and Y be arbitrary spaces and $h : X \to Y$ a continuous bijection. Then the following are equivalent:
(a) h is a homeomorphism.
(b) h is an open map.
(c) h is a closed map.

Exercise 1-20. Prove Theorem 1-21.   Q.E.D.

Example 1-2. (a) All open intervals in $\mathbf{R}$ are homeomorphic to $\mathbf{R}$. To see this, first observe that the map $x \to e^x$ is a homeomorphism of $\mathbf{R}$ onto the interval $(0, \infty)$ and that $x \to -x$ maps $(0, \infty)$ homeomorphically onto $(-\infty, 0)$. The affine map $x \to x + a$ is a homeomorphism that carries $(0, \infty)$ onto (a, ∞) and $(-\infty, 0)$ onto $(-\infty, a)$. All that remains is to show that each bounded open interval (a, b) is homeomorphic to $\mathbf{R}$. For this we first let c be a positive real number and define $h:(-c, c) \to \mathbf{R}$ by $y = h(x) = cx/(c^2 - x^2)^{1/2}$. Then h is continuous and, solving for x, we find that $x = h^{-1}(y) = cy/(c^2 + y^2)^{1/2}$, which is defined and continuous on all of $\mathbf{R}$. Thus, h is a homeomorphism. Finally, if $a < b$ and d is the midpoint of (a, b), then $x \to x - d$ maps (a, b) homeomorphically onto $(-c, c)$, where $c = \frac{1}{2}(b - a)$. Thus, (a, b) is also homeomorphic to $\mathbf{R}$.

(b) Each open ball $U_r(x_0)$ in $\mathbf{R}^n$ is homeomorphic to $\mathbf{R}^n$.

Exercise 1-21. Generalize the constructions in (a) to prove (b).

(c) Each sphere $\{x \in \mathbf{R}^n : \|x\| = r\}$ in $\mathbf{R}^n$ is homeomorphic to S^{n-1}. Indeed, the map h from $\{x \in \mathbf{R}^n : \|x\| = r\}$ to S^{n-1} defined by $h(x) = x/\|x\| = (x_1/(x_1^2 + \cdots + x_n^2)^{1/2}, \ldots, x_n/(x_1^2 + \cdots + x_n^2)^{1/2})$ is the required homeomorphism since it is continuous and has inverse $x \to rx$.

(d) Each closed ball $B_r(x_0)$ in $\mathbf{R}^n$ is homeomorphic to B^n. To see this, first note that $B_r(x_0)$ is homeomorphic to $B_r(0)$ by translation. To show that $B_r(0)$ is homeomorphic to B^n, we let h be the homeomorphism constructed in (c) from bdy $B_r(0)$ onto bdy $B^n = S^{n-1}$. Now observe that each point of $B_r(0)$ can be expressed as tx for some $x \in$ bdy $B_r(0)$ and some $t \in [0, 1]$. Moreover, these representations are unique for all x in $B_r(0)$ except 0 for which $t = 0$ and x is arbitrary. Define $\bar{h}(tx) = th(x) = t(x/\|x\|)$ for each t in $[0, 1]$ and x in bdy $B_r(0)$.

Exercise 1-22. Show that $\bar{h} : B_r(0) \to B^n$ is a homeomorphism.

(e) In general, we can show that two spaces are not homeomorphic by exhibiting a topological property of one that is not possessed by the other. As a particularly simple example we show that the closed unit interval $I = [0, 1]$ is not homeomorphic to $\mathbf{R}$. Recall from the calculus that any continuous real-valued function on I assumes its maximum value. Since this is not true of every continuous real-valued function on $\mathbf{R}$, we need only show that if X is a space with the property that each of its continuous real-valued functions assumes its maximum value, then any space homeomorphic to X also has this property. Thus, suppose h is a homeomorphism of X onto a space Y, and let $f : Y \to \mathbf{R}$ be a continuous function. Then $f \circ h : X \to \mathbf{R}$ is continuous and therefore assumes its maximum value at some point $x_0 \in X$. Thus, $(f \circ h)(x) \leq (f \circ h)(x_0)$ for every $x \in X$. Since h is surjective, we find that $f(y) \leq f(h(x_0))$ for every y in Y so that, since $h(x_0) \in Y$, f assumes its maximum value at $h(x_0)$ and the proof is complete.

Remark: The property of I that accounts for the success of the argument in part (e) is its "compactness." A general investigation of this property is begun in Section 1-4. Another important topological property of I is its "connectedness," which we consider in Section 1-5 and which has one consequence already known to the reader:

Exercise 1-23. Use the Intermediate Value Theorem from the calculus and arguments analogous to those in (e) to show that I is not homeomorphic to the subspace $[0, \frac{1}{2}] \cup [1, \frac{3}{2}]$ of $\mathbf{R}$.

Before turning to a serious investigation of the most important topological properties, we simplify our terminology by agreeing that henceforth the term "map" unmodified will always mean "continuous map."

1-4   Compact spaces

A *cover* of a space X is a collection $\mathscr{C}$ of subsets of X whose union is X; $\mathscr{C}$ is an *open* (respectively, *closed*) *cover* of X if each element of $\mathscr{C}$ is an open (respectively, closed) subset of X. A *subcover* for $\mathscr{C}$ is a subcollection $\mathscr{C}'$ of $\mathscr{C}$ whose union is also all of X.

Lemma 1-22. Let $\{A_1, \ldots, A_k\}$ be an open (or closed) cover of a space X. For each $i = 1, \ldots, k$ let $f_i : A_i \to Y$ be a continuous map and assume that $f_i|A_i \cap A_j = f_j|A_i \cap A_j$ for all i and j. Then the map $f : X \to Y$ defined by $f|A_i = f_i$ for each $i = 1, \ldots, k$ is continuous.

Proof: Suppose first that $\{A_1, \ldots, A_k\}$ is an open cover of X. To show that f is continuous, let V be an open subset of Y. Since $f^{-1}(V) \cap A_i = f_i^{-1}(V)$ and each f_i is continuous on A_i, we find that $f^{-1}(V) \cap A_i$ is open in A_i and thus also in X since A_i is open in X. Now, $f^{-1}(V) = \cup_{i=1}^{k} f^{-1}(V) \cap A_i$ since the A_i cover X, so $f^{-1}(V)$ is open in X and f is continuous. If $\{A_1, \ldots, A_k\}$ is a closed cover of X, then we can show in the same way that $f^{-1}(C)$ is closed in X for every closed subset C of Y.   Q.E.D.

Theorem 1-23. Let X be an arbitrary space and $\mathscr{U}$ an open cover of X. Then $\mathscr{U}$ has a countable subcover.

Proof: By Theorem 1-18 we may select a countable open base $\mathscr{B}$ for X. For each $U \in \mathscr{U}$ and $x \in U$ there is a $B_{x,U} \in \mathscr{B}$ with $x \in B_{x,U} \subseteq U$. Now, $\mathscr{B}' = \{B_{x,U} : U \in \mathscr{U}, x \in U\}$ is a countable set since $\mathscr{B}' \subseteq \mathscr{B}$, so we may write $\mathscr{B}' = \{B_{x_1,U_1}, B_{x_2,U_2}, \ldots\}$. Thus, $\{U_1, U_2, \ldots\}$ is a countable subcover for $\mathscr{U}$.   Q.E.D.

It is certainly not true that every open cover of an arbitrary space has a subcover that is finite. For example, the open cover $\{(-n, n) : n \in \mathbf{N}\}$ of $\mathbf{R}$ has no finite subcover. A space with the property that each of its open covers has a finite subcover is said to be *compact*.

Remark: A subset A of a space X is called a *compact subset* of X if A is compact as a subspace of X, that is, if every cover of A by sets that are open *in* A has a finite subcover. Since the open sets in A are precisely the intersections with A of open sets in X, it follows that A is a compact subset of X iff for every collection $\mathscr{U}$ of open sets *in* X whose union contains A there exist finitely many elements $U_1, \ldots, U_k$ of $\mathscr{U}$ with $A \subseteq U_1 \cup \cdots \cup U_k$. In particular, a subspace X of $\mathbf{R}^n$ is compact iff for every collection $\mathscr{U}$ of open sets in $\mathbf{R}^n$ whose union contains X there exist finitely many elements $U_1, \ldots, U_k$ of $\mathscr{U}$ with $X \subseteq U_1 \cup \cdots \cup U_k$. This "extrinsic" characterization of compactness is quite common in advanced calculus and is rather useful since the compact subsets of $\mathbf{R}^n$ are relatively easy to describe. We assume the reader is familiar with the following fundamental result (see Apostol, Theorem 3-40, or Buck, Theorems 4 and 5 of Chapter 1):

Theorem 1-24. Let X be a subspace of $\mathbf{R}^n$. Then the following are equivalent:

(a) X is compact.
(b) X is closed and bounded in $\mathbf{R}^n$.
(c)   Every infinite subset of X has an accumulation point in X.

From (b) of Theorem 1-24 it follows that every closed subset of a compact space is compact and that a compact subset of any space X is closed in X. Next we show that compactness is a topological property. In fact, we will show more generally that compactness is preserved under arbitrary continuous maps.

Theorem 1-25. The continuous image of a compact space is compact.

Proof: Suppose X is compact and f is a continuous map of X onto Y. Let $\mathcal{U}$ be an open cover of Y. Then $\{f^{-1}(U) : U \in \mathcal{U}\}$ is an open cover of X. Select a finite subcover, say, $\{f^{-1}(U_1), \ldots, f^{-1}(U_k)\}$. Then, since f is surjective, $\{U_1, \ldots, U_k\}$ covers Y so Y is compact.   Q.E.D.

Exercise 1-24. Show that a continuous map defined on a compact space is a closed map and conclude that a continuous bijection on a compact space is a homeomorphism.

Exercise 1-25. Let X be a compact space and $f: X \to \mathbf{R}$ a continuous real-valued function. Show that f attains its supremum and infimum, that is, that there is at least one $x_0 \in X$ with $f(x_0) = \sup\{f(x) : x \in X\}$ and at least one $x_1 \in X$ with $f(x_1) = \inf\{f(x) : x \in X\}$.

Corollary 1-26. A product space $X_1 \times \cdots \times X_k$ is compact iff each of the factor spaces $X_1, \ldots, X_k$ is compact.

Proof: Suppose first that $X_1 \times \cdots \times X_k$ is compact. For each $i = 1, \ldots, k$ the projection map π_i maps $X_1 \times \cdots \times X_k$ continuously onto X_i so by Theorem 1-25 each X_i is compact. Next suppose X_i is a compact subspace of $\mathbf{R}^n$ for each i. By (b) of Theorem 1-24 X_i is closed and bounded in $\mathbf{R}^{n_i}$. Thus, $X_1 \times \cdots \times X_k$ is closed and bounded in $\mathbf{R}^{(n_1 + \cdots + n_k)}$ so $X_1 \times \cdots \times X_k$ is compact.   Q.E.D.

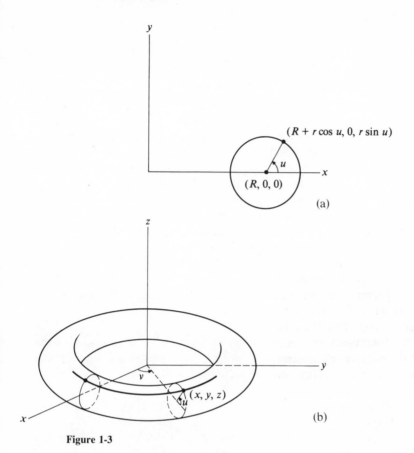

$(R + r \cos u, 0, r \sin u)$

u

$(R, 0, 0)$

(a)

v

(x, y, z)

u

(b)

Figure 1-3

Example 1-3. (a) The closed ball B^n and the sphere are both closed and bounded in $\mathbf{R}^n$ and are therefore compact. In particular, the closed unit interval I is compact.

(b) The Cantor set C, being an intersection of closed subsets of I (see Example 1-1 (f)), is closed in I and therefore compact.

(c) By Corollary 1-26 the square $I \times I$ and, more generally, the n-dimensional cube $I \times I \times \cdots \times I$ (n-factors) is compact, as is the cylinder $S^1 \times I$.

(d) Consider a circle in the xz-plane of radius $r > 0$ about a point $(R, 0, 0)$, where $R > r$ (see Figure 1-3 (a)). Any point on this circle has coordinates $(R + r \cos u, 0, r \sin u)$ for some $0 \leqslant u \leqslant 2\pi$. Now rotate this circle about the z-axis to obtain a surface called the *torus* T (see Figure 1-3(b)).

Notice that, for each point on the circle, $x^2 + y^2$ and z remain con-

Figure 1-4

stant during the rotation. Thus, if (x, y, z) is any point on the torus and v denotes the angle through which the circle was rotated to arrive at this point, then $x = (R + r \cos u)\cos v$, $y = (R + r \cos u)\sin v$ and $z = r \sin u$, where $(R + r \cos u, 0, r \sin u)$ is the point on the original circle which, when rotated through the angle v, arrives at (x, y, z). Defining a map $f: \mathbf{R}^2 \to \mathbf{R}^3$ by $f(u, v) = ((R + r \cos u)\cos v, (R + r \cos u)\sin v, r \sin u)$, we find that the torus is the image under f of the square $[0, 2\pi] \times [0, 2\pi] = \{(u, v) : 0 \leqslant u, v \leqslant 2\pi\}$. Since f is continuous, it follows that T is compact.

Exercise 1-26. Show that the torus T is homeomorphic to the subspace $S^1 \times S^1$ of $\mathbf{R}^4$.

(e) The surface obtained from a long rectangular strip by fastening the ends of the strip together after rotating one end through an angle of 180 degrees from its original position is called the *Möbius strip* (see Figure 1-4).

We obtain an analytic representation of this surface as follows: Consider first the circle $x = 2 \cos u$, $y = 2 \sin u$, $0 \leqslant u \leqslant 2\pi$, in the xy-plane. At the point on this circle corresponding to the value u of the parameter we construct a unit vector j that starts at the point of the circle and makes an angle of $u/2$ with the positive z-axis. At the same point we also construct the vector $-j$. Thus, we have a line segment of length 2 composed of two vectors that has its midpoint on the circle (see Figure 1-5).

As u goes from 0 to 2π this line segment travels with u, turning through an angle of π so that j finally comes to the original position of $-j$. Clearly, then, this line segment traces out the Möbius strip. For each value of u, $0 \leqslant u \leqslant 2\pi$, the point on this line segment at a distance $v, -1 \leqslant v \leqslant 1$, from the circumference in the direction of j has coordinates $x = 2 \cos u + v \sin(u/2) \cos u$, $y = 2 \sin u + v \sin(u/2) \sin u$ and $z = v \cos(u/2)$. Thus, the Möbius strip is the image of the rectangle $[0, 2\pi] \times [-1, 1]$ under the continuous map $f: \mathbf{R}^2 \to \mathbf{R}^3$ given by $f(u, v) = (2 \cos u + v \sin(u/2) \cos u$, $2 \sin u + v \sin(u/2) \sin u, v \cos(u/2))$. In particular, the Möbius strip is compact.

We will say that a space is *locally compact* if each point in that space has a nbd base consisting of compact sets.

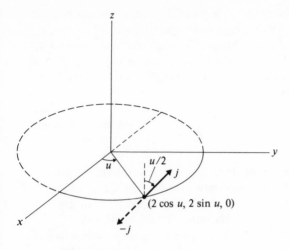

Figure 1-5

Exercise 1-27. Show that a space X is locally compact iff each point in X has a compact nbd in X.

In particular, it follows from Exercise 1-27 that every compact space is locally compact. The converse, however, is false since $\mathbf{R}^n$ is locally compact, but not compact.

Exercise 1-28. Show that the subspace $\mathbf{Q}$ of $\mathbf{R}$ is not locally compact.

1-5   Connectivity properties

Among the simplest and yet most fundamental topological properties are those related to various notions of "connectedness." In this section we begin an investigation of several such notions, but others must await the development of more sophisticated machinery (in Chapters 3 and 4). Suffice it to say that the theme of connectivity is a recurrent one and that all of the rather elementary material of this section will serve us well in later chapters.

We shall say that a space X is *connected* if it cannot be expressed as the union of two disjoint nonempty open subsets of X; if, on the other hand, we can find nonempty open subsets H and K of X with $H \cap K = \varnothing$ and $H \cup K = X$, then X is *disconnected* and the sets $\{H, K\}$ are said to constitute a *disconnection* of X. Notice that, since $H = X - K$ and $K = X - H$, the sets in a disconnection are closed as well as open. It follows that a space X is disconnected iff there exist two nonempty, disjoint, closed subsets of X whose union is all of X. Connected spaces are therefore precisely those in which only $\varnothing$ and the space itself are both closed and open. We say that a subset A of a space X is a *connected subset* of X if A is connected as a subspace of X; this is the case iff A is not contained in the union of two open subsets of X whose intersections with A are disjoint and nonempty.

Theorem 1-27. A subspace X of $\mathbf{R}$ is connected iff it is an interval (open, closed or half-open).

Proof: Suppose first that X is connected. We may assume that X contains more than one point, since if $X = \{x_0\}$, then $X = [x_0, x_0]$. Consequently, if X were not an interval, there would exist three numbers x, y, and z with x and z in X, $x < y < z$ and $y \notin X$. But this immediately implies that $X = [X \cap (-\infty, y)] \cup [X \cap (y, \infty)]$ is a disconnection of X, which is a contradiction. Thus, X is an interval.

Now, let X be an interval. Again, if X consists of a single point, then X is connected, so we may assume that X contains more than one point. Suppose X were disconnected. Then there exist nonempty, disjoint subsets H and K of X that are both open and closed and whose union is all of X. Choose a point $x \in H$ and a point $z \in K$. Since H and K are disjoint, $x \neq z$ and, by relabeling H and K if necessary, we may assume that $x < z$. Since X is an interval, $[x, z] \subseteq X$ and each point in $[x, z]$ is in either H or K. Let $y = \sup\{t \in [x, z] : t \in H\}$. Then $x \leq y \leq z$ so $y \in X$. Since H is closed, it follows that $y \in H$. Thus, $y < z$. Now, by the definition of y, $y + \epsilon \in K$ for every $\epsilon > 0$ such that $y + \epsilon \leq z$. Since K is closed in X, it follows that $y \in K$. Thus, $y \in H \cap K$ and this contradicts the assumption that these sets are disjoint. Consequently, X is connected.   Q.E.D.

Remark: It follows from Theorem 1-27 and Example 1-2 (a) that all connected open subsets of $\mathbf{R}$ are homeomorphic. A connected open subset of $\mathbf{R}^2$ is generally called a *domain*. Not all domains are homeomorphic since, as we shall see later, the open disc $\{(x, y) \in \mathbf{R}^2 : (x^2 + y^2)^{1/2} < 1\}$ and the annulus $\{(x, y) \in \mathbf{R}^2 : \frac{1}{2} < (x^2 + y^2)^{1/2} < \frac{3}{2}\}$ are not homeomorphic, although both are connected and open in $\mathbf{R}^2$.

Theorem 1-28. The continuous image of a connected space is connected.

Proof: Suppose X is connected and f is a continuous map of X onto Y. If Y were disconnected by H and K, then X would be disconnected by $f^{-1}(H)$ and $f^{-1}(K)$. Thus, Y must be connected.   Q.E.D.

In particular, it follows from Theorem 1-28 that connectedness is a topological property.

Exercise 1-29. Show that a space X is connected iff the only continuous maps of X into the two-point discrete space $\{-1, 1\}$ are the constant maps.

Lemma 1-29. Suppose $X = \cup X_\alpha$, where each X_α is a connected subset of X and $\cap X_\alpha \neq \varnothing$. Then X is connected.

Proof: Suppose $X = H \cup K$, where H and K are open in X and $H \cap K = \varnothing$. Since X_α is connected and $X_\alpha \subseteq H \cup K$ for every α, each X_α is contained entirely in either H or K. Since $\cap X_\alpha \neq \varnothing$ and $H \cap K = \varnothing$, all of the X_α are contained in one of these sets, say, $X_\alpha \subseteq H$ for each α. Thus, $X = \cup X_\alpha \subseteq H$, so $K = \varnothing$; this is a contradiction, so it follows that X is connected.   Q.E.D.

Theorem 1-30. A product space $X_1 \times \cdots \times X_k$ is connected iff each of the factor spaces $X_1, \ldots, X_k$ is connected.

Proof: Suppose first that $X_1 \times \cdots \times X_k$ is connected. For each $i = 1, \ldots, k$ the projection π_i maps $X_1 \times \cdots \times X_k$ onto X_i, so by Theorem 1-28 each X_i is connected. For the converse it suffices to show that $X_1 \times X_2$ is connected whenever X_1 and X_2 are connected, and this we shall prove by contradiction. Thus, we assume that X_1 and X_2 are connected, but that $X_1 \times X_2 = H \cup K$, where H and K are nonempty disjoint open subsets of $X_1 \times X_2$. Choose points $(a_1, b_1) \in H$ and $(a_2, b_2) \in K$. The subspaces $\{a_1\} \times X_2$ and $X_1 \times \{b_2\}$ are homeomorphic to X_2 and X_1, respectively, and are therefore connected. Moreover, $(\{a_1\} \times X_2) \cup (X_1 \times \{b_2\})$ is nonempty since it contains the point (a_1, b_2). By Lemma 1-29 the subspace $X = (\{a_1\} \times X_2) \cup (X_1 \times \{b_2\})$ of $X_1 \times X_2$ is connected. This, however, is impossible since X intersects both H and K and is therefore disconnected by $\{X \cap H, X \cap K\}$.   Q.E.D.

Example 1-4. (a) All intervals in $\mathbf{R}$ are connected by Theorem 1-27. By Theorem 1-30, then, all open or closed rectangles in $\mathbf{R}^n$ are connected. In particular, $\mathbf{R}^n$ itself is connected.

(b) All convex subsets of $\mathbf{R}^n$ are connected. To see this, suppose $X \subseteq \mathbf{R}^n$ is convex and let x_0 be an arbitrary point of X. Then X is the union of all the line segments $[x_0, x]$ for $x \in X$. Since each $[x_0, x]$ is homeomorphic to a closed interval in $\mathbf{R}$ and is therefore connected, the connectedness of X follows from Lemma 1-29.

(c) The torus T and the Möbius strip M, being continuous images of closed rectangles in $\mathbf{R}^2$, are both connected.

(d) The reader will be asked to show in Exercise 3-15 that the sphere S^n has a property ("pathwise connectivity") that implies connectivity (see Theorem 1-37).

Exercise 1-30. Show that the *punctured plane* $\mathbf{R}^2 - \{(0, 0)\}$ is connected.

Connectivity properties are particularly useful for showing that various spaces are not homeomorphic. For example, we show that the line and the plane are not homeomorphic by observing that if $h : \mathbf{R}^2 \to \mathbf{R}$ were a homeomorphism, then $h|\mathbf{R}^2 - \{(0, 0)\}$ would map the punctured plane homeomorphically onto $\mathbf{R} - \{h(0, 0)\}$. This, however, is impossible since $\mathbf{R}^2 - \{0, 0)\}$ is connected by Exercise 1-30 and $\{(-\infty, h(0, 0)), (h(0, 0), \infty)\}$ is a disconnection of $\mathbf{R} - \{h(0, 0)\}$.

Remark: The argument just given to show that $\mathbf{R}$ and $\mathbf{R}^2$ are not homeomorphic will not distinguish $\mathbf{R}^n$ and $\mathbf{R}^m$ topologically for n, $m > 1$ since neither $\mathbf{R}^n$ nor $\mathbf{R}^m$ can be disconnected by removing a point. We return to the problem of showing that $\mathbf{R}^n$ and $\mathbf{R}^m$ are not homeomorphic for $n \neq m$ in Sections 1-8 and 2-10.

Exercise 1-31. Show that the circle S^1 and the interval I are not homeomorphic.

Another area of vital concern to topology in which connectedness plays a central role is the proof of Existence Theorems such as the familiar Intermediate Value Theorem:

Exercise 1-32. Let $g : [a, b] \to \mathbf{R}$ be continuous and assume that $g(a) \neq g(b)$. Show that for each real number r between $g(a)$ and $g(b)$ there exists an x_0 in $[a, b]$ with $g(x_0) = r$.

One of our major concerns will be with that particular type of existence theorem known as a Fixed Point Theorem, the simplest of which follows next.

Theorem 1-31. Let $f : I \to I$ be continuous. Then there exists a point x_0 in I such that $f(x_0) = x_0$.

Proof: If either $f(0) = 0$ or $f(1) = 1$, then there is nothing to prove, so we will assume that $f(0) > 0$ and $f(1) < 1$. Define $g : I \to \mathbf{R}$ by $g(x) = x - f(x)$ for each x in I. Then g is continuous. Moreover,

$g(0) = -f(0) < 0$ and $g(1) = 1 - f(1) > 0$. Thus, by Exercise 1-32, there is an x_0 in I with $g(x_0) = 0$. But then $x_0 - f(x_0) = 0$, so $f(x_0) = x_0$ as required.   Q.E.D.

If X is an arbitrary space and $f : X \to X$ is a map, then a point $x_0 \in X$ for which $f(x_0) = x_0$ is called a *fixed point* of f. If a space X is such that every continuous map of X into itself has a fixed point, then X is said to have the *fixed point property*. Theorem 1-31 then simply says that the closed unit interval I has the fixed point property. We show next that the fixed point property is a topological property.

Proposition 1-32. Suppose that the space X has the fixed point property and that Y is homeomorphic to X. Then Y has the fixed point property.

Proof: Let $h : X \to Y$ be a homeomorphism and suppose $f : Y \to Y$ is a continuous map. Then $h^{-1} \circ f \circ h : X \to X$ is continuous, so by assumption there is an $x_0 \in X$ with $(h^{-1} \circ f \circ h)(x_0) = x_0$, that is, $h^{-1}(f(h(x_0))) = x_0$. But then $f(h(x_0)) = h(x_0)$, so $h(x_0)$ is a fixed point of f.   Q.E.D.

It now follows from Proposition 1-32 and Example 1-2 (d) that every closed interval $[a, b]$ in $\mathbf{R}$ has the fixed point property. In particular, $B^1 = [-1, 1]$ has the fixed point property, and this simple fact is the one-dimensional case of the

Brouwer Fixed Point Theorem. For each integer $n \geq 1$ the closed ball B^n has the fixed point property.

Remark: This remarkable and extremely powerful result was first proved by L. E. J. Brouwer in 1910 and will occupy quite a bit of our time. In the remaining chapters we will develop a wide range of rather diverse ideas and techniques, some combinatorial, some algebraic, and some differential, and there is a danger that this overabundance of trees will obscure the reader's view of the forest. Perhaps there is no more convincing demonstration of the essential unity of our subject than the fact that we will eventually be able to give three entirely different proofs of the Brouwer Theorem, one combinatorial (Section 2-9), one algebraic (Sections 3-6 and 4-13) and one differential (Section 5-11).

Next we show that if a space X is not itself connected, then at least it can be written as the union of a disjoint collection of maximal connected subspaces. For each $x \in X$ we define the *component* C_x of x in X to be the union of all connected subsets of X containing x. By Lemma 1-29 the component of any point is a connected subset of X. Moreover, if x and y are distinct points of X, then either $C_x = C_y$ or $C_x \cap C_y = \varnothing$, for otherwise $C_x \cup C_y$ would be a connected set containing x and y and larger than C_x or C_y and this is impossible. Thus, $\{C_x : x \in X\}$ partitions X into maximal connected subspaces.

Lemma 1-33. Let X be an arbitrary space and A a connected subset of X. If B is a subset of X with $A \subseteq B \subseteq \bar{A}$, then B is connected. In particular, $\bar{A}$ is connected.

Proof: Assume that B is not connected. Then there exist two closed subsets H and K of X whose union contains B and whose intersections with B are nonempty and disjoint. Since A is connected and contained in $H \cup K$, A is contained in either H or K and is disjoint from the other, say, $A \subseteq H$ and $A \cap K = \varnothing$. Since H is closed, $\bar{A} \subseteq H$ and it follows that $\bar{A} \cap K = \varnothing$. But then $B \cap K = \varnothing$ and this is a contradiction, so B must be connected.   Q.E.D.

Theorem 1-34. Let X be an arbitrary space. Then each component of X is closed in X.

Proof: If C is the component of $x \in X$, then by Lemma 1-33 $\bar{C}$ is a connected set containing x so $\bar{C} \subseteq C$, showing that C is closed.   Q.E.D.

The components of a space need not be open however. Consider, for example, the subspace **Q** of **R**. We claim that no subset of **Q** containing more than one point is connected. Indeed, if $S \subseteq \mathbf{Q}$ contains two distinct points x and z with, say, $x < z$, then we may select an irrational number y with $x < y < z$ and $\{S \cap (-\infty, y), S \cap (y, \infty)\}$ is a disconnection of S. Consequently, if x is any point in **Q**, then $C_x = \{x\}$, that is, the components of **Q** are its points. Since no point of **Q** is isolated in **Q**, these components are not open. A space X such as **Q** with the property that, for each $x \in X$, $C_x = \{x\}$ is said to be *totally disconnected*.

Exercise 1-33. Show that the Cantor space is totally disconnected.

There is, however, an important class of spaces in which components are open as well as closed. Let us say that a space X is *locally connected* if each $x \in X$ has a nbd base consisting of connected open sets.

Theorem 1-35. A space X is locally connected iff each component of each open subspace of X is open in X.

Proof: Suppose X is locally connected, Y is an open subspace of X, C is a component of Y and $x \in C$. Since Y is open in X there is a connected nbd

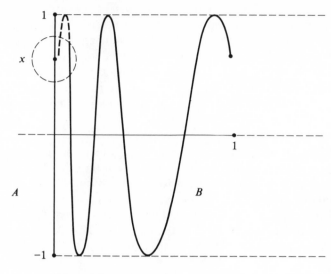

Figure 1-6

U of x in X that is contained in Y. Thus, U is a nbd of x in Y and, since it is connected, $x \in U \subseteq C$ so C is open in X.

Conversely, suppose each component of each open subspace of X is open in X. Let $x \in X$ and U be an open nbd of x in X. The component of x in U is then open in X, connected and contained in U, so it follows that X is locally connected.   Q.E.D.

We claim that local connectedness neither implies nor is implied by connectedness. Since $(0, 1) \cup (2, 3)$ is locally connected, but not connected, our first contention is trivial. For the second we define a space X called the *topologist's sine curve* as follows: X is the subspace of $\mathbf{R}^2$ given by $X = A \cup B$, where $A = \{(x, y) : x = 0, -1 \le y \le 1\}$ and $B = \{(x, y) : 0 < x \le 1, y = \sin(1/x)\}$ (see Figure 1-6).

Since B is the image of the interval $(0, 1]$ under the continuous map $f : (0, 1] \to \mathbf{R}^2$ given by $f(x) = (x, \sin(1/x))$, it is connected by Theorem 1-28. But $X = \mathrm{cl}_{\mathbf{R}^2}B$, so by Lemma 1-33 X is also connected. However, X is not locally connected since each $x \in A$ has a nbd in X that contains no connected nbd of x (again, see Figure 1-6).

Local connectivity is certainly a topological property. More generally, we have the next proposition.

Proposition 1-36. Let X be locally connected and $f : X \to Y$ a continuous map onto Y that is either open or closed. Then Y is locally connected.

Proof: Suppose U is an open subset of Y and C is a component of U. For each $x \in f^{-1}(C)$ let C_x be the component of x in the open set $f^{-1}(U)$. Then each C_x is open in X by Theorem 1-35. Now, $f(C_x)$ is connected and contains $f(x) \in C$ so $f(C_x) \subseteq C$. Thus, $x \in C_x \subseteq f^{-1}(C)$. Since C_x is open in X, $f^{-1}(C)$ is open in X. Now, if f is an open map, $f(f^{-1}(C)) = C$ is open in Y. If f is a closed map, then $f(X - f^{-1}(C)) = Y - f(f^{-1}(C)) = Y - C$ is closed in Y, so again C is open in Y and the result follows from Theorem 1-35.   Q.E.D.

From the point of view of analysis there is a somewhat more natural notion of "connectivity" than the one we have considered thus far. Let us say that a space X is *pathwise connected* if for any two points x and y in X there is a continuous map $\alpha : I \to X$ with $\alpha(0) = x$ and $\alpha(1) = y$; such a map α is called a *path* in X from x to y. For example, if $A \subseteq \mathbf{R}^n$ is convex and x and y are two points of A, then the map $\alpha : I \to A$ defined by $\alpha(t) = ty + (1 - t)x$ is a path in A from $\alpha(0) = x$ to $\alpha(1) = y$.

Exercise 1-34. Prove that a continuous image of a pathwise connected space is pathwise connected.

Theorem 1-37. A pathwise connected space is connected.

Proof: Suppose $\{H, K\}$ is a disconnection of the pathwise connected space X. Let $\alpha : I \to X$ be any path in X from $x \in H$ to $y \in K$. Then H and K disconnect $\alpha(I)$, which is impossible by Theorems 1-27 and 1-28.
    Q.E.D.

Exercise 1-35. Show that the converse of Theorem 1-37 is false. Hint: The topologist's sine curve is not pathwise connected.

Nevertheless, there is a useful partial converse for Theorem 1-37. We say that a subset A of an arbitrary space X is a *pathwise connected subset* of X if A is pathwise connected as a subspace of X, that is, if any two points in A can be joined by a path whose image lies in A. Now a space X is said to be *locally pathwise connected* if each point in X has a nbd base consisting of pathwise connected subsets of X.

Theorem 1-38. A connected, locally pathwise connected space is pathwise connected.

Proof: Let X be connected and locally pathwise connected. Fix some point $x_0 \in X$ and denote by H the set of all points in X that can be joined to x_0 by a path. Then $H \neq \varnothing$ since $x_0 \in H$. Thus, if H is both open and closed, it must be all of X by connectedness. To prove that this is indeed the case it is convenient to introduce at this point a concept that will play a central role in our discussion of homotopy theory in Chapter 3. Let α, $\beta : I \to X$ be paths in a space X with $\alpha(0) = x_0$, $\alpha(1) = \beta(0) = x_1$, and $\beta(1) = x_2$. Define $\alpha\beta : I \to X$ by

$$(\alpha\beta)(t) = \alpha(2t), \qquad 0 \leqslant t \leqslant \tfrac{1}{2}$$
$$= \beta(2t - 1), \qquad \tfrac{1}{2} \leqslant t \leqslant 1.$$

Then $\alpha\beta$ is a path in X by Lemma 1-22. Moreover, $(\alpha\beta)(0) = \alpha(0) = x_0$ and $(\alpha\beta)(1) = \beta(1) = x_2$. The path $\alpha\beta$ is called the *product* of α and β (see Section 3-4 for more details).

Now, to show that H is open in X, we let x_1 be an arbitrary point of H and U a pathwise connected nbd of x_1 in X. We claim that $U \subseteq H$. To see this, let x_2 be an arbitrary point of U. Let α, $\beta : I \to X$ be paths with $\alpha(0) = x_0$, $\alpha(1) = \beta(0) = x_1$, and $\beta(1) = x_2$. Then the product $\alpha\beta : I \to X$ is a path in X from x_0 to x_2, so $x_2 \in H$. Thus, $U \subseteq H$, so H is open. Next we show that H is closed. Let $x_2 \in \bar{H}$, and let U be any pathwise connected nbd of x_2 in X. Then $U \cap H \neq \varnothing$. Choose some point x_1 in $U \cap H$. Since $x_1 \in H$ there is a path α from x_0 to x_1, and since $x_1 \in U$ there is a path β from x_1 to x_2. Again, the product $\alpha\beta$ is a path from x_0 to x_2, so $x_2 \in H$, that is, $\bar{H} \subseteq H$ so H is closed. Thus, $H = X$.

Exercise 1-36. Complete the proof by showing that $H = X$ implies that X is pathwise connected.   Q.E.D.

It follows, in particular, that any connected open subset of $\mathbf{R}^n$ is pathwise connected.

Now let X be an arbitrary space. Define a relation $\sim$ on X as follows: If x and y are two points in X, then $x \sim y$ iff there is a path in X from x to y. Then $x \sim x$ for every $x \in X$. Moreover, if $x \sim y$ and $y \sim z$, then there exist paths α, $\beta : I \to X$ with $\alpha(0) = x$, $\alpha(1) = \beta(0) = y$ and $\beta(1) = z$, so $\alpha\beta : I \to X$ is a path in X from x to z and it follows that $x \sim z$. Finally, if $\alpha : I \to X$ is a path from x to y, then $\alpha^{-1} : I \to X$ defined by $\alpha^{-1}(t) = \alpha(1 - t)$ for each $t \in I$ is a path from y to x so $y \sim x$. Thus, $\sim$ is an equivalence relation on X. The equivalence classes in X under this equivalence relation are called the *path components* of X. Each path component of X is pathwise connected and therefore contained in some component of X by Theorem 1-37.

Exercise 1-37. Show that the path components of a space X need not be closed in X.

1-6   Real-valued continuous functions

For any space X we denote by $C(X)$ the set $C(X, \mathbf{R})$ of all continuous maps from X to $\mathbf{R}$. We define addition, multiplication, and scalar multiplication in $C(X)$ pointwise:

$$(f + g)(x) = f(x) + g(x)$$
$$(fg)(x) = f(x)g(x)$$
$$(af)(x) = af(x)$$

or all $f, g \in C(X)$, $a \in \mathbf{R}$ and $x \in X$.

Exercise 1-38. Show that if f and g are in $C(X)$ and $a \in \mathbf{R}$, then $f + g$, fg, and af are all in $C(X)$ and that, moreover, under these operations $C(X)$ is an algebra with unit element over the field of real numbers (see Herstein, p. 218, for the definition of an algebra).

If $f_1, \ldots, f_k$ are elements of $C(X)$, we define two elements $\max(f_1, \ldots, f_k)$ and $\min(f_1, \ldots, f_k)$ of $C(X)$ by $\max(f_1, \ldots, f_k)(x) = \max\{f_1(x), \ldots, f_k(x)\}$ and by $\min(f_1, \ldots, f_k)(x) = \min\{f_1(x), \ldots, f_k(x)\}$ for each $x \in X$.

Lemma 1-39. Let X be an arbitrary space and A an arbitrary subset of X. Then the function $\varphi : X \to \mathbf{R}$ defined by $\varphi(x) = \mathrm{dist}(x, A)$ is continuous. Moreover, if A is closed in X, then φ is strictly positive on $X - A$.

Proof: Let x and y be two points of X. Then, for each $a \in A$, $d(x, a) \leq d(x, y) + d(y, a)$, so we have $\mathrm{dist}(x, A) = \inf_{a \in A} d(x, a) \leq d(x, y) + \inf_{a \in A} d(y, a) = d(x, y) + \mathrm{dist}(y, A)$. Thus, $\mathrm{dist}(x, A) - \mathrm{dist}(y, A) \leq d(x, y)$. Interchanging the roles of x and y we obtain $\mathrm{dist}(y, A) - \mathrm{dist}(x, A) \leq d(y, x) = d(x, y)$ so that $|\mathrm{dist}(x, A) - \mathrm{dist}(y, A)| \leq d(x, y)$, that is, $d(\varphi(x), \varphi(y)) \leq d(x, y)$, and from this the continuity of φ is immediate. Observe that $\varphi(x) \geq 0$ for every $x \in X$. Suppose $x_0 \in X$ is such that $\varphi(x_0) = \mathrm{dist}(x_0, A) = 0$. Then x_0 must be in $\bar{A}$. In particular, if A is closed in X, then $\varphi(x_0) = 0$ implies $x_0 \in A$, so φ is strictly positive on $X - A$.   Q.E.D.

Proposition 1-40. Let X be an arbitrary space, A a closed subset of X, and C a compact subset of X. Then there exists a point $c_0 \in C$ such that $\mathrm{dist}(C, A) = \mathrm{dist}(c_0, A)$. If A is also compact, then there is an $a_0 \in A$ such that $\mathrm{dist}(C, A) = d(c_0, a_0)$.

Proof: By Lemma 1-39 $\varphi(x) = \text{dist}(x, A)$ is continuous on X and therefore on C. Thus, by Exercise 1-25 $\varphi|C$ assumes its infimum, that is, there is a point c_0 in C such that $\varphi(c_0) = \inf\{\varphi(c) : c \in C\} = \inf\{\text{dist}(c, A) : c \in C\} = \text{dist}(C, A)$. If A is also compact, the same arguments apply to the restriction of $\psi(x) = \text{dist}(c_0, x) = d(c_0, x)$ to A, so there is a point $a_0 \in A$ such that $d(c_0, a_0) = \text{dist}(C, A)$.   Q.E.D.

Our next result, while it may appear rather technical, describes one of the most important characteristics of compact spaces and will prove to be indispensible in Chapters 2, 3, and 4.

Theorem 1-41 (Lebesgue Covering Lemma). Let $\mathcal{U}$ be an open cover of the compact space X. Then there exists a real number $\delta > 0$ (called a *Lebesgue number* for $\mathcal{U}$) such that any subset of X of diameter less than δ is contained in some element of $\mathcal{U}$.

Proof: Let $\{U_1, \ldots, U_k\}$ be a finite subcover for $\mathcal{U}$. For each $i = 1, \ldots, k$ let $\varphi_i(x) = \text{dist}(x, X - U_i)$ and let $\varphi = \max(\varphi_1, \ldots, \varphi_k)$. Then φ is continuous and, since each $x \in X$ is in some U_i and each $X - U_i$ is closed, $\varphi(x) \geq \varphi_i(x) > 0$. Thus, $\varphi(X)$ is a compact subset of **R** that does not contain 0. Consequently, $\text{dist}(0, \varphi(X)) > 0$, so there is a $\delta > 0$ such that $\varphi(x) > \delta$ for each $x \in X$. Thus, the open ball of radius δ about any point in X is contained entirely in some U_i and from this the theorem is immediate.   Q.E.D.

Exercise 1-39. Use Theorem 1-41 to show that any continuous map $f : X \to Y$, where X is compact and Y is arbitrary, is uniformly continuous. Hint: Recall that f is uniformly continuous iff for every $\epsilon > 0$ there is a $\delta > 0$ (which depends only on ϵ) such that $d(f(x_1), f(x_2)) < \epsilon$ for all x_1, $x_2 \in X$ with $d(x_1, x_2) < \delta$. (See Apostol, Section 4-11, or Buck, Section 2.3.)

Theorem 1-42 is also of considerable importance in that it assures us that any subspace of $\mathbf{R}^n$ has a sufficiently rich supply of continuous real-valued functions to "separate" all of its disjoint closed subsets.

Theorem 1-42 (Urysohn's Lemma). Let X be an arbitrary space and A and B disjoint closed subsets of X. Then, for any real numbers a and b with $a < b$ there exists a continuous function $g : X \to [a, b]$ with $g(A) = \{a\}$ and $g(B) = \{b\}$.

Proof: It will suffice to prove the result when $a = -1$ and $b = 1$, for then we need only compose with a homeomorphism of $[-1, 1]$ onto $[a, b]$ (see Example 1-2 (d)). Now define $g : X \to [-1, 1]$ by

$$g(x) = \frac{\text{dist}(x, A) - \text{dist}(x, B)}{\text{dist}(x, A) + \text{dist}(x, B)}$$

for each $x \in X$. By Lemma 1-39, $\text{dist}(x, A) + \text{dist}(x, B)$ is never zero since $A \cap B = \emptyset$, and therefore g is continuous. Moreover, $g(A) = \{-1\}$ and $g(B) = \{1\}$, so the proof is complete.   Q.E.D.

Corollary 1-43. Let X be an arbitrary space and A and B disjoint closed subsets of X. Then there exist disjoint open subsets U and V of X with $A \subseteq U$ and $B \subseteq V$.

Proof: Let $g : X \to [-1, 1]$ be a continuous map with $g(A) = \{-1\}$ and $g(B) = \{1\}$. Then we need only set $U = g^{-1}([-1, \frac{1}{2}))$ and $V = g^{-1}((\frac{1}{2}, 1])$.
    Q.E.D.

Another interesting class of topological problems we will consider in detail are the so-called Extension Problems. Suppose X, Y, and Z are arbitrary spaces with Y a subspace of X. If $F : X \to Z$ is a continuous map on X, then $F|Y$ is continuous on Y. On the other hand, we have already observed that a continuous map $f : Y \to Z$ need not be the restriction to Y of any continuous map on X (see the remarks following Theorem 1-12). If, in fact, there is a continuous map $F : X \to Z$ such that $f = F|Y$, then we call F a *continuous extension* of f to X. The most basic facts about the existence of continuous extensions are summarized in the following sequence of classical results. First we consider the problem of extending a continuous *real-valued* function f that is *bounded*, that is, for which there exists a positive constant M such that $|f(y)| \leq M$ for every y, and defined on a closed subspace.

Theorem 1-44 (Tietze Extension Theorem). Let X be an arbitrary space, Y a closed subspace of X and f a bounded continuous real-valued function on Y. Then there exists a bounded continuous real-valued function F on X such that $F|Y = f$.

Proof: Let $M > 0$ be such that $|f(y)| \leq M$ for every $y \in Y$. For each $n = 1, 2, \ldots$, define $r_n = (M/2)(2/3)^n$. Then $|f(y)| \leq 3r_1$ for every $y \in Y$. We define a sequence $f_1, f_2, \ldots$ of continuous functions on Y inductively such that $|f_n(y)| \leq 3r_n$ for every $y \in Y$. Set $f_1 = f$ and suppose that $f_1, \ldots, f_n$ have been defined. Let $A_n = \{y \in Y : f_n(y) \leq -r_n\}$ and $B_n = \{y \in Y : f_n(y) \geq r_n\}$. Then A_n and B_n are disjoint closed subsets of Y

and are therefore closed in X also since Y is closed in X. By Theorem 1-42 there is a continuous map $g_n : X \rightarrow [-r_n, r_n]$ with $g_n(A_n) = \{-r_n\}$ and $g_n(B_n) = \{r_n\}$. Now, the values of f_n and g_n on A_n lie between $-3r_n$ and $-r_n$; on B_n they lie between r_n and $3r_m$; elsewhere on Y they lie between $-r_n$ and r_n. Now define

$$f_{n+1} = f_n - g_n | Y.$$

Then f_{n+1} is continuous on Y and $|f_{n+1}(y)| \leq 2r_n$, so $|f_{n+1}(y)| \leq 3r_{n+1}$ for every $y \in Y$.

Now consider the sequence $g_1, g_2, \ldots$ of continuous functions on X just constructed. Since $|g_n(x)| \leq r_n$ for each $x \in X$ and since the series $\sum_{n=1}^{\infty}(M/2)(2/3)^n$ converges, the Weierstrass M-test (Apostol, Theorem 13-7, or Buck, Theorem 16, Chapter 4) implies that $\sum_{n=1}^{\infty} g_n(x)$ converges uniformly on X and thus represents a continuous function $F(x) = \sum_{n=1}^{\infty} g_n(x)$ on X (see Apostol, Theorem 13-3, or Buck, Theorem 17, Chapter 4). Note that, for any n, $(g_1 + \cdots + g_n)|Y = (f_1 - f_2) + \cdots (f_n - f_{n+1}) = f_1 - f_{n+1} = f - f_{n+1}$. Since the sequence $\{f_{n+1}(y)\}_{n=1}^{\infty}$ converges to zero for every $y \in Y$, we conclude that $F(y) = f(y)$ for each $y \in Y$, that is, $F|Y = f$. Finally, for any $x \in X$

$$|F(x)| \leq \sum_{n=1}^{\infty} |g_n(x)| \leq \frac{M}{3} \sum_{n=0}^{\infty} \left(\frac{2}{3}\right)^n = \frac{M}{3} \left[\frac{1}{1 - (\frac{2}{3})}\right] = M.$$

Q.E.D.

It is a relatively simple matter to remove the boundedness hypothesis in Theorem 1-44, but then, of course, the extension will not be bounded.

Corollary 1-45. Let X be an arbitrary space, Y a closed subspace of X, and f a continuous real-valued function on Y. Then there exists a continuous real-valued function F on X such that $F|Y = f$.

Proof: Arctan $\circ f$ is a bounded continuous real-valued function on Y and thus has an extension G to X. Let $A = \{x \in X : |G(x)| \geq \pi/2\}$. Then A is closed in C and disjoint from Y. By Theorem 1-42 there is a continuous function $g : X \rightarrow [0, 1]$ with $g(A) = \{0\}$ and $g(Y) = \{1\}$. Thus, the product Gg agrees with arctan $\circ f$ on Y and satisfies $|Gg(x)| \leq \pi/2$ for all $x \in X$. Consequently, $F = \tan \circ (Gg)$ is a continuous extension of f to X.
    Q.E.D.

Corollary 1-46. Let X be an arbitrary space, Y a closed subspace of X, and $f : Y \rightarrow \mathbf{R}^k$ a continuous map. Then there exists a continuous map $F : X \rightarrow \mathbf{R}^k$ such that $F|Y = f$.

Exercise 1-40. Prove Corollary 1-46.   Q.E.D.

Corollary 1-47. Let X be an arbitrary space, Y a closed subspace of X, and $f : Y \to I \times I \times \cdots \times I$ (k-factors) a continuous map. Then there exists a continuous map $F : X \to I \times I \times \cdots \times I$ such that $F|Y = f$.

Exercise 1-41. Prove Corollary 1-47.   Q.E.D.

At this point the reader may be tempted to conjecture that continuous maps on closed subspaces always extend, regardless of the range. To see why this is not true, we consider a space X, a closed subspace Y of X, and a continuous map f of Y into the sphere S^n. Since $S^n \subseteq \mathbf{R}^{n+1}$ we may, of course, regard f as a map into $\mathbf{R}^{n+1}$. By Corollary 1-46 there is then a continuous map F of X into $\mathbf{R}^{n+1}$ that extends f. The difficulty is that the map F will, in general, not take all of its values in S^n and therefore cannot be considered an extension of the map $f : Y \to S^n$. In some cases it may be possible to correct this difficulty. For example, if F does not map any point of X onto the origin in $\mathbf{R}^{n+1}$, then the map $\hat{F}(x) = F(x)/\|F(x)\|$ is continuous, maps into S^n, and extends F. Naturally, this rather fortuitous property of F is a bit too much to expect in general. We can, of course, let U be the open subset of X defined by $U = \{x \in X : F(x) \neq 0\}$ and note that $Y \subseteq U$ and $\hat{F} : U \to S^n$ is a continuous extension of f to U, thereby obtaining the next theorem.

Theorem 1-48. Let X be an arbitrary space, Y a closed subspace of X, and $f : Y \to S^n$ a continuous map. Then there exists an open nbd U of Y in X and a continuous map $\hat{F} : U \to S^n$ such that $\hat{F}|Y = f$.

In general, we can do no better. For example, suppose $n = 0$ ($S^0 = \{-1, 1\}$), X is connected, and Y is the union of two nonempty, disjoint, closed subsets A and B of X. The map $f : Y \to S^0$ defined by $f(A) = \{-1\}$ and $f(B) = \{1\}$ is continuous, but it cannot extend to X by Exercise 1-29. In particular, the identity map $\mathrm{id}_{S^0} : S^0 \to S^0$ does not extend to a continuous map of $B^1 = [-1, 1]$ to S^0. We show somewhat later (Section 2-9) that, for any $n \geq 0$, the map $\mathrm{id}_{S^n} : S^n \to S^n$ fails to extend continuously to B^{n+1}.

There are, however, certain closed subsets with the property that continuous maps on them always extend, and we now turn to a study of these subsets.

1-7   Retracts

A subset A of a space X is called a *retract* of X if there exists a continuous map $r : X \to A$ such that $r|A = \mathrm{id}_A$; the map r is called a *retraction* of X onto A.

Exercise 1-42. Show that a retract of a space X is necessarily closed in X.

Proposition 1-49. Let X be an arbitrary space and A a subset of X. Then A is a retract of X iff for every space Y each continuous map $f : A \to Y$ has a continuous extension to X.

Proof: If $r : X \to A$ is a retraction and $f : A \to Y$ is continuous, then $f \circ r : X \to Y$ is a continuous extension of f to x. On the other hand, if every continuous map $f : A \to Y$ extends to X, then, in particular, $\text{id}_A : A \to A$ admits a continuous extension to a map $r : X \to A$, which is a retraction of X onto A.   Q.E.D.

Proposition 1-50. A retract of a space with the fixed point property also has the fixed point property.

Proof: Suppose X has the fixed point property and A is a retract of X. Let $f : A \to A$ be a continuous map. By Proposition 1-49 f extends to a continuous map $F : X \to A \subseteq X$. By assumption there is an $x_0 \in X$ such that $F(x_0) = x_0$. But since F maps into A, $x_0 \in A$, so $f(x_0) = F(x_0) = x_0$ and f also has a fixed point at x_0.   Q.E.D.

In order to prove the major result of this section (and, indeed, of this chapter), we require the following rather remarkable property of the Cantor set.

Lemma 1-51. Every nonempty closed subset A of the Cantor space C is a retract of C.

Proof: We construct a retraction $r : C \to A$ as follows: For each point in C there is at least one point q in A such that $\text{dist}(p, A) = d(p, q)$ (Proposition 1-40). Suppose $p \in C$ is such that this point q is not unique. Let a_p and b_p be points in A such that $d(p, a_p) = d(p, b_p) = \text{dist}(p, A)$ and assume $a_p < b_p$. Then $a_p < p < b_p$. Since $[0, 1] - C$ is dense in $[0, 1]$, we may select a y_p in $[0, 1] - C$ such that $a_p < y_p < p < b_p$. For each $x \in C \cap [a_p, y_p]$ let $r(x) = a_p$, and for each $x \in C \cap (y_p, b_p]$ let $r(x) = b_p$ (in particular, $r(p) = b_p$). Now let p be a point of C for which $r(p)$ has not yet been determined. Then there exists a unique point $q_p \in A$ such that $d(p, q_p) = \text{dist}(p, A)$. Let $r(p) = q_p$ for every such p. In particular, if $p \in A$, then $\text{dist}(p, A) = 0$ and $q_p = p$ is unique, so $r(p) = p$, that is, $r|A = \text{id}_A$.

Exercise 1-43. Complete the proof by showing that r is continuous.
      Q.E.D.

Theorem 1-52. Every nonempty compact space is a continuous image of the Cantor space.

Proof: Let X be a nonempty compact space. By Theorem 1-18 we may let $\mathscr{B} = \{B_n\}_{n=1}^{\infty}$ be a countable open base for X. For each $n = 1, 2, \ldots$ let f_n be the set-valued map defined on $\{0, 2\}$ by

$$f_n(0) = \bar{B}_n$$
$$f_n(2) = X - B_n.$$

For each $x \in C$ let $:x_1 x_2 x_3 \ldots$ be a triadic expansion of x that does not contain the digit 1. Since $\mathscr{B}$ is a open base for X, the intersection $\cap_{n=1}^{\infty} f_n(x_n)$ must either be empty or a singleton $\{p_x\}$. To see this, note that if p and q are any two distinct points of X, then there exists a $B_n \in \mathscr{B}$ with $p \in B_n$ and $q \notin \bar{B}_n$. Since $f_n(x_n)$ is either $\bar{B}_n$ or $X - B_n$, not both of p and q can lie in $f_n(x_n)$. Thus, $\cap_{n=1}^{\infty} f_n(x_n)$ can contain at most one point. If $\cap_{n=1}^{\infty} f_n(x_n) = \{p_x\}$, we define $g(x) = p_x$; if $\cap_{n=1}^{\infty} f_n(x_n) = \varnothing$, then $g(x)$ is not defined. We thus have a map

$$g : A \to X, \qquad \text{where } A = \{x \in C : \cap_{n=1}^{\infty} f_n(x_n) \neq \varnothing\}.$$

Observe first that this map g is surjective. Indeed, for any $p \in X$, $\cap \{\bar{B}_i : p \in B_i \in \mathscr{B}\} = \{p\}$ and if we define $x = :x_1 x_2 x_3 \ldots$, where $x_i = 0$ if $p \in B_i$ and $x_i = 2$ if $p \notin B_i$, then $x \in C$ and $g(x) = p$. Moreover, g is continuous. To see this, let $x = :x_1 x_2 x_3 \ldots$ (without 1's) in A and $\epsilon > 0$ be given. Let $B_k \in \mathscr{B}$ be such that $g(x) \in B_k$ and diam $\bar{B}_k < \epsilon$. Now let y be any point of A with $d(x, y) < 3^{-2k}$ and let $y = :y_1 y_2 y_3 \ldots$ (without 1's). Since $d(x, y) < 3^{-2k}$, $y_k = x_k$ and we have $g(y) \in f_k(y_k) = f_k(x_k) = \bar{U}_k$. Thus, $d(g(x), g(y)) < \epsilon$ and g is continuous at x. Note also that A is closed in C: Let $x \in C - A$. Then $\cap_{n=1}^{\infty} f_n(x_n) = \varnothing$, that is, $\cup_{n=1}^{\infty}(X - f_n(x_n)) = X$. Since each $X - f_n(x_n)$ is open in X and X is compact, there is an $m \geq 1$ such that $\cup_{n=1}^{m}(X - f_n(x_n)) = X$, that is, $\cap_{n=1}^{m} f_n(x_n) = \varnothing$. Let y be any point of C with $d(x, y) < 3^{-2m}$. Then $x_n = y_n$ for $n = 1, \ldots, m$, so $\cap_{n=1}^{m} f_n(y_n) = \varnothing$ and we conclude that $y \in C - A$. Thus, $C - A$ is open and A is closed in C.

In summary, we now have a continuous map g of the closed subset A of C onto X. But by Lemma 1-51 A is a retract of C, so by Lemma 1-49 g has a continuous extension $G : C \to X$ that must also be surjective.
     Q.E.D.

This extraordinary theorem was first proved by Alexandroff and Urysohn in 1929 and has some rather startling consequences. For example, we have the following corollary.

Corollary 1-53. For each $k \in \mathbf{N}$ there exists a continuous map of the closed unit interval I onto the k-dimensional cube $I^k = I \times \cdots \times I$ (k-factors).

Proof: Let $f : C \to I^k$ be a continuous map of the Cantor space C onto I^k. Since C is closed in I, Corollary 1-47 yields an extension $F : I \to I^k$ of f to I that is also surjective.   Q.E.D.

The interval $[0, 1]$ can be "continuously deformed" to cover any cube $[0, 1] \times \cdots \times [0, 1]$! It would require a great deal of sophistication indeed not to be taken aback at first exposure to a result of this sort, which so clearly shatters our intuitive ideas about continuity and "dimension." The mathematical community of the late nineteenth century was indeed taken aback when Peano first constructed a continuous mapping of the interval I onto the square $I \times I$ (see Apostol, Section 13-8), and if we today are less startled by Corollary 1-53, it is perhaps because we know the situation to be even more dramatic than Peano had imagined. Consider, for example, what we know must be true of a space if it has any chance at all of being a continuous image of I. Certainly, it must be compact and connected, and by Proposition 1-36 it must also be locally connected. Believe it or not, this is enough:

Hahn–Mazurkiewicz Theorem. A subspace X of $\mathbf{R}^n$ is a continuous image of the closed unit interval I iff X is compact, connected, and locally connected.

To prove that a compact, connected, locally connected space X is a continuous image of I, we begin, as in Corollary 1-53, with a continuous map $f : C \to X$ of the Cantor space C onto X, and the problem is to extend f continuously to I. In this case, however, Corollary 1-47 is no longer applicable and the construction of this extension requires a rather delicate analysis that we do not intend to carry out here. We include a statement of the Hahn–Mazurkiewicz Theorem only to reinforce a point that should already be clear from Corollary 1-53, namely, that the status of our intuitive conception of "dimension" in a topological setting is far from clear and warrants a more careful examination.

1-8   Topological dimension

Classical dimension theory arose in response to the problem of distinguishing topologically between the Euclidean spaces $\mathbf{R}^n$ and $\mathbf{R}^m$ for $n \neq m$. Prior to the emergence of topology as an independent discipline, mathematicians, being primarily concerned with relatively nonpathological

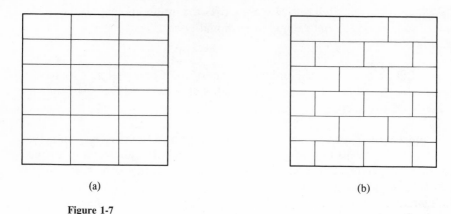

(a)   (b)

Figure 1-7

subsets of $\mathbf{R}^n$, where the context clearly indicated the proper "dimension" to be assigned a given figure, tended to use the term in a rather vague sense. The inadvisability of such an attitude became abundantly clear when Peano exhibited a continuous mapping of the interval I onto the square $I \times I$ and thus raised the possibility that the infant notion of a topological mapping was not sufficiently sensitive to distinguish objects of different dimension such as $\mathbf{R}^n$ and $\mathbf{R}^m$, which it is so obviously desirable to distinguish.

We have seen (Section 1-5) that a relatively simple connectivity argument suffices to show that $\mathbf{R}$ and $\mathbf{R}^2$ are not homeomorphic, and it is clear that the same argument will distinguish $\mathbf{R}$ and $\mathbf{R}^n$ for any $n > 1$. We also observed, however, that if both m and n are greater than 1, then the problem of showing that $\mathbf{R}^n$ and $\mathbf{R}^m$ are not homeomorphic when $n \neq m$ is less trivial. What is required, of course, is some topological property of $\mathbf{R}^n$ that distinguishes it from $\mathbf{R}^m$. Such a property can be described in a number of ways (see Hurewicz and Wallman, Introduction), but perhaps the most easily motivated approach is the following modification of ideas we owe to H. Lebesgue.

Let us consider how we might go about saying, in a topologically invariant way, that the square $I \times I$ is "two-dimensional." Suppose we were assigned the task of covering $I \times I$ with finitely many ("small") closed rectangles in such a way that as few of the rectangles as possible have points in common. A first attempt might be the covering shown in Figure 1-7 (a), where no more than four of the rectangles share a common point. However, we can clearly do better than this since the covering in Figure 1-7 (b) has the property that no more than three rectangles have a common point.

Intuitively, it seems clear that, however small the rectangles are, an arrangement such as that shown in Figure 1-7 (b) is possible and that, moreover, if the rectangles are sufficiently small, then this arrangement cannot be improved. We conclude then that a square can be covered by arbitrarily small closed rectangles in such a way that no point of the square is contained in more than three of the rectangles, but that if the rectangles are sufficiently small, then at least three have a common point. Similarly, a cube in $\mathbf{R}^3$ can be covered by arbitrarily small rectangular solids in such a way that no point in the cube is contained in more than four of the solids, but if the solids are sufficiently small, at least four have a common point. In the same way we conjecture that the cube I^n can be decomposed into arbitrarily small closed rectangles in such a way that no more than $n + 1$ of the rectangles have a common point and that $n + 1$ is the least such number; that is, in any decomposition of I^n into sufficiently small rectangles there must be a point common to at least $n + 1$ of them. With these examples as motivation we now venture the following definitions:

Let X be a subspace of $\mathbf{R}^n$, $\mathscr{C}$ a cover of X by finitely many sets, and $\epsilon > 0$ a real number. Then $\mathscr{C}$ is said to be an ϵ-*cover* of X if each element of $\mathscr{C}$ has diameter less than ϵ. The *order* of $\mathscr{C}$, denoted ord $\mathscr{C}$, is m if there is at least one point of X that belongs to m elements of $\mathscr{C}$, but no point of X belongs to $m + 1$ elements of $\mathscr{C}$. Finally, a compact space X is said to have (*topological*) *dimension* k, written dim $X = k$, if k is the least nonnegative integer for which, for every $\epsilon > 0$, there exists a finite, closed ϵ-cover of X of order $k + 1$.

Theorem 1-54. Let X and Y be homeomorphic compact spaces and suppose dim $X = k$. Then dim $Y = k$.

Proof: Let h be a homeomorphism of X onto Y. We show first that dim $Y \leq k$, that is, that for every $\epsilon > 0$ there exists a finite, closed ϵ-cover of Y of order $k + 1$. Let $\epsilon > 0$ be given and let $\mathscr{U}$ be the cover of Y by all open ϵ-balls in Y. Then $\{h^{-1}(U) : U \in \mathscr{U}\}$ is an open cover of X. Let δ be a Lebesgue number (Theorem 1-41) for this cover. Since dim $X = k$, there exists a finite, closed δ-cover $\{C_1, \ldots, C_m\}$ of X of order $k + 1$. Then $\{h(C_1), \ldots, h(C_m)\}$ is a finite, closed cover of Y of order $k + 1$ and, since each C_i is contained in some $h^{-1}(U)$ for $U \in \mathscr{U}$, it is also an ϵ-cover.

Exercise 1-44. Complete the proof by showing that dim $Y \geq k$, that is, that there exists an $\epsilon > 0$ such that every finite, closed ϵ-cover of Y has order at least $k + 1$.   Q.E.D.

Thus, the dimension of a compact space as defined above is a topological invariant. Note, however, that this definition applies only to compact spaces and, in particular, not to $\mathbf{R}^n$. We will, nevertheless, be able to use the dimension function dim to show that $\mathbf{R}^n$ and $\mathbf{R}^m$ are not homeomorphic if $n \neq m$. Roughly, the procedure is to observe that, if we assume $n > m$, $\mathbf{R}^n$ is distinguished topologically from $\mathbf{R}^m$ by the fact that it contains compact subsets of higher dimension than can exist in $\mathbf{R}^m$ (see Section 2-10). These and other results of a dimension-theoretic nature are, however, beyond the capabilities of the purely set-theoretic techniques developed thus far (with the one important exception noted in Exercise 1-45). Indeed, it should come as no surprise to the reader that a deeper analysis of topological dimension makes considerable use of the linear structure of $\mathbf{R}^n$, since it is this structure that most effectively captures our intuitive notion of dimension. It is to these matters that we direct our attention in Chapter 2.

Exercise 1-45. Show that the following are equivalent for a compact space X:
(a)   dim $X = 0$.
(b) X has a base consisting of sets that are both open and closed in X.
(c) X is totally disconnected.

Supplementary exercises

1-46   Find two nonhomeomorphic spaces X and Y such that X is homeomorphic to a subspace of Y and Y is homeomorphic to a subspace of X.

1-47   Show that if A and B are retracts of a space X, $A \cap B$ need not be a retract of X.

1-48   Prove that the unit disc B^2 is a retract of $\mathbf{R}^2$.

1-49   Find spaces X, Y, and Z such that $X \times Y \cong X \times Z$, but $Y \ncong Z$.

1-50   Show that a connected subspace of $\mathbf{R}^n$ that contains more than one point is necessarily uncountable.

1-51   Let $A \times B$ be a compact subset of $X \times Y$ and W an open subset of $X \times Y$ with $A \times B \subseteq W$. Show that there exist open sets U in X and V in Y such that $A \times B \subseteq U \times V \subseteq W$.

1-52   For any space X let $H(X)$ denote the collection of all homeomorphisms of X onto itself.
(a) Show that $H(X)$ is a group under composition.
(b) Find two nonhomeomorphic spaces X and Y for which $H(X)$ and $H(Y)$ are isomorphic.

1-53   Show that if A is any countable subset of $\mathbf{R}^n$, $n > 1$, then $\mathbf{R}^n - A$ is connected.

1-54 Show that a space X is compact iff every continuous real-valued function on X is bounded.

1-55 Let X be compact and $f : X \to X$ a map for which $d(f(x), f(y)) = d(x, y)$ for all x and y in X. Show that f is surjective. Hint: Suppose there is a $y \in X - f(X)$, and consider the sequence y, $f(y), f(f(y)), \ldots$.

1-56 Let $f : S^1 \to \mathbf{R}$ be an arbitrary continuous map. Show that there exists a pair of antipodal points $z, -z \in S^1$ such that $f(z) = f(-z)$. Hint: Let $F(x) = f(x) - f(-x)$ for each $x \in S^1$ and define $g : \mathbf{R} \to S^1$ by $g(t) = (\cos t, \sin t)$. Show that $(F \circ g)(t) = -(F \circ g)(t + \pi)$ and apply Exercise 1-32.

Chapter 2
Elementary combinatorial techniques

2-1 Introduction

The previous chapter concluded by noting the inadequacy of the purely set-theoretic techniques developed thus far for the resolution of certain very basic topological problems, particularly those concerned with topological dimension. We observed that we might reasonably expect the linear structure of $\mathbf{R}^n$ to play a rather prominant role in such considerations. In this chapter we shall begin to exploit some of the rich structure of $\mathbf{R}^n$ that was essentially ignored previously. We will find that by restricting attention to a class of spaces ("polyhedra") that can be constructed in a very natural way from certain simple building blocks ("simplexes") a wealth of new and powerful techniques become available to us. The utility of these "combinatorial" techniques is demonstrated by a number of applications, including a complete elementary proof of the Brouwer Fixed Point Theorem and a theorem to the effect that dim $B^n = n$ from which we are able to conclude that $\mathbf{R}^n \cong \mathbf{R}^m$ iff $n = m$.

2-2 Hyperplanes in $\mathbf{R}^n$

The subspaces of $\mathbf{R}^n$ that we intend to investigate here are those that can be constructed by piecing together, in a manner we shall describe later, a finite number of elementary "building blocks," each of which is constrained to lie in a "plane" of some dimension in $\mathbf{R}^n$. Our first goal then is to accumulate a certain amount of basic information about such planes.

Intuitively, a "hyperplane" (or "affine subspace") in $\mathbf{R}^n$ is nothing other than a "translation" of a linear subspace of $\mathbf{R}^n$. More precisely, a subset H^p of $\mathbf{R}^n$ is a *p-dimensional hyperplane* in $\mathbf{R}^n$ if there exists a linearly independent set $\{a_1, \ldots, a_p\}$ of vectors (points) in $\mathbf{R}^n$ and a vector a_0 in $\mathbf{R}^n$ such that H^p is precisely the set of all x in $\mathbf{R}^n$ that can be written in the form $x = a_0 + \Sigma_{i=1}^p t_i a_i$, where $t_1, \ldots, t_p$ are real numbers. Thus, the hyperplane H^p is determined by $p + 1$ points a_0, $a_1, \ldots, a_p$; $a_1, \ldots, a_p$ are linearly independent and thus span a linear subspace V of $\mathbf{R}^n$, which we translate by a_0 to obtain H^p. We often write $H^p = a_0 + V$ to indicate that H^p consists of all vectors of the form $a_0 + v$ for $v \in V$. Reversing our point of view, suppose we are given a set $s = \{a_0, a_1, \ldots, a_p\}$ of $p + 1$ points in $\mathbf{R}^n$, and suppose that the vectors $a_1 - a_0, \ldots, a_p - a_0$ are linearly independent (so that $p \le n$). In this case we say that the set s is *geometrically independent*, and we claim that there is a unique p-dimensional hyperplane in $\mathbf{R}^n$ that contains s. Before proving this, we obtain a more useful formulation of the concept of geometrical independence to make it clear that the dependence of our definition on which of the vectors we choose to call a_0 is only apparent.

Lemma 2-1. A set $s = \{a_0, a_1, \ldots, a_p\}$ of points in $\mathbf{R}^n$ is geometrically independent iff $\Sigma_{i=0}^p \lambda_i a_i = 0$ and $\Sigma_{i=0}^p \lambda_i = 0$ together imply that $\lambda_i = 0$ for each $i = 0, 1, \ldots, p$.

Proof: First suppose that $\Sigma_{i=0}^p \lambda_i a_i = 0$ and $\Sigma_{i=0}^p \lambda_i = 0$ imply that $\lambda_i = 0$ for $i = 0, 1, \ldots, p$. We must show that the vectors $a_1 - a_0, \ldots,$ $a_p - a_0$ are linearly independent. Thus, suppose that $\Sigma_{i=1}^p \alpha_i (a_i - a_0) = 0$. Then $0 = \Sigma_{i=1}^p \alpha_i (a_i - a_0) = \Sigma_{i=1}^p \alpha_i a_i - (\Sigma_{i=1}^p \alpha_i) a_0 = \Sigma_{i=0}^p \lambda_i a_i$, where $\lambda_i = \alpha_i$ for $i = 1, \ldots, p$ and $\lambda_0 = -\Sigma_{i=1}^p \alpha_i$. Thus, $\Sigma_{i=0}^p \lambda_i = 0$ so, by assumption, $\lambda_i = 0$ for $i = 0, 1, \ldots, p$. Consequently, $\alpha_i = 0$ for $i = 1, \ldots,$ p as required.

Exercise 2-1. Prove the converse.   Q.E.D.

Proposition 2-2. Let the set $s = \{a_0, a_1, \ldots, a_p\} \subseteq \mathbf{R}^n$ be geometrically independent. Then there is a unique p-dimensional hyperplane in $\mathbf{R}^n$, denoted $\pi(s)$, which contains s. Furthermore, $\pi(s)$ is precisely the set of all points x in $\mathbf{R}^n$ that can be expressed in the form

$$(1) \qquad x = \sum_{i=0}^p \lambda_i a_i, \qquad \text{where } \sum_{i=0}^p \lambda_i = 1.$$

For each $x \in \pi(s)$, the constants $\lambda_0, \ldots, \lambda_p$ in the expression (1), called the *barycentric coordinates* of x relative to $\{a_0, a_1, \ldots, a_p\}$, are unique.

Proof: Since $s = \{a_0, a_1, \ldots, a_p\}$ is geometrically independent, the set $\pi(s)$ of all $x \in \mathbf{R}^n$ of the form $x = a_0 + \Sigma_{i=1}^p t_i (a_i - a_0)$, where $t_i \in \mathbf{R}$ for $i = 1, \ldots, p$, is a p-dimensional hyperplane in $\mathbf{R}^n$ that contains each a_i. Observe that for each such x we may write $x = \Sigma_{i=1}^p t_i a_i + (1 - \Sigma_{i=1}^p t_i) a_0 = \Sigma_{i=0}^p \lambda_i a_i$, where $\lambda_i = t_i$ for $i = 1, \ldots, p$ and $\lambda_0 = 1 - \Sigma_{i=1}^p t_i$. Thus, $\Sigma_{i=0}^p \lambda_i = 1$, and we have shown that any $x \in \pi(s)$ has the required form (1). Moreover, for any such x the t_i in $x = a_0 + \Sigma_{i=1}^p t_i (a_i - a_0)$ are unique by the linear independence of the vectors $a_1 - a_0, \ldots, a_p - a_0$, so the barycentric coordinates λ_i are unique as well. Conversely, if $x = \Sigma_{i=0}^p \lambda_i a_i$, where $\Sigma_{i=0}^p \lambda_i = 1$, then $\lambda_0 = 1 - \Sigma_{i=1}^p \lambda_i$, so $x = a_0 - \Sigma_{i=1}^p \lambda_i a_0 + \Sigma_{i=1}^p \lambda_i a_i = a_0 + \Sigma_{i=1}^p \lambda_i (a_i - a_0)$ and x is in $\pi(s)$.

Exercise 2-2. Complete the proof by showing that $\pi(s)$ is the only p-dimensional hyperplane in $\mathbf{R}^n$ that contains s.   Q.E.D.

Again, let $s = \{a_0, a_1, \ldots, a_p\} \subseteq \mathbf{R}^n$ be geometrically independent. Then any x in $\pi(s)$ has unique barycentric coordinates $\lambda_i(x)$ such that $x = \Sigma_{i=0}^p \lambda_i(x) a_i$ and $\Sigma_{i=0}^p \lambda_i(x) = 1$. Thus, each $\lambda_i(x)$ is a real-valued function on $\pi(s)$; we claim that these functions are continuous. For the proof we first consider the case $n = p$. In this case there is a particularly simple reformulation of the concept of geometrical independence that follows immediately from Lemma 2-1.

Lemma 2-3. Let $s = \{a_0, a_1, \ldots, a_p\}$ be a subset of $\mathbf{R}^p$ where, for each $i = 0, 1, \ldots, p$, $a_i = (a_i^1, \ldots, a_i^p)$. Then the set s is geometrically independent iff the determinant

$$
d(s) = \begin{vmatrix}
a_0^1 & a_0^2 & a_0^3 & \ldots & a_0^p & 1 \\
a_1^1 & a_1^2 & a_1^3 & \ldots & a_1^p & 1 \\
\cdot & \cdot & \cdot & & \cdot & \cdot \\
\cdot & \cdot & \cdot & & \cdot & \cdot \\
\cdot & \cdot & \cdot & & \cdot & \cdot \\
a_p^1 & a_p^2 & a_p^3 & \ldots & a_p^p & 1
\end{vmatrix}
$$

is nonzero.

A simple application of Cramer's Rule now yields the next proposition.

Proposition 2-4. Let $s = \{a_0, a_1, \ldots, a_p\}$ be a geometrically independent subset of $\mathbf{R}^p$. For each $x = (x_1, \ldots, x_p)$ in $\pi(s) = \mathbf{R}^p$ denote by $\lambda_i(x)$, $i = 0, 1, \ldots, p$, the barycentric coordinates of x relative to s. Then $\lambda_i(x) = d_i(x)/d(s)$, where $d_i(x)$ is the determinant obtained from $d(s)$ by replacing the ith row by the vector $(x_1, \ldots, x_p, 1)$.

Exercise 2-3. Prove Proposition 2-4.   Q.E.D.

In particular, each barycentric coordinate function $\lambda_i(x)$ relative to a geometrically independent set $\{a_0, a_1, \ldots, a_p\}$ in $\mathbf{R}^p$ is a linear function of the Euclidean coordinates $x_1, \ldots, x_p$ on $\mathbf{R}^p$ and is therefore certainly continuous on $\pi(s) = \mathbf{R}^p$.

For the general case we now assume that $s = \{a_0, a_1, \ldots, a_p\}$ is a

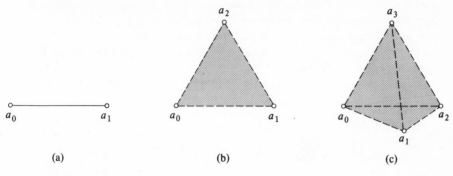

Figure 2-1

geometrically independent subset of $\mathbf{R}^n$, where $n > p$. Since the vectors $a_1 - a_0, \ldots, a_p - a_0$ are linearly independent in $\mathbf{R}^n$, there exist points $b_{p+1}, \ldots, b_n$ such that the set $\{a_1 - a_0, \ldots, a_p - a_0, b_{p+1}, \ldots, b_n\}$ is a basis for $\mathbf{R}^n$. Let $a_i = b_i + a_0$ for $i = p + 1, \ldots, n$. Then $s' = \{a_0, a_1, \ldots, a_p, a_{p+1}, \ldots, a_n\}$ is a geometrically independent set in $\mathbf{R}^n$. For each x in $\pi(s)$, let $\mu_i(x)$, $i = 0, 1, \ldots, p$, be the barycentric coordinates of x relative to s. For each x in $\pi(s') = \mathbf{R}^n$ let $\lambda_i(x)$, $i = 0, 1, \ldots, n$, denote the barycentric coordinates of x relative to s'. By Proposition 2-4 each $\lambda_i(x)$ is continuous on $\mathbf{R}^n$. But by uniqueness of barycentric coordinates $\mu_i = \lambda_i|\pi(s)$, so μ_i is continuous on $\pi(s)$ and we have proved the following theorem.

Theorem 2-5. Let $s = \{a_0, a_1, \ldots, a_p\}$ be a geometrically independent set in $\mathbf{R}^n$, and for each x in the hyperplane $\pi(s)$ determined by s let $\lambda_i(x)$, $i = 0, 1, \ldots, p$ be the barycentric coordinates of x relative to s. Then each $\lambda_i(x)$ is a continuous real-valued function on $\pi(s)$.

2-3   Simplexes and complexes

We are now in a position to define the elementary building blocks from which the spaces of interest to us here will be constructed. Observe first that any set $s = \{a_0, a_1\}$ consisting of two distinct points is geometrically independent in $\mathbf{R}^n$; the open line segment joining a_0 and a_1 is a typical "1-dimensional simplex." A set $s = \{a_0, a_1, a_2\}$ of three points in $\mathbf{R}^n$, $n \geq 2$, is geometrically independent iff they are the vertices of a triangle; the $\pi(s)$-interior of such a triangle is a "2-dimensional simplex." Similarly, a set $s = \{a_0, a_1, a_2, a_3\}$ of four points in $\mathbf{R}^n$, $n \geq 3$, is geometrically independent iff these points are the vertices of a tetrahedron; the $\pi(s)$-interior of such a tetrahedron is a "3-dimensional simplex" (see Figure 2-1).

Note that in each case the set of interest is just the set of those points in $\pi(s)$ with positive barycentric coordinates, for example, the open line segment joining a_0 and a_1 is the set of all points $\lambda_0 a_0 + \lambda_1 a_1$, where $\lambda_0 + \lambda_1 = 1$ and $\lambda_0, \lambda_1 > 0$. We generalize as follows:

Let $s = \{a_0, a_1, \ldots, a_p\}$ be a geometrically independent set in $\mathbf{R}^n$. Then the *(open) p-dimensional geometric simplex* spanned by s, denoted s_p or $(a_0, a_1, \ldots, a_p)$, is the set of all points in the hyperplane $\pi(s)$ with positive barycentric coordinates, that is,

$$s_p = (a_0, a_1, \ldots, a_p)$$
$$= \left\{ \sum_{i=0}^{p} \lambda_i a_i : \sum_{i=0}^{p} \lambda_i = 1, \quad \lambda_i > 0 \text{ for } i = 0, \ldots, p \right\}.$$

The points $a_0, a_1, \ldots, a_p$ are called the *vertices* of s_p and p is the *algebraic dimension* of s_p. A p-dimensional geometric simplex is generally referred to simply as a "p-simplex." We claim that s_p is an open subset of the hyperplane $\pi(s)$. To see this, define, for each $i = 0, 1, \ldots, p$, the set $U_i = \{x \in \pi(s) : \lambda_i(x) > 0\} = \lambda_i^{-1}(0, \infty)$. By Theorem 2-5 each U_i is open in $\pi(s)$. But $s_p = U_0 \cap U_1 \cap \cdots \cap U_p$, so s_p is open as well. Note, however, that s_p is, in general, not open in the ambient Euclidean space $\mathbf{R}^n$. Indeed, this is the case iff $n = p$: If $n = p$, $\pi(s) = \mathbf{R}^n = \mathbf{R}^p$, and the result follows at once. On the other hand, we shall prove later (see the Remark following Theorem 2-31) that if $n \neq p$, then an open subset of $\mathbf{R}^p$ cannot be homeomorphic to an open subset of $\mathbf{R}^n$, so that, since $\pi(s)$ is homeomorphic to $\mathbf{R}^p$ and s_p is open in $\pi(s)$, the converse will follow. One of the most important properties of simplexes is stated in the lemma:

Lemma 2-6. If $s = \{a_0, a_1, \ldots, a_p\}$ is a geometrically independent subset of $\mathbf{R}^n$, then the simplex s_p is a convex subset of $\mathbf{R}^n$.

Exercise 2-4. Prove Lemma 2-6.   Q.E.D.

Exercise 2-5. Prove that if $s = \{a_0, a_1, \ldots, a_p\}$ is a geometrically independent subset of $\mathbf{R}^n$, then the hyperplane $\pi(s)$ is closed in $\mathbf{R}^n$.

By virtue of Exercise 2-5 the closure of s_p in $\pi(s)$ coincides with its closure in $\mathbf{R}^n$. This closure is denoted $\bar{s}_p$ or $[a_0, a_1, \ldots, a_p]$ and is called the *closed p-dimensional geometric simplex* spanned by s, or simply the "closed p-simplex" spanned by s. The points $a_0, a_1, \ldots, a_p$ are the *vertices* of $\bar{s}_p$ and p is its *algebraic dimension*. By continuity of the barycentric coordinates

$$\bar{s}_p = [a_0, a_1, \ldots, a_p]$$
$$= \left\{ \sum_{i=0}^{p} \lambda_i a_i : \sum_{i=0}^{p} \lambda_i = 1, \quad \lambda_i \geq 0 \text{ for } i = 0, \ldots, p \right\}.$$

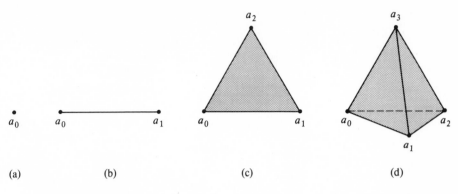

Figure 2-2

Note that if a_0 is any point in $\mathbf{R}^n$, then $s_0 = (a_0) = [a_0] = \bar{s}_0 = \{a_0\}$, that is, a 0-simplex (closed or open) is a point (vertex). If $s = \{a_0, a_1\}$ is geometrically independent, then $\bar{s}_1 = [a_0, a_1]$ is the closed line segment joining a_0 and a_1. If $s = \{a_0, a_1, a_2\}$ is geometrically independent, then $\bar{s}_2 = [a_0, a_1, a_2]$ is the closed triangle ($\pi(s)$-interior and boundary) whose vertices are a_0, a_1, and a_2. Finally, if $s = \{a_0, a_1, a_2, a_3\}$ is geometrically independent, then $\bar{s}_3 = [a_0, a_1, a_2, a_3]$ is the closed solid tetrahedron with vertices a_0, a_1, a_2, and a_3 (see Figure 2-2).

By Proposition 1-14 any closed simplex $\bar{s}_p$ is convex. More generally, we have this proposition:

Proposition 2-7. If $s = \{a_0, a_1, \ldots, a_p\} \subseteq \mathbf{R}^n$ is geometrically independent, then $\bar{s}_p$ is the closed convex hull of s.

Exercise 2-6. Prove Proposition 2-7. Hint: One need only show that $\bar{s}_p$ is contained in any convex set C containing s. Let $\Sigma_{i=0}^{p}\lambda_i a_i$ be any point in $\bar{s}_p$, and proceed by induction on the number of nonzero λ_i.

The *boundary* $\dot{s}_p$ of a p-simplex s_p is defined by $\dot{s}_p = \bar{s}_p - s_p$. If $s_p = (a_0, a_1, \ldots, a_p)$ and $\{i_0, i_1, \ldots, i_k\}$ is a subset of $\{0, 1, \ldots, p\}$, then the simplex $s_k = (a_{i_0}, a_{i_1}, \ldots, a_{i_k})$ is called a *k-face* of s_p and we write $s_k \leqslant s_p$. s_k is a *proper face* of s_p, written $s_k < s_p$, if $s_k \leqslant s_p$ and $s_k \neq s_p$. The *face opposite the vertex* a_i is denoted $(a_0, a_1, \ldots, \hat{a}_i, \ldots, a_p)$ and is the $(p - 1)$-face $(a_0, a_1, \ldots, a_{i-1}, a_{i+1}, \ldots, a_p)$. The 0-faces of any simplex are its vertices, while the 1-faces are its "edges" and the 2-faces are its "plane sides"; for the simplex $s_3 = (a_0, a_1, a_2, a_3)$, see Figure 2-3.

Proposition 2-8. Let $s_p = (a_0, a_1, \ldots, a_p)$ be a p-simplex. Then $\bar{s}_p$ is the union of all of the faces of s_p and $\dot{s}_p$ is the union of all the proper

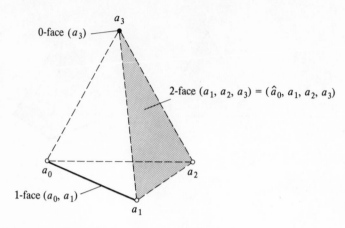

0-face (a_3)

2-face $(a_1, a_2, a_3) = (\hat{a}_0, a_1, a_2, a_3)$

a_0

a_2

1-face (a_0, a_1)

a_1

Figure 2-3

faces of s_p. Moreover, each point of $\bar{s}_p$ is contained in a unique face of s_p.

Proof: Let x be a point of $\bar{s}_p$. Then $x = \sum_{i=0}^{p}\lambda_i a_i$, where $\sum_{i=0}^{p}\lambda_i = 1$ and $\lambda_i \geq 0$ for all $i = 0, 1, \ldots, p$. If $i_0, \ldots, i_k$ are the values of i for which $\lambda_i > 0$, then x is in $s_k = (a_{i_0}, \ldots, a_{i_k}) \leq s_p$ and s_k is the only face of s_p containing x by uniqueness of the barycentric coordinates. Moreover, s_k is a proper face of s_p iff $\{i_0, \ldots, i_k\}$ is a proper subset of $\{0, \ldots, p\}$, that is, iff some barycentric coordinate of x is zero, in which case $x \in \dot{s}_p$. Conversely, if $s_k = (a_{i_0}, \ldots, a_{i_k})$ is a face of s_p and x is in s_k, then we can write $x = \sum_{j=0}^{k}\lambda_{i_j} a_{i_j}$, where $\sum_{j=0}^{k}\lambda_{i_j} = 1$ and $\lambda_{i_j} > 0$ for $j = 0, \ldots, k$. If $\{i_0, \ldots, i_k\} = \{0, \ldots, p\}$, then $x \in s_k = s_p \subseteq \bar{s}_p$. If $\{i_0, \ldots, i_k\}$ is a proper subset of $\{0, \ldots, p\}$, then s_k is a proper face of s_p, and we may write $x = \sum_{j=0}^{k}\lambda_{i_j} a_{i_j} + \sum_{i \neq i_j}\lambda_i a_i$, where $\lambda_i = 0$ for $i \neq i_j$, so $x \in \dot{s}_p \subseteq \bar{s}_p$.   Q.E.D.

It will be useful to observe that the "size" of a simplex is rather easy to compute.

Lemma 2-9. The diameter of a simplex $s_p = (a_0, \ldots, a_p)$ is the length of its longest edge, that is, diam $s_p = d(a_i, a_j)$ for some pair of vertices a_i and a_j of s_p.

Proof: Let D be the length of the longest edge in s_p and let x and y be arbitrary points of s_p. We claim that $d(x, y) \leq D$. Let a_{i_0} be the vertex of s_p farthest from x. The closed ball of radius $d(x, a_{i_0})$ about x contains all of the vertices of s_p and is convex, so by Proposition 2-7 it must contain all

Figure 2-4

of s_p. Thus, $d(x, y) \le d(x, a_{i_0})$. Now let a_{i_1} be the vertex of s_p farthest from a_{i_0}. The same argument shows that $d(x, a_{i_0}) \le d(a_{i_0}, a_{i_1})$. Since $d(a_{i_0}, a_{i_1}) \le D$ by definition of D, we obtain $d(x, y) \le d(x, a_{i_0}) \le d(a_{i_0}, a_{i_1}) \le D$.   Q.E.D.

Observe that by Proposition 1-15 diam $\bar{s}_p$ = diam s_p.

With the information we have accumulated thus far about simplexes, we can describe a canonical procedure for "piecing together" these elementary building blocks to obtain the class of spaces that is our main concern here ("polyhedra").

A *geometric complex K* in $\mathbf{R}^n$ is a finite collection of open geometric simplexes in $\mathbf{R}^n$ that satisfies the following conditions:

(K1)   If s_p is a simplex in K and $s_k < s_p$, then s_k is in K.

(K2)   If s_p and s_q are in K and $s_p \ne s_q$, then $s_p \cap s_q = \varnothing$.

The *algebraic dimension* of K is the highest dimension of any of the simplexes in K. The *polyhedron of K*, denoted $|K|$, is the point-set union of the simplexes in K. The complex K is called a *triangulation* of the polyhedron $|K|$. If x is a point in $|K|$, then the unique simplex in K which contains x is called the *carrier* of x. A subcollection K^* of K that satisfies (K1) is called a *subcomplex* of K. If s_p is any simplex, then s_p together with all of its faces forms a complex that we denote by $K(s_p)$ and call a *standard p-dimensional complex*. From Proposition 2-8 it follows that $|K(s_p)| = \bar{s}_p$, so $K(s_p)$ is a triangulation of the closed simplex $\bar{s}_p$. Further examples are easily obtained by simply piecing together various simplexes in the manner prescribed by (K1) and (K2). (The reader should pause to carry out several such constructions.) Observe that two different geometric complexes can have the same polyhedron, for example, the complexes K and L shown in Figure 2-4, where $K = K(s_1)$, $s_1 = (a_0, a_1)$, and the simplexes of L are (a_0, a_2), (a_2, a_1), (a_0), (a_1), and (a_2).

If K is an arbitrary geometric complex and r is a nonnegative integer less than or equal to the dimension of K, then the *r-skeleton K_r* of K is the subcomplex of K consisting of all the simplexes of K of dimension at most r. In particular, K_0 is the set of vertices of K.

Observe that thus far we have defined only the polyhedron *of a geometric complex K*. Since we are interested only in the topological properties of the spaces under consideration, it seems appropriate to define a *topological polyhedron* to be a space that is homeomorphic to the polyhe-

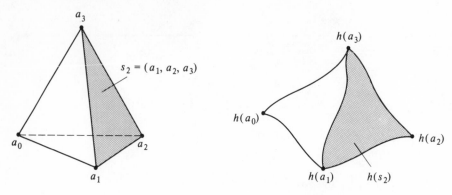

Figure 2-5

dron of some geometric complex. Observe that at this point it is not yet clear that a topological polyhedron has a uniquely defined algebraic dimension since it is at least conceivable that such a space might be homeomorphic to polyhedra of geometric complexes having different algebraic dimensions; that this cannot happen follows from Theorem 2-29. Note also that if K is a geometric complex and $h : |K| \to X$ is a homeomorphism of $|K|$ onto a space X, then h carries each simplex $s_p = (a_0, \ldots, a_p)$ of K onto a subspace $h(s_p)$ of X that is homeomorphic to s_p and thus might reasonably be called a "curved simplex" with "vertices" $h(a_0), \ldots, h(a_p)$; see Figure 2-5.

Defining the "faces" of such a curved simplex in the obvious way (i.e., $(h(a_{i_0}), \ldots, h(a_{i_k})) \leqslant (h(a_0), \ldots, h(a_p))$ iff $\{i_o, \ldots, i_k\} \subseteq \{0, \ldots, p\}$), we find that the collection of all $h(s_p)$ for s_p in K has all the properties required of a complex except "rectilinearity" and thus might be referred to as a "curvilinear complex" and regarded as providing a "triangulation" of X into curved simplexes. More precisely, a *triangulation* of a topological polyhedron X is a pair (K, h), where K is a geometric complex and h is a homeomorphism of $|K|$ onto X. Although we deal almost exclusively with geometric complexes and their polyhedra (noting that any topological result we obtain about such polyhedra is true of any homeomorph), the "curvilinear" point of view is also quite useful at times, particularly in determining whether or not a given space is a topological polyhedron. (We need only decide whether or not the space can be "dissected" into a complex of curved simplexes.)

Observe that every topological polyhedron is compact but that not every compact subspace of $\mathbf{R}^n$ is a topological polyhedron, for example, the Cantor space or the subspace $\{1/n\}_{n=1}^{\infty} \cup \{0\}$ of $\mathbf{R}$. On the other hand, we shall now show that a great many of the spaces we have encountered

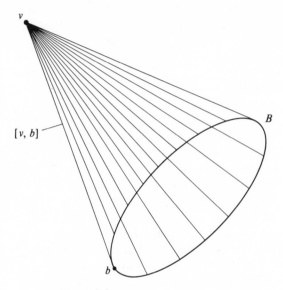

Figure 2-6

previously, for example, the sphere S^n, the ball B^{n+1}, the torus, and the Möbius strip, are topological polyhedra.

2-4 Sample triangulations

First we let $K = K_n(K(s_{n+1}))$ denote the n-skeleton of a standard $(n + 1)$-complex $K(s_{n+1})$. We shall exhibit a homeomorphism h of $|K|$ onto the n-sphere S^n. Geometrically, we view s_{n+1} as being inscribed in S^n and h as the map that "pushes" K out onto S^n. Specifically, let $s_{n+1} = (a_0, \ldots, a_{n+1})$ be a simplex in $\mathbf{R}^{n+1}$ that contains the origin 0 and is contained in the open ball $\{x \in \mathbf{R}^{n+1} : \|x\| < 1\}$. For each $x = (x_1, \ldots, x_{n+1})$ in $|K| = |K_n(K(s_{n+1}))|$ we let $h(x)$ be the point of inter-section of S^n and the ray from 0 through x. Explicitly, $h(x) = (1/\|x\|) \cdot (x_1, \ldots, x_{n+1})$. Then h is continuous and maps into S^n, and by Exercise 1-24 we need only show that h is bijective to complete the construction. Although geometrically obvious, this fact is not quite trivial and to prove it we must introduce two concepts that will play an important role at several points in the sequel as well.

Let v be a point in $\mathbf{R}^n$ and B a subset of $\mathbf{R}^n$. The pair (v, B) is said to be in *general position* if v is not in B and if, for every pair b_1 and b_2 of dis-tinct points in B, $[v, b_1] \cap [v, b_2] = \{v\}$. If (v, B) is in general position, then the *cone* with vertex v and base B, denoted $v * B$, is defined by $v * B = \cup_{b \in B}[v, b]$; see, for example, Figure 2-6.

Lemma 2-10. Let $s_p = (a_0, \ldots, a_p)$ be a p-simplex in $\mathbf{R}^n$, $v \in s_p$, $K(s_p)$ the standard triangulation of $\bar{s}_p$ by the faces of s_p and $K_{p-1}(s_p)$ the $(p - 1)$-skeleton of $K(s_p)$. Then $(v, |K_{p-1}(s_p)|)$ is in general position and $v * |K_{p-1}(s_p)| = |K(s_p)| = \bar{s}_p$.

Proof: Exercise 2-7. Prove that $(v, |K_{p-1}(s_p)|)$ is in general position. Hint: Let $b_1, b_2 \in |K_{p-1}(s_p)|$, and suppose there is a w in $[v, b_1] \cap [v, b_2]$ with $w \neq v$. Write $b_1 = \Sigma\lambda_{1i}a_i$, $b_2 = \Sigma\lambda_{2i}a_i$ and $v = \Sigma\mu_i a_i$. Note that $\lambda_{1i_1} = 0$ and $\lambda_{2i_2} = 0$ for some i_1 and some i_2, but $\mu_i > 0$ for all i. Write $w = t_1 v + (1 - t_1)b_1$ and $w = t_2 v + (1 - t_2)b_2$ for some $0 \leqslant t_1, t_2 \leqslant 1$, show that $t_1 = t_2$ and conclude that $b_1 = b_2$.

We conclude the proof by showing that $v * |K_{p-1}(s_p)| = |K(s_p)|$. Since $|K(s_p)| = \bar{s}_p$ is convex, it is clear that $v * |K_{p-1}(s_p)| \subseteq |K(s_p)|$. Conversely, suppose $w \in |K(s_p)|$. If $w = v$ or $w \in |K_{p-1}(s_p)|$, then w is in $v * |K_{p-1}(s_p)|$, so we may assume that $w \in s_p$ and $w \neq v$. We show that there exists a b in $|K_{p-1}(s_p)| = \dot{s}_p$ such that $w \in [v, b]$. Let $x = w - v$ and consider the set of all y in $\mathbf{R}^n$ of the form $y = v + tx$ for $t \geqslant 0$ (the ray from v through w). If t is sufficiently small, $y \in s_p$. Moreover, since $|K(s_p)|$ is compact, y must lie outside of $|K(s_p)|$ for sufficiently large t. Thus, the set of all $t \geqslant 0$ for which $y = v + tx$ is in $|K(s_p)|$ is bounded above and therefore has a least upper bound $t_0 > 0$. By compactness of $|K(s_p)|$, $v + t_0 x \in |K(s_p)|$ so t_0 is the maximum value of t for which $y = v + tx$ is in $|K(s_p)|$. Let $b = v + t_0 x$. Then b must be in $\dot{s}_p = |K_{p-1}(s_p)|$ for otherwise t_0 would not be maximal. Now, since $w = v + x$ is in s_p, it follows (again by maximality of t_0) that $t_0 > 1$ so $0 < 1/t_0 < 1$. Solving $b = v + t_0 x = v + t_0(w - v)$ for w, we find that $w = (1/t_0)b + (1 - (1/t_0))v$ and thus $w \in [v, b] \subseteq v * |K_{p-1}(s_p)|$ as required.   Q.E.D.

Now we return to the map $h : |K| \to S^n$ defined above. Since $0 \in s_{n+1}$ and $K = K_n(K(s_{n+1})) = K_n(s_{n+1})$, Lemma 2-10 implies that $(0, |K|)$ is in general position and from this it follows at once that h is one-to-one. Moreover, $0 * |K| = |K(s_{n+1})|$, so we find that h is surjective. Thus, h is a homeomorphism.

Next we use this result to produce a triangulation of the closed ball B^{n+1}; indeed, we show that the homeomorphism $h : |K_n(s_{n+1})| \to S^n$ extends to a homeomorphism $\bar{h} : |K(s_{n+1})| \to B^{n+1}$. Intuitively, the procedure is quite clear. For any $x \in |K_n(s_{n+1})|$ we "stretch" the line segment joining 0 and x linearly until it covers the line segment joining 0 and $h(x)$, that is, until it has unit length. Specifically, we find from Lemma 2-10 that for any $y \in |K(s_{n+1})| = 0 * |K_n(s_{n+1})|$ there is a unique $x \in |K_n(s_{n+1})|$ and a unique t in $[0, 1]$ such that $y = tx$, and we define $\bar{h}(y) = \bar{h}(tx) = th(x)$. Then $\bar{h}(0) = 0$ and $\bar{h}(x) = h(x)$ for every $x \in |K_n(s_{n+1})|$. Since $\bar{h}$ is continuous, one-to-one, and onto B^{n+1}, it is the required homeomorphism by Exercise 1-24.

Before considering additional examples, we pause to record the following lemma, which we shall have occasion to call upon somewhat later.

Lemma 2-11. If $s_p = (a_0, \ldots, a_p)$ is a p-simplex in $\mathbf{R}^n$ and $v \in \mathbf{R}^n$, then $(v, |K(s_p)|)$ is in general position iff the set $\{a_0, \ldots, a_p, v\}$ is geometrically independent and, in this case, $v * |K(s_p)| = [a_0, \ldots, a_p, v]$.

Proof: Exercise 2-8. Prove that if $\{a_0, \ldots, a_p, v\}$ is geometrically independent, then $(v, |K(s_p)|)$ is in general position.

Conversely, we show that if $(v, |K(s_p)|)$ is in general position, then $\{a_0, \ldots, a_p, v\}$ is geometrically independent. Assume that $\{a_0, \ldots, a_p, v\}$ is not geometrically independent. Then there exist constants λ, $\lambda_0, \ldots, \lambda_p$ such that

$$(2) \qquad v + \sum_{i=0}^{p} \lambda_i a_i = 0, \qquad \lambda + \sum_{i=0}^{p} \lambda_i = 0 \quad \text{and} \quad \lambda \neq 0.$$

Since this relation remains true when multiplied by an arbitrary nonzero number, the constants $\lambda, \lambda_0, \ldots, \lambda_p$ may be chosen with arbitrarily small absolute values. Let b_1 be any point of s_p with barycentric coordinates λ_{1i}, $i = 0, \ldots, p$, let t_1 be in $(0, 1)$ and set $w = t_1 v + (1 - t_1)b_1$, that is,

$$(3) \qquad w = t_1 v + \sum_{i=0}^{p} \lambda_{1i}(1 - t_1)a_i.$$

Note that all of the coefficients in (3) are positive. Thus, by taking the coefficients in (2) with sufficiently small absolute values and adding (2) and (3) we obtain

$$(4) \qquad w = (\lambda + t_1)v + \sum_{i=0}^{p} (\lambda_i + \lambda_{1i}(1 - t_1))a_i$$

with all coefficients positive and $\lambda + t_1 < 1$. Letting $t_2 = \lambda + t_1$ and $\lambda_{2i} = (\lambda_i + \lambda_{1i}(1 - t_1))/(1 - t_2)$, equation (4) becomes

$$(5) \qquad w = t_2 v + (1 - t_2) \sum_{i=0}^{p} \lambda_{2i} a_i$$

with $t_2 \in (0, 1)$,

$$\sum_{i=0}^{p} \lambda_{2i} = \frac{1}{1 - t_2} \left(\sum_{i=0}^{p} \lambda_i + (1 - t_1) \sum_{i=0}^{p} \lambda_{1i} \right)$$

$$= \frac{1}{1 - \lambda - t_1} (-\lambda + (1 - t_1)) = 1$$

and $\lambda_{2i} > 0$ for all i. Let b_2 be the point of s_p with barycentric coordinates λ_{2i}.

Then (5) becomes

(6) $w = t_2 v + (1 - t_2)b_2.$

But $w = t_1 v + (1 - t_1)b_1$ also, so we can prove that $(v, |K(s_p)|)$ is not in general position by showing that $b_1 \neq b_2$. If $b_1 = b_2 = b$, then $w = t_1 v + (1 - t_1)b$ and $w = t_2 v + (1 - t_2)b$. Since $t_2 = \lambda + t_1$ and $\lambda \neq 0$, $t_2 \neq t_1$ and we have two distinct representations of w in the 1-simplex (v, b) and this is impossible. Thus, $b_1 \neq b_2$.

Exercise 2-9. Show that if $(v, |K(s_p)|)$ is in general position, then $v * |K(s_p)| = [a_0, \ldots, a_p, v]$.

Many of the most interesting examples of topological polyhedra are "surfaces" in $\mathbf{R}^3$, that is, homeomorphs of polyhedra of 2-dimensional complexes. The problem of producing triangulations for such polyhedra is often most easily visualized by "cutting" the surface a sufficient number of times, "flattening" the result out onto a plane, triangulating this plane figure, and "pasting" the surface back together again. As an example we shall consider the torus. In order to flatten the torus out onto a plane, two cuts are required. First we cut along the curve c_1 in Figure 2-7 (a) and straighten out the resulting tube to obtain a cylinder as shown in Figure 2-7 (b) and (c). Next we cut along c_2 and flatten to obtain a rectangle (see Figure 2-7 (c), (d), and (e)).

Now the object is to find a suitable triangulation of this rectangle and paste the torus back together to obtain a triangulation of it. In order to keep clear how the torus is to be recovered from the rectangle, it is customary to label the vertices and provide the edges with directions (indicated by arrows) and to specify pairs of edges to be identified (pasted together) with coinciding directions in the reconstruction of the torus. Thus, in Figure 2-8 (a) we identify the edges BA and CD as well as BC and AD in such a way that the direction of the arrows is the same on identified edges.

Now, the vertices of the rectangle in Figure 2-8 (a) have been labeled in the "obvious" way, but this natural labeling is quite misleading since all of these vertices arise from the same point on the torus (the intersection of the curves c_1 and c_2) and thus will all be identified to the same point when the torus is reconstructed. We therefore prefer the more accurate diagram in Figure 2-8 (b).

The labeling of the vertices in Figure 2-8 (b) also has the advantage of making it clear that not every triangulation of the rectangle will do for our purposes. For example, the most obvious triangulation of the rectangle (Figure 2-9 (a)) is inadequate since it is determined by two simplexes spanned by $\{A, A, A\}$ and $\{A, A, A\}$, that is, by $\{A\}$, and thus it collapses

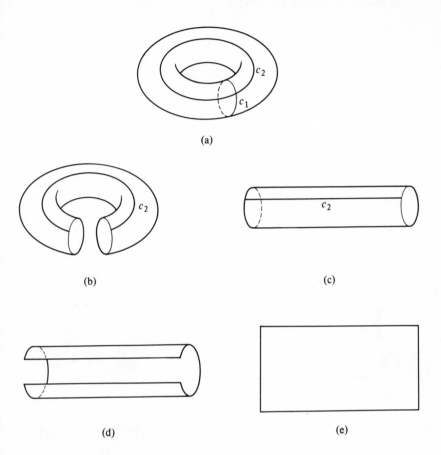

(a)

(b)   (c)

(d)   (e)

Figure 2-7

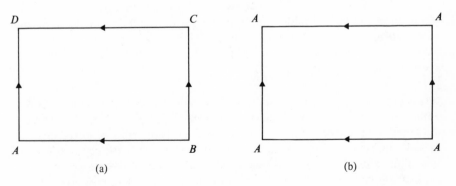

(a)   (b)

Figure 2-8

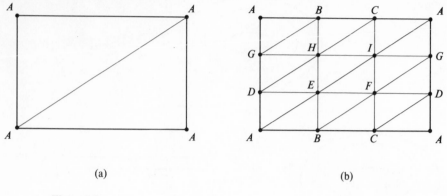

(a)   (b)

Figure 2-9

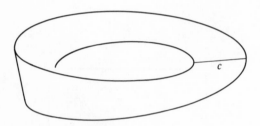

Figure 2-10

to a point. A satisfactory triangulation of the torus is shown in Figure 2-9 (b). Identifying appropriate edges to reconstruct the torus spreads the simplexes in Figure 2-9 (b) out over the surface of the torus thus yielding a triangulation of it.

Exercise 2-10. Triangulate the Möbius strip by cutting along a curve such as *c* in Figure 2-10 and finding an appropriate triangulation of the resulting plane figure.

2-5   Simplicial maps

Let us emphasize once again that, since any topological polyhedron is (by definition) homeomorphic to the polyhedron of some geometric complex and since we are interested only in topological properties of polyhedra, we will almost invariably be able to work exclusively with geometric complexes and their polyhedra. In particular, since one of the major objectives of topology is the analysis of continuous maps, it seems appropriate at this point to observe that the simplicial structure of such a polyhedron determines one particularly natural class of continuous maps, that is, those that are piecewise linear.

Let K and L be two geometric complexes. A map $\varphi:|K| \to |L|$ is a *simplicial map* (relative to the triangulations of $|K|$ and $|L|$ by K and L, respectively) if

(a)   For each vertex a of K, $\varphi(a)$ is a vertex of L.

(b)   For each simplex $(a_0, \ldots, a_p)$ of K, the vertices $\varphi(a_0), \ldots,$ $\varphi(a_p)$, which need not all be distinct, span a simplex of L.

(c) φ is linear on each simplex of K, that is, if $(a_0, \ldots, a_p) \in K$ and if $x = \Sigma_{i=0}^{p}\lambda_i(x)a_i$, where $\Sigma_{i=0}^{p}\lambda_i(x) = 1$ and $\lambda_i(x) > 0$ for each $i = 0, \ldots, p$, then $\varphi(x)$ is given by

(7) $$\varphi(x) = \varphi\left(\sum_{i=0}^{p} \lambda_i(x)a_i\right) = \sum_{i=0}^{p} \lambda_i(x)\varphi(a_i).$$

A map $\varphi_0:K_0 \to L_0$ that satisfies (a) and (b) is called a *vertex map*. Any vertex map extends uniquely to a simplicial map by linearity, that is, by (c). By virtue of the obvious one-to-one correspondence between vertex maps and the simplicial maps they induce it is generally unnecessary to distinguish between them and we shall often use the same symbol for both.

Remark: Whether or not a given map from one polyhedron to another is simplicial depends entirely on the particular triangulations of these polyhedra that are under consideration. In particular, if K_1, K_2, and L are geometric complexes and K_1 and K_2 have the same polyhedron, then a given map φ from this polyhedron to $|L|$ may be simplicial relative to K_1 and not relative to K_2, that is, $\varphi:|K_1| \to |L|$ may be simplicial even though $\varphi:|K_2| \to |L|$ is not, despite the fact that, as point sets, $|K_1| = |K_2|$. Nevertheless, it follows from Theorem 2-5 and equation (7) that any map between two polyhedra that is simplicial relative to some triangulations of the domain and range is necessarily continuous.

It is often the case in mathematics that the particular situation in which we find ourselves suggests a "natural" class of mappings to be studied and used in the analysis of more complex phenomena. Perhaps the most notable example is the use of orthogonal sequences of functions in the study of boundary value problems for ordinary and partial differential equations. In particular, in the study of oscillatory phenomena the "simple cases" are those that arise when all functions involved are linear combinations of trigonometric functions. In order to handle the general situation we then attempt to approximate all relevant functions by elements of this distinguished class and apply results and techniques developed in the special cases. It is important to note that the meaning of "approximate" varies with the situation: Pointwise, uniform, and mean approximation and even more general notions of "summability" all play a role. Our situation here is similar. We have just isolated a "natural" class of continuous maps between polyhedra and, with a little thought, an appropriate notion of "approximation" also emerges:

Let K and L be geometric complexes and $f : |K| \to |L|$ a continuous map. A simplicial map $\varphi : |K| \to |L|$ is called a *simplicial approximation* to f if, for each x in $|K|$, $f(x)$ and $\varphi(x)$ lie in the same closed simplex of L, that is, if

(8) $f(x) \in s \in L$ implies $\varphi(x) \in \bar{s}.$

To obtain another useful formulation of this condition, we introduce several new concepts which turn out to be of extreme importance in themselves. Let K be an arbitrary geometric complex and $s_p \in K$. The *star* of s_p in K, denoted $\mathrm{st}(s_p)$, is the union of all the simplexes in K of which s_p is a face, that is, $\mathrm{st}(s_p) = \cup \{s_q \in K : s_p \leq s_q\}$.

Lemma 2-12. If s_p is any simplex in a geometric complex K, then $\mathrm{st}(s_p)$ is an open subset of $|K|$.

Proof: Let K^* denote the set of all simplexes in K which do not have s_p as a face. Then K^* is a subcomplex of K, so $|K^*|$ is compact and thus closed in $|K|$. But $\mathrm{st}(s_p) = |K| - |K^*|$, so $\mathrm{st}(s_p)$ is open in $|K|$.   Q.E.D.

In particular, this is true for any 0-simplex in K, that is, for each vertex a of K, $\mathrm{st}(a) = \cup \{s_q \in K : a$ is a vertex of $s_q\}$ is open in $|K|$. Note that $\{\mathrm{st}(a) : a \in K_0\}$ is, in fact, an open cover of $|K|$ since, if $x \in |K|$ and s_q is the carrier of x in K, then $x \in \mathrm{st}(a)$ for any vertex a of s_q. Also observe that a is the only vertex of K contained in $\mathrm{st}(a)$ since the only open simplex containing a vertex is the 0-simplex consisting of that vertex alone.

Lemma 2-13. Let $\{a_0, \ldots, a_p\}$ be a set of vertices of a geometric complex K. Then $a_0, \ldots, a_p$ span a simplex of K iff $\cap_{i=0}^{p} \mathrm{st}(a_i) \neq \emptyset$.

Exercise 2-11. Prove Lemma 2-13.   Q.E.D.

Proposition 2-14. Let K and L be geometric complexes and $f : |K| \to |L|$ a continuous map. A simplicial map $\varphi : |K| \to |L|$ is a simplicial approximation to f iff for each $a \in K_0$

(9) $f(\mathrm{st}(a)) \subseteq \mathrm{st}(\varphi(a)).$

Proof: Exercise 2-12. Show that if φ is a simplicial approximation to f, then (9) is satisfied for each $a \in K_0$.

Conversely, suppose (9) holds for each $a \in K_0$. Let $x \in |K|$, and suppose $f(x) \in s \in L$. Let $t = (a_0, \ldots, a_p)$ be the carrier of x in K. For each a_i, $f(x) \in f(t) \subseteq f(\text{st}(a_i)) \subseteq \text{st}(\varphi(a_i))$, so $s \cap \text{st}(\varphi(a_i))$ is nonempty for every i. Since L is a complex and $\text{st}(\varphi(a_i))$ is a union of open simplexes in L, $s \subseteq \text{st}(\varphi(a_i))$ for each i. Thus, each (a_i) is a vertex of s, so $\varphi(a_0), \ldots, \varphi(a_p)$ span a face of s. In terms of barycentric coordinates in t, $x = \Sigma \lambda_i a_i$, so since φ is simplicial $\varphi(x) = \Sigma \lambda_i \varphi(a_i)$. Thus, $\varphi(x)$ is in a face of s and by Proposition 2-8 $\varphi(x) \in \bar{s}$ as required.   Q.E.D.

2-6   Barycentric subdivision

If K and L are arbitrary geometric complexes, then a continuous map $f:|K| \to |L|$ need not have a simplicial approximation $\varphi:|K| \to |L|$. For example, let $K = K(s_1)$, $s_1 = (a_0, a_1)$, where $a_0 = (0, 0)$ and $a_1 = (1, 0)$ in $\mathbf{R}^2$, and let L be the one-dimensional complex in $\mathbf{R}^2$ with vertices $b_0 = (0, 0)$, $b_1 = (0, \frac{1}{2})$ and $b_2 = (0, 1)$ and 1-simplexes (b_0, b_1) and (b_1, b_2). Then the continuous map $f : |K| \to |L|$ defined by $f(x, 0) = (0, x^2)$ for every x in $[0, 1]$ can be pictured graphically as in Figure 2-11 (a). Now suppose $\varphi:|K| \to |L|$ is a simplicial approximation to f. Then $\varphi(a_0) = f(a_0) = b_0$ and $\varphi(a_1) = f(a_1) = b_2$. Since φ is simplicial, we must therefore have $\varphi(x, 0) = (0, x)$ for every x in $[0, 1]$. But then, for any w in $|K|$ between $(\frac{1}{2}, 0)$ and $f^{-1}(b_1) = (\sqrt{2}/2, 0)$, $f(w) \in (b_0, b_1)$ whereas $\varphi(w) \in (b_1, b_2)$, so equation (8) is not satisfied at w. Of course, this fact is scarcely surprising since we are dealing here with two quite distinct levels of structure. Continuity is a purely topological notion, while the existence of simplicial approximations to maps between $|K|$ and $|L|$ depends entirely on the particular triangulations of $|K|$ and $|L|$ with which we happen to be dealing. It is essential to observe that the failure of our map f to have a simplicial approximation is not an intrinsic property of f, but rather is the result of the particular manner in which $|K|$ is triangulated by K. Indeed, suppose we construct a new complex K' by inserting a vertex a_2 at $f^{-1}(b_1)$ and thus "subdividing" K (the 1-simplexes of K' are, of course, (a_0, a_2) and (a_2, a_1)). Then $|K'| = |K|$, so we may regard f as a map from $|K'|$ to $|L|$. Moreover, the simplicial map $\psi : |K'| \to |L|$ induced by the assignments $\psi(a_0) = b_0$, $\psi(a_2) = b_1$, and $\psi(a_1) = b_2$ is easily seen to be a simplicial approximation to f (see Figure 2-11 (b)).

Now, it should be kept in mind that our object is to analyze general, continuous maps between polyhedra in terms of appropriate simplicial maps. This example indicates that the "appropriate" simplicial structure will, in general, depend on the map. This is, of course, perfectly satisfac-

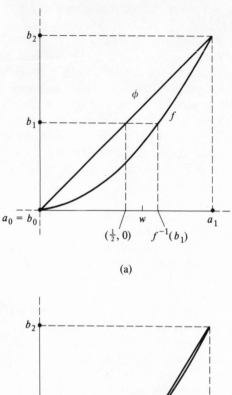

(a)

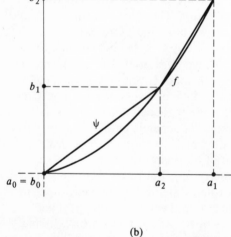

(b)

Figure 2-11

tory from our point of view since our primary concern is with topological questions and the dissection of polyhedra into simplexes is merely a useful technique for their study and thus may be carried out in any convenient way. With this in mind we begin an investigation of a general proce-

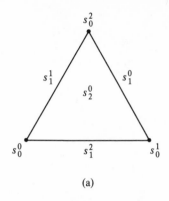

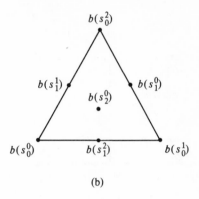

(a)   (b)

Figure 2-12

dure for "subdividing" an arbitrary geometric complex in the hope that any continuous map between polyhedra will have a simplicial approximation relative to a sufficiently "fine" triangulation of its domain.

Let K and L be two geometric complexes. We say that L is a *subdivision* of K if

(T1) $|L| = |K|$.

(T2)   For each simplex s' in L there is a simplex s in K such that
$s' \subseteq s$.

Naturally, there are any number of ways to produce subdivisions of a given complex. What we require is some canonical procedure for generating "finer" subdivisions of arbitrary complexes. One particularly useful technique is the method of *barycentric subdivision*. If $s_p = (a_0, \ldots, a_p)$ is a simplex, the *barycenter* $b(s_p)$ of s_p is the point of s_p all of whose barycentric coordinates are equal. Thus, $b(s_p) = (1/(p + 1))\Sigma_{i=0}^{p}a_i$. The barycentric subdivision K' of a complex K will have as its vertices the barycenters of all the simplexes of K. The higher dimensional simplexes of K' are those that arise naturally, but the precise definition requires some discussion. We motivate the procedure we have in mind by first constructing K' for a standard, two-dimensional complex.

Example 2-1. The complex $K = K(s_2)$ and its barycentric subdivision K'. Let $s_2 = s_2^0 = (s_0^0, s_0^1, s_0^2)$ denote a standard 2-simplex with vertices s_0^0, s_0^1, and s_0^2, and let its faces be denoted as in Figure 2-12 (a). The barycenters $b(s_i^j)$ of all the simplexes s_i^j in $K(s_2^0)$ are shown in Figure 2-12 (b).

The barycenters $b(s_i^j)$ are the 0-simplexes (vertices) of K'. Note that the vertices of K are also vertices of K'. The 1-simplexes of K' will be determined by certain pairs of these vertices. Of course, not all such pairs can determine a 1-simplex; for example, if $b(s_0^2)$ and $b(s_2^0)$ determine a 1-

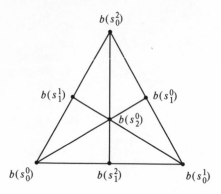

Figure 2-13

simplex, then $b(s_1^1)$ and $b(s_1^0)$ cannot, for 1-simplexes in a complex must either be disjoint or identical. The shorter line segments between vertices shown in Figure 2-13 are the 1-simplexes we propose to include in K'.

Although this choice will certainly appear natural to the reader, it may not be clear precisely how it was made. A careful examination of the 1-simplexes shown in Figure 2-13 will, however, reveal that two barycenters $b(s)$ and $b(t)$ span a 1-simplex in K' iff either $s < t$ or $t < s$; for example, $(b(s_2^0), b(s_0^2)) \in K'$ because $s_0^2 < s_2^0$, but $b(s_1^1)$ and $b(s_1^0)$ do not determine a 1-simplex of K' because s_1^1 is not a face of s_1^0 and s_1^0 is not a face of s_1^1. As to the 2-simplexes of K', the situation is similar. Each such simplex is determined by a triple $(b(r), b(s), b(t))$ of vertices, but not all such triples can span a 2-simplex. Indeed, it is clear that only the smaller triangles in Figure 2-13 can be taken as 2-simplexes in K' since K' is to be a complex. Thus, for example, $(b(s_0^2), b(s_1^1), b(s_2^0))$ is in K', but $b(s_0^2)$, $b(s_1^2)$ and $b(s_0^1)$ will not determine a 2-simplex of K'. Again, the difference between these triples is easily expressed in terms of the defining simplexes. Note that, in the first case $s_0^2 < s_1^1 < s_2^0$, but in the second case the defining simplexes s_0^2, s_1^2, and s_0^1 do not form such an "ascending sequence" in any order. Generalizing these ideas, we arrive at the following definitions.

If K is a geometric complex, then an *ascending sequence* of simplexes in K is a collection $\{s^0, \ldots, s^k\}$ of simplexes in K such that $s^0 < s^1 < \cdots < s^k$. (We use superscripts rather than subscripts since s^i need not have dimension i.) The *barycentric subdivision* K' of K can now be defined as follows: The vertices of K' are the barycenters $b(s)$ for $s \in K$, while the simplexes of K' of dimension greater than zero are defined by those sets of barycenters whose defining simplexes form an ascending sequence (when taken in some appropriate order). We must, of course, prove that K' thus defined is indeed a subdivision of K (it is not even clear that K' is a complex). In our next proposition we make use of Lemmas 2-10 and 2-11 to obtain a procedure for producing a subdivision of $K(s_p)$ from a subdivision of $K_{p-1}(s_p)$ and a point inside s_p thus opening the way for an inductive proof of our major result (Theorem 2-16).

Proposition 2-15. Let $s_p = (a_0, \ldots, a_p)$ be a p-simplex, $v \in s_p$, $K = K(s_p)$ and K_{p-1} the $(p-1)$-skeleton of K. Suppose K_{p-1}^- is a subdivision of K_{p-1}. Let

$$\tilde{K} = K_{p-1}^- \cup \left(\bigcup_{s^- \in K_{p-1}^-} (s^-, v) \right) \cup (v),$$

where $(s^-, v) = (v_0, \ldots, v_r, v)$ if $s^- = (v_0, \ldots, v_r)$. Then $\tilde{K}$ is a subdivision of K.

Proof: Exercise 2-13. Show that the elements of $\tilde{K}$ are open simplexes. Hint: Use both Lemma 2-10 and Lemma 2-11.

Now, it is clear that $\tilde{K}$ satisfies condition (K1). To prove (K2) we must show that distinct open simplexes in $\tilde{K}$ are disjoint. Of course, this is clear for pairs of simplexes in K_{p-1}^- and, moreover, (v) meets no other simplex of $\tilde{K}$. In addition if $s^- \in K_{p-1}^-$, then $s^- \cap (t^-, v) = \emptyset$ for every $t^- \in K_{p-1}^-$ since $(t^-, v) \subseteq s_p$ and $s^- \cap s_p = \emptyset$. Thus, we need only show that if $(s^-, v) \cap (t^-, v) \neq \emptyset$, then $(s^-, v) = (t^-, v)$. Suppose $w \in (s^-, v) \cap (t^-, v)$. Then $w \neq v$. By Lemma 2-10 there is a unique x in $|K_{p-1}| = |K_{p-1}^-|$ such that $w \in [v, x]$. By Lemma 2-11 $|K(s^-, v)| = [s^-, v] = v * |K(s^-)|$, so $w \in (s^-, v)$ implies $x \in s^-$. Similarly, $x \in t^-$ so $s^- \cap t^- \neq \emptyset$, and thus $s^- = t^-$ so $(s^-, v) = (t^-, v)$. Thus, we have shown that $\tilde{K}$ is a complex.

Exercise 2-14. Show that $\tilde{K}$ is a subdivision of K. Hint: To prove that $|\tilde{K}| = |K|$, use Lemmas 2-10 and 2-11.   Q.E.D.

With this we can prove that the barycentric subdivision K' of a complex K is actually a subdivision of K.

Theorem 2-16. Let K be a geometric complex and

$$K' = \{(b(s^0), \ldots, b(s^k)) : s^0, \ldots,$$
$$s^k \in K \quad \text{and} \quad s^0 < \cdots < s^k\}.$$

Then K' is a subdivision of K. Furthermore, for each ascending sequence $s^0 < \cdots < s^k$ of simplexes in K, $(b(s^0), \ldots, b(s^k)) \subseteq s^k$.

Proof: The proof is by induction on the dimension of K. If K has dimension zero, then $K' = K$ so the result is trivial. Now assume the theorem is true for all complexes of dimension at most $n - 1$ and suppose K has dimension n. Thus, the result is true for the $(n - 1)$-skeleton K_{n-1} of K; that is, if $s^0 < s^1 < \cdots < s^k$ is an ascending sequence in K with the dimension of s^k at most $n - 1$, then $(b(s^0), \ldots, b(s^k))$ is an open simplex in the complex K'_{n-1} with $(b(s^0), \ldots, b(s^k)) \subseteq s^k$.

Now suppose $s^0 < \cdots < s^{k-1} < s^k$ is an ascending sequence in K for which the dimension of s^k is n. Since $s^{k-1} < s^k$, the inductive hypothesis implies that $(b(s^0), \ldots, b(s^{k-1}))$ is a simplex contained in s^{k-1}. Since $b(s^k) \in s^k$, Lemma 2-10 implies that $(b(s^k), \dot{s}^k)$ is in general position, so by Lemma 2-11 $\{b(s^0), \ldots, b(s^{k-1}), b(s^k)\}$ is geometrically independent. Thus, $(b(s^0), \ldots, b(s^k))$ is an open simplex. Moreover, $(b(s^0), \ldots, b(s^k))$ is the interior of $[b(s^0), \ldots, b(s^k)]$, which by Lemma 2-11 is $b(s^k) * [b(s^0), \ldots, b(s^{k-1})] \subseteq \bar{s}^k$. Thus, $(b(s^0), \ldots, b(s^k)) \subseteq s^k$ as required.

Thus far we know that, for any ascending sequence $s^0 < \cdots < s^k$ in K, $(b(s^0), \ldots, b(s^k))$ is an open simplex contained in s^k. It remains to be shown that these simplexes together form a complex that is a subdivision of K. First we show that K' is a complex. (K1) is satisfied because K' is defined in terms of ascending sequences. To prove (K2), suppose $s^0 < \cdots < s^k$ and $t^0 < \cdots < t^l$ are ascending sequences in K and that there is some w in $(b(s^0), \ldots, b(s^k))$ $\cap (b(t^0), \ldots, b(t^l))$. Then $w \in s^k \cap t^l$ so, since K is a complex $s^k = t^l$ and, in particular, $b(s^k) = b(t^l)$. Moreover, we have $(b(s^0), \ldots, b(s^{k-1})) \subseteq s^{k-1}$ and $(b(t^0), \ldots, b(t^{l-1})) \subseteq t^{l-1}$ and both s^{k-1} and t^{l-1} are faces of s^k $(= t^l)$. Let L be the complex consisting of the proper faces of s^k. By the induction hypothesis L' exists and is a subdivision of L. Both $(b(s^0), \ldots, b(s^{k-1}))$ and $(b(t^0), \ldots, b(t^{l-1}))$ are in L'. Constructing the subdivision $\tilde{K}$ of $K(s^k)$ from L' and $v = b(s^k) = b(t^l)$ as in Proposition 2-15, we find that both $(b(s^0), \ldots, b(s^{k-1}), b(s^k))$ and $(b(t^0), \ldots, b(t^{l-1}), b(t^l))$ are in $\tilde{K}$. Since they intersect and $\tilde{K}$ is a complex, they are identical.

Exercise 2-15. Complete the proof by showing that $|K'| = |K|$. Hint: $|K'| \subseteq |K|$ is obvious. Also, $|K'_{n-1}| = |K_{n-1}|$ by the induction hypothesis, so $|K'| \supseteq |K_{n-1}|$. Show that $|K'| \supseteq |K| - |K_{n-1}|$ by using Lemma 2-10 and the induction hypothesis again.   Q.E.D.

Remark: If $s^0 < \cdots < s^k$ is an ascending sequence in K, then $b(s^k)$ is called the *leading vertex* of the simplex $(b(s^0), \ldots, b(s^k))$ in K'.

Exercise 2-16. Show that the algebraic dimension of K' is the same as that of K.

The *mesh* $\mu(K)$ of a geometric complex K is defined by $\mu(K) =$ max$\{$diam $s : s \in K\}$. The real advantage to the barycentric method of subdivision lies in the following lemma and the theorem we prove from it.

Lemma 2-17. If the dimension of the geometric complex K is m, then $\mu(K') \le (m/m + 1)\mu(K)$.

Proof: Let s' be an edge (1-simplex) of K' with vertices $b(s_p)$ and $b(s_q)$ and assume that $s_p < s_q$. Then $p < q \le m$. Let $s_p = (a_0, \ldots, a_p)$ and $s_q = (a_0, \ldots, a_p, a_{p+1}, \ldots, a_q)$ and compute

$$b(s_p) - b(s_q) = \frac{1}{p+1} \sum_{i=0}^{p} a_i - \frac{1}{q+1} \sum_{i=0}^{q} a_i$$

$$= \left(\frac{1}{p+1} - \frac{1}{q+1} \right) \sum_{i=0}^{p} a_i - \frac{1}{q+1} \sum_{i=p+1}^{q} a_i$$

$$= \frac{q-p}{q+1} \left(\frac{1}{p+1} \sum_{i=0}^{p} a_i - \frac{1}{q-p} \sum_{i=p+1}^{q} a_i \right).$$

Now, $\dfrac{1}{p+1} \displaystyle\sum_{i=0}^{p} a_i$ and $\dfrac{1}{q-p} \displaystyle\sum_{i=p+1}^{q} a_i$

are both in $\bar{s}_q$ so, since diam $\bar{s}_q =$ diam s_q, their distance apart cannot exceed $\mu(K)$. Thus,

$$\|b(s_p) - b(s_q)\| \le \frac{q-p}{q+1} \mu(K) \le \frac{q}{q+1} \mu(K) \le \frac{m}{m+1} \mu(K).$$

It follows from Lemma 2-9 that $\mu(K')$ is the length of the longest edge in K'. Since we have just shown that the length of any edge in K' is less than or equal to $(m/m + 1)\mu(K)$, the result follows.   Q.E.D.

By repeatedly forming successive barycentric subdivisions, we obtain a sequence $K^{(0)} = K$, $K^{(1)} = K'$, $K^{(2)} = (K^{(1)})' = (K')'$, $\ldots$, $K^{(n)} = (K^{(n-1)})'$, $\ldots$ of subdivisions of K that we now prove become arbitrarily "fine."

Theorem 2-18. Let K be a geometric complex. For any $\epsilon > 0$, there exists an $n \ge 0$ such that $\mu(K^{(n)}) < \epsilon$.

Proof: If the dimension of K is m, then by Lemma 2-17 $\mu(K^{(n)}) \le (m/m + 1) \mu(K^{(n-1)}) \le (m/m + 1)^2 \mu(K^{(n-2)}) \le \cdots \le (m/m + 1)^n \mu(K)$. Since $(m/m + 1) < 1$, $(m/m + 1)^n \to 0$ as $n \to \infty$ so the result follows.   Q.E.D.

2-7   The Simplicial Approximation Theorem

With Theorem 2-18 we can now prove the result that initially prompted our interest in the construction of subdivisions.

Theorem 2-19 (Simplicial Approximation Theorem). Let K and L be geometric complexes and $f : |K| \to |L|$ a continuous map. Then there exists an $n \geq 0$ and a simplicial map $\varphi : |K^{(n)}| \to |L|$ that is a simplicial approximation to $f : |K^{(n)}| = |K| \to |L|$.

Proof: $\{\text{st}(b) : b \in L_0\}$ is an open cover of $|L|$ so $\mathcal{U} = \{f^{-1}(\text{st}(b)) : b \in L_0\}$ is an open cover of $|K|$. Let δ be a Lebesgue number for $\mathcal{U}$. Choose $n \geq 0$ sufficiently large that $\mu(K^{(n)}) < \delta/2$.

Claim: $\mu(K^{(n)}) < \delta/2$ implies that diam $\overline{\text{st}(a)} < \delta$ for each vertex a of $K^{(n)}$.

  To prove the claim, let x and y be two points in $\overline{\text{st}(a)} = \cup\{s_q : a < s_q \in K^{(n)}\} = \cup\{\bar{s}_q : a < s_q \in K^{(n)}\}$. Then x is in a face of some simplex in $K^{(n)}$ of which a is a vertex. $\mu(K^{(n)}) < \delta/2$ implies that $d(x, a) < \delta/2$. Similarly, $d(a, y) < \delta/2$, so $d(x, y) \leq d(x, a) + d(a, y) < \delta$. Since st$(a)$ is compact, diam $\overline{\text{st}(a)} < \delta$. In particular, diam st$(a) < \delta$.

  Thus, for each $a \in K_0^{(n)}$ we may select some vertex $\varphi_0(a) \in L_0$ such that

(10) $\text{st}(a) \subseteq f^{-1}(\text{st}(\varphi_0(a)))$.

Exercise 2-17. Show that the map $\varphi_0 : K_0^{(n)} \to L_0$ thus defined is a vertex map and therefore induces a simplicial map $\varphi : |K^{(n)}| \to |L|$. Hint: Use Lemma 2-13.

  The simplicial map $\varphi : |K^{(n)}| \to |L|$ is a simplicial approximation to f since by statement (10) $f(\text{st}(a)) \subseteq \text{st}(\varphi(a))$ for each $a \in K_0^{(n)}$ and this is condition (9) of Proposition 2-14.   Q.E.D.

  It is natural to seek some quantitative measure of the degree to which the simplicial approximations guaranteed by Theorem 2-19 actually approximate continuous functions. Although there are several viable possibilities, we shall content ourselves with the most obvious and define the *distance* $d(g, h)$ between the continuous maps $g, h : X \to Y$, where X and Y are arbitrary subspaces of some Euclidean spaces, by $d(g, h) = \sup_{x \in X} d(g(x), h(x)) = \sup_{x \in X} \|g(x) - h(x)\|$. In general, $d(g, h)$ may be infinite, for example, if $g : \mathbf{R} \to \mathbf{R}$ is the identity function and $h : \mathbf{R} \to \mathbf{R}$ is identically zero. However, if X is compact, then $d(g, h)$ must be finite since the image of $g - h$ is compact and thus bounded. Moreover, in this case,

$\sup_{x \in X} \|g(x) - h(x)\|$ is actually assumed at some point of X, so $d(g, h)$ $= \max_{x \in X} \|g(x) - h(x)\|$. This is true, in particular, for polyhedra. Now observe that if $f : |K| \to |L|$ is continuous and φ is a simplicial approximation to f, then (8) implies that

(11) $\quad d(f, \varphi) \leq \mu(L)$.

This estimate is generally of little value, however, particularly if L is a "coarse" triangulation of $|L|$ so that $\mu(L)$ is relatively large. Ideally, one would like to be assured of the existence of simplicial approximations to f with $d(f, \varphi)$ arbitrarily small. In the next exercise you will show that this is indeed possible.

Exercise 2-18. Let K and L be geometric complexes and $f : |K| \to |L|$ a continuous map. Show that for each $\epsilon > 0$ there exist integers $n \geq 0$ and $m \geq 0$ and a simplicial map $\varphi : |K^{(n)}| \to |L^{(m)}|$ that is a simplicial approximation to f with $d(f, \varphi) < \epsilon$.

More important for us, however, is the fact that sufficiently "close" continuous maps between polyhedra have common simplicial approximations.

Proposition 2-20. Let K and L be geometric complexes, δ a Lebesgue number for the open cover $\{st(b_j) : b_j \in L_0\}$ of $|L|$, and $f,\ g : |K| \to |L|$ continuous maps with $d(f, g) < \delta/3$. Then there is an $n \geq 0$ and a simplicial map $\varphi : |K^{(n)}| \to |L|$ that is a simplicial approximation to both f and g.

Proof: For each $b_j \in L_0$ set $B_j = \{x \in |L| : \text{dist}(x, |L| - st(b_j)) > \delta/3\}$. The sets B_j are open since the distance function is continuous. Moreover, we claim that the B_j cover $|L|$. To see this, let $x \in |L|$ and denote by V the open $\delta/3$-ball about x. Since diam $\bar{V} < \delta$, $\bar{V} \subseteq st(b_j)$ for some b_j. If y is a point in $|L| - st(b_j)$, then, in particular, $y \notin \bar{V}$ so $d(x, y) > \delta/3$. Since $|L| - st(b_j)$ is compact, $\text{dist}(x, |L| - st(b_j))$ is greater than $\delta/3$ so $x \in B_j$ as required. Observe also that $b_k \in B_j$ iff $k = j$.

Since the B_j cover $|L|$, the sets $f^{-1}(B_j)$ cover $|K|$. Let ϵ be a Lebesgue number for this cover and choose $n \geq 0$ such that $\mu(K^{(n)}) < \epsilon/2$. Just as in the proof of Theorem 2-19, to every vertex a of $K^{(n)}$ there corresponds a vertex $b_j = \varphi_0(a)$ of L with $st(a) \subseteq f^{-1}(B_j)$, and this correspondence defines a vertex map $\varphi_0 : K_0^{(n)} \to L_0$, which induces a simplicial map $\varphi : |K^{(n)}| \to |L|$. By construction φ is a simplicial approximation to f. To see that φ is also a simplicial approximation to g, let $x \in |K^{(n)}|$. Then $x \in st(a)$ for some $a \in K_0^{(n)}$. Let $\varphi(a) = b_j$. By definition of φ, $f(x) \in B_j$ so that, since $d(f, g) < \delta/3$, $\text{dist}(g(x), B_j) < \delta/3$. We thus have

$g(x) \in \text{st}(b_j) = \text{st}(\varphi(a))$ so that $g(\text{st}(a)) \subseteq \text{st}(\varphi(a))$ for each $a \in K_0^{(n)}$. By Proposition 2-14 φ is a simplicial approximation to g.   Q.E.D.

Before leaving the subject of simplicial approximations, a few remarks on uniqueness are in order. Observe first that, since altering the triangulation of a polyhedron changes the very notion of "simplicial," it would be unreasonable to expect any but the most trivial sort of uniqueness result in general. Indeed, the following Lemma is essentially the best result available.

Lemma 2-21. Let K and L be geometric complexes, $f : |K| \to |L|$ a simplicial map, and $\varphi : |K| \to |L|$ a simplicial approximation to f. Then $\varphi = f$.

Proof: Let a be a vertex of K. By condition (9), $f(a) \in f(\text{st}(a)) \subseteq \text{st}(\varphi(a))$. Since f is simplicial, $f(a)$ is a vertex. But $\text{st}(\varphi(a))$ contains only one vertex, namely, $\varphi(a)$. Thus, $f(a) = \varphi(a)$ and f and φ agree on the vertices of K. Since f and φ are simplicial, they agree everywhere.   Q.E.D.

Nevertheless, distinct simplicial approximations to a given continuous map are "close" in a sense made precise by the following definition. Let K and L be geometric complexes, n and m integers with $n \geqslant m \geqslant 0$, and $\varphi : |K^{(n)}| \to |L|$ and $\psi : |K^{(m)}| \to |L|$ simplicial maps. Let $s^{(n)}$ be a simplex in $K^{(n)}$ and $s^{(m)}$ the unique simplex in $K^{(m)}$ containing $s^{(n)}$ (the "carrier" of $s^{(n)}$). We say that φ and ψ are *contiguous* if for every such $s^{(n)}$ there is a simplex $t \in L$ such that $\varphi(s^{(n)}) < t$ and $\psi(s^{(m)}) < t$. In particular, if φ, $\psi : |K| \to |L|$, then φ and ψ are contiguous if for each $s \in K$ there is a $t \in L$ such that $\varphi(s) < t$ and $\psi(s) < t$.

Proposition 2-22. Let K and L be geometric complexes, $f : |K| \to |L|$ a continuous map and $\varphi : |K^{(n)}| \to |L|$ and $\psi : |K^{(m)}| \to |L|$ two simplicial approximations to f. Then φ and ψ are contiguous.

Proof: Assume $n \geqslant m$. Let $s^{(n)}$ be a simplex in $K^{(n)}$ and $s^{(m)}$ its carrier in $K^{(m)}$. Let $x \in s^{(n)}$ and let t be the carrier of $f(x)$ in L. Then $\varphi(x)$ and $\psi(x)$ both lie in $\bar{t}$. Since φ and ψ are simplicial, both of the simplexes $\varphi(s^{(n)})$ and $\psi(s^{(m)})$ lie in $\bar{t}$. By Proposition 2-8 $\varphi(s^{(n)}) < t$ and $\psi(s^{(m)}) < t$.   Q.E.D.

Unfortunately, even when $n = m = 0$ the property of being contiguous is not an equivalence relation since, although it is reflexive and symmetric, it is not transitive.

Exercise 2-19. Construct an example consisting of two geometric complexes K and L and three simplicial maps φ_1, φ_2, $\varphi_3 : |K| \to |L|$ such that φ_1 and φ_2 are contiguous and φ_2 and φ_3 are contiguous, but φ_1 and φ_3 are not contiguous.

To obtain an equivalence relation, we proceed as follows: Let K and L be geometric complexes, n and m integers with $n \geqslant m \geqslant 0$ and $\varphi : |K^{(n)}| \to |L|$, and $\psi : |K^{(m)}| \to |L|$ simplicial maps. We say that φ and ψ are *contiguous equivalent* or in the same *contiguity class* if there is a sequence of integers $n = n_0 \geqslant n_1 \geqslant \cdots \geqslant n_k = m$ and a sequence $\varphi_0, \ldots, \varphi_k$ of simplicial maps with $\varphi_i : |K^{(n_i)}| \to |L|$ for each $i = 0,$ $\ldots, k$ such that $\varphi_0 = \varphi$, $\varphi_k = \psi$ and φ_i and φ_{i+1} are contiguous for each $i = 0, \ldots, k - 1$.

2-8 Sperner's lemma

All of this material on simplicial approximations will be put to good use in Chapter 4 where we study the simplicial homology theory of polyhedra. For the present, however, we shall focus our attention on one particularly important class of simplicial maps.

Lemma 2-23. Let K be a geometric complex. Any map $\omega : K_0' \to K_0$ defined by

$$\omega(b(s_p)) = \text{one of the vertices of } s_p$$

is a vertex map, called a *Sperner map*.

Exercise 2-20. Prove Lemma 2-23.   Q.E.D.

For each $n \geqslant 0$ we can define generalized Sperner maps $\omega^{(n)} : K_0^{(n)} \to K_0$ by

$$\omega^0 = \mathrm{id}_{K_0}$$
$$\omega^{(n)} = \omega \circ \omega^{(n-1)}.$$

Any such map is called a *standard map*.

Remark: This last definition is somewhat abbreviated, but easily interpreted. For example, each of the maps $\omega^{(2)} : K_0^{(2)} \to K_0$ is defined as follows: Each vertex of $K^{(2)}$ is a barycenter $b(s')$ of some simplex s' in K'. A Sperner map $\omega : K_0^{(2)} \to K_0'$ carries $b(s')$ onto one of the vertices of s', which in turn is a barycenter $b(s)$ of some $s \in K$. Thus, another Sperner map $\omega : K_0' \to K_0$ (note the use of the same symbol for another map) carries $b(s)$ onto some vertex of s. The composition of these two Sperner maps is a typical $\omega^{(2)}$.

Any standard map is a vertex map. (Verify this!) By virtue of the one-to-one correspondence between vertex maps and simplicial maps we shall not distinguish between a standard map $\omega^{(n)} : K_0^{(n)} \to K_0$ and the simplicial map $\omega^{(n)} : |K^{(n)}| \to |K|$ induced by it.

Lemma 2-24 (Sperner's Lemma). Let $s_p = (a_0, \ldots, a_p)$ be a p-simplex and $K = K(s_p)$. If $\omega^{(n)} : K_0^{(n)} \to K_0$ is a standard map, then there is a unique simplex s in $K^{(n)}$ that $\omega^{(n)} : |K^{(n)}| \to |K|$ maps onto s_p.

Proof: By induction it suffices to prove the result for a Sperner map $\omega : K_0' \to K_0$. The proof in this case is by induction on the dimension of s_p. For $p = 0$, $K' = K$ and $\omega = \mathrm{id}_{K_0}$, so the result is trivial. Now assume the result for simplexes of dimension less than p. By relabeling the vertices if necessary, we may assume that $\omega(b(s_p)) = a_p$. Now let $L = K(s_{p-1}) = K(a_0, \ldots, a_{p-1})$ be the triangulation of $\bar{s}_{p-1} = [a_0, \ldots, a_{p-1}]$ by its faces. Then, by restriction, ω induces a Sperner map $\eta : L_0' \to L_0$. By the induction hypothesis there is a unique $s_{p-1}' \in L'$ such that $\eta(s_{p-1}') = s_{p-1}$. This s_{p-1}' is a face of the simplex s_p' in K' that has as its vertices those of s_{p-1}' together with $b(s_p)$. Since $\omega(b(s_p)) = a_p$, $\omega(s_p') = s_p$ and all that remains is to show that $s = s_p'$ is unique.

To this end suppose t_p' is a simplex in K' with $\omega(t_p') = s_p$. We show that $t_p' = s_p'$. Note that $b(s_p)$ must be among the vertices of t_p' for otherwise $\omega(t_p')$ would be contained in $\dot{s}_p$. Let t_{p-1}' be the face of t_p' opposite $b(s_p)$. Note that a_p cannot be a vertex of t_p' for, if it were, then $a_p = \omega(a_p) = \omega(b(s_p))$ and thus, since $a_p \neq b(s_p)$, $\omega(t_p')$ would be a simplex of dimension at most $p - 1$, contradicting $\omega(t_p') = s_p$. Thus, t_{p-1}' lies in s_{p-1} so $t_{p-1}' \in L'$. Therefore, $s_{p-1} = \omega(t_{p-1}') = \eta(t_{p-1}')$. But, by the induction hypothesis, s_{p-1}' is the unique simplex in L' with $\eta(s_{p-1}') = s_{p-1}$, so $t_{p-1}' = s_{p-1}'$ and thus $t_p' = s_p'$.   Q.E.D.

Now we collect a few consequences of this extremely powerful lemma that will facilitate the discussion of the rather deep applications we propose to consider in the next two sections.

Lemma 2-25. Let $s_p = (a_0, \ldots, a_p)$ be a p-simplex and let $\mathscr{B} = \{B_i : i = 0, \ldots, p\}$ be a family of closed sets in $|K(s_p)| = \bar{s}_p$ such that $(a_{i_0}, \ldots, a_{i_k}) \subseteq B_{i_0} \cup \cdots \cup B_{i_k}$ for each subset $\{i_0, \ldots, i_k\}$ of $\{0, \ldots, p\}$. Then $\cap_{i=0}^p B_i \neq \varnothing$.

Proof: Suppose $\cap_{i=0}^p B_i = \varnothing$. Then $\{\bar{s}_p - B_i : i = 0, \ldots, p\}$ is a finite open cover of the compact space $\bar{s}_p$. Let δ be a Lebesgue number for this cover. Thus, any subset of $\bar{s}_p$ of diameter less than δ misses some B_i. Find an $n \geq 0$ such that $\mu(K^{(n)}(s_p)) < \delta$. We define a standard map $\omega^{(n)} : K_0^{(n)}(s_p) \to K_0(s_p)$ by selecting for any $b \in K_0^{(n)}(s_p)$ a B_i containing b and setting $\omega^{(n)}(b) = a_i$. (Verify that this does indeed define a standard map!) By Sperner's Lemma there is a unique simplex $s_p' = (b_0, \ldots, b_p)$ in $K^{(n)}(s_p)$ such that $\omega^{(n)}(s_p') = s_p$. After a possible rearrangement of the indices we thus have $b_i \in B_i$ for each $i = 0, \ldots, p$. Since $b_i \in \bar{s}_p'$

for each i, we find that $\bar{s}'_p \cap B_i \neq \varnothing$ for each i. But then we must have diam $\bar{s}'_p \geq \delta$, which contradicts $\mu(K^{(n)}(s_p)) < \delta$.   Q.E.D.

Lemma 2-26. Let $s_p = (a_0, \ldots, a_p)$ be a p-simplex and $\mathscr{B} = \{B_i : i = 0, \ldots, p\}$ a closed cover of $\bar{s}_p$ such that $a_i \in B_i$ and $(a_0, \ldots, \hat{a}_i, \ldots, a_p) \cap B_i = \varnothing$ for each $i = 0, \ldots, p$. Then $\cap_{i=0}^p B_i \neq \varnothing$.

Exercise 2-21. Prove Lemma 2-26. Hint: Apply Lemma 2-25.   Q.E.D.

2-9   The Brouwer Fixed Point Theorem

We are now in a position to apply the techniques we have developed thus far to obtain proofs of several of the most important theorems in topology.

Theorem 2-27 (Brouwer Fixed Point Theorem). For any integer $n \geq 1$, the closed ball B^n has the fixed point property.

Proof: Since B^n is homeomorphic to $\bar{s}_n$ (Section 2-4), we need only show that every continuous map $f : \bar{s}_n \to \bar{s}_n$ has a fixed point (Proposition 1-32). Let $s_n = (a_0, \ldots, a_n)$, and let x be any point in $\bar{s}_n$. Then $x = \Sigma_{i=0}^n \lambda_i(x) a_i$, where $\Sigma_{i=0}^n \lambda_i(x) = 1$ and $0 \leq \lambda_i(x) \leq 1$ for each $i = 0, \ldots, n$. Since $f(x) \in \bar{s}_n$ as well, we may also write $f(x) = \Sigma_{i=0}^n \mu_i(x) a_i$, where $\Sigma_{i=0}^n \mu_i(x) = 1$ and $0 \leq \mu_i(x) \leq 1$ for each $i = 0, \ldots, n$. Define sets $B_i = \{x \in \bar{s}_n : \mu_i(x) \leq \lambda_i(x)\}$ for each i and observe that each B_i is closed by continuity of f and Theorem 2-5. We show that $\mathscr{B} = \{B_i : i = 0, \ldots, n\}$ satisfies the conditions of Lemma 2-25. Suppose $\{i_0, \ldots, i_k\} \subseteq \{0, \ldots, p\}$ and $x \in (a_{i_0}, \ldots, a_{i_k})$. Then $x = \Sigma_{j=0}^k \lambda_{i_j}(x) a_{i_j}$, where $\Sigma_{j=0}^k \lambda_{i_j}(x) = 1$ and $\lambda_{i_j}(x) > 0$ for each j. Now, $f(x) = \Sigma_{i=0}^n \mu_i(x) a_i$, where $\Sigma_{i=0}^n \mu_i(x) = 1$ and $\mu_i(x) \geq 0$ for each i. It follows that for at least one $j \leq k$, say r, we must have $\mu_{i_r}(x) \leq \lambda_{i_r}(x)$ so $x \in B_{i_r}$. Therefore, $(a_{i_0}, \ldots, a_{i_k}) \subseteq B_{i_0} \cup \cdots \cup B_{i_k}$. By Lemma 2-25 there is some y in $\cap_{i=0}^n B_i$. Then $y \in \bar{s}_n$ and $\mu_i(y) \leq \lambda_i(y)$ for each $i = 0, \ldots, n$. But $\Sigma_{i=0}^n \lambda_i(y) = \Sigma_{i=0}^n \mu_i(y) = 1$, so we must have $\mu_i(y) = \lambda_i(y)$ for each i and thus $f(y) = y$ as required.   Q.E.D.

This renowned theorem of Brouwer has an important reformulation generally known as the No Retraction Theorem. First observe that it follows easily from Theorem 2-27 that the sphere S^{n-1} is not a retract of B^n since a retract of a space with the fixed point property also has the fixed point property (Proposition 1-50), but the "antipodal map" $f : S^{n-1} \to S^{n-1}$ defined by $f(x) = -x$ is continuous and has no fixed points. We show next that, conversely, the Brouwer Fixed Point Theorem

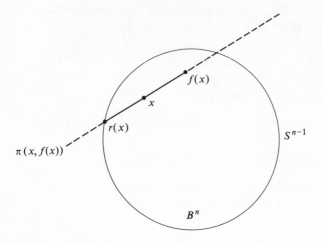

Figure 2-14

follows from the fact that S^{n-1} is not a retract of B^n, so that these two statements are logically equivalent.

Theorem 2-28. The following (valid) statements are equivalent:
(a) S^{n-1} is not a retract of B^n.
(b) B^n has the fixed point property.

Proof: All that remains is to show that (a) implies (b). For this we assume the existence of a continuous map $f: B^n \to B^n$ without fixed points and construct from it a retraction $r: B^n \to S^{n-1}$. Now, since $f(x) \neq x$ for each $x \in B^n$, the points x and $f(x)$ determine a unique line $\pi(x, f(x))$. We intend to let $r(x)$ be the point where the ray from $f(x)$ through x intersects $S^{n-1} = $ bdy B^n; see Figure 2-14.

More precisely, we observe that any y in $\pi(x, f(x))$ can be written in the form

(12) $y = tx + (1 - t)f(x)$

for $t \in \mathbf{R}$. We claim that there is exactly one such y with $\|y\| = 1$ and $t \geqslant 1$. To see this, take the dot product of both sides of equation (12) with itself and use $\|y\| = 1$ to obtain

(13) $t^2 x \cdot x + 2(t - t^2)x \cdot f(x) + (1 - t)^2 f(x) \cdot f(x) = 1$

or, equivalently,

(14) $t^2\|x - f(x)\|^2 + 2tf(x) \cdot (x - f(x)) + (\|f(x)\|^2 - 1) = 0.$

Now, (14) is a quadratic in t and an elementary calculation reveals that it has precisely one root greater than or equal to 1. We define $r(x)$ to be the unique y determined by (12) with this root as the value of t. Thus, we have a map $r : B^n \to S^{n-1}$. Note that if $x \in S^{n-1}$, then $x \cdot x = 1$, so it follows that $t = 1$ is a solution of (13) and thus $r(x) = x$. Consequently, $r|S^{n-1} = \mathrm{id}_{S^{n-1}}$ and all that remains is to prove the continuity of r. Applying the quadratic formula to (14) and choosing the root greater than or equal to 1, we obtain an expression for t in terms of continuous functions of x. Substituting this expression for t into $r(x) = tx + (1 - t)f(x)$, we find that $r(x)$ is continuous.   Q.E.D.

Exercise 2-22. Show that the open ball $\{x \in \mathbf{R}^n : \|x\| < 1\}$ does not have the fixed point property.

Exercise 2-23. Find two examples of compact spaces that do not have the fixed point property.

2-10 Topological dimension of compact subsets of $\mathbf{R}^n$

Next we bring our combinatorial techniques to bear on the problem of determining the topological dimension of certain compact subsets of $\mathbf{R}^n$. Recall that a compact subspace X of $\mathbf{R}^n$ is said to have topological dimension k, written $\dim X = k$, iff k is the least nonnegative integer for which, for each $\epsilon > 0$, there is a finite, closed ϵ-cover of X of order $k + 1$ (see Section 1-8). Our major result is the following theorem.

Theorem 2-29. Let $s_p = (a_0, \ldots, a_p)$ be a p-simplex. Then $\dim |K(s_p)| = \dim \bar{s}_p = p$.

Proof: We must prove
(a) $\dim \bar{s}_p \leqslant p$, that is, for every $\epsilon > 0$ there is a finite, closed ϵ-cover of $\bar{s}_p$ of order $k + 1$.
(b) $\dim \bar{s}_p \geqslant p$, that is, p is the least integer for which such a cover exists for every $\epsilon > 0$; in other words there exists an $\epsilon > 0$ such that all finite, closed ϵ-covers of $\bar{s}_p$ have order at least $p + 1$.
For (a) we suppose $\epsilon > 0$ is given and choose $n \geqslant 0$ such that $\mu(K^{(n)}(s_p)) < \epsilon/2$. Let $c_0, \ldots, c_k$ be the vertices of $K^{(n-1)}(s_p)$ and $\mathcal{B} = \{\,\mathrm{st}(c_i) : i = 0, \ldots, k\}$, where $\mathrm{st}(c_i)$ means the star of c_i in $K^{(n)}(s_p)$. We claim that $\mathcal{B}$ is the required cover of $\bar{s}_p$. Each $\overline{\mathrm{st}(c_i)}$ is certainly closed and $\mu(K^{(n)}(s_p)) < \epsilon/2$ implies diam $\overline{\mathrm{st}(c_i)} < \epsilon$ (see the Claim in the proof of Theorem 2-19).

Exercise 2-24. Prove that $\mathcal{B}$ is a cover of $\bar{s}_p$. Hint: Show that every simplex s in $K^{(n)}(s_p)$ is contained in some $\text{st}(c_i)$.

Thus, $\mathcal{B}$ is a finite, closed ϵ-cover of $\bar{s}_p$ and we need only show that ord $\mathcal{B} = p + 1$. First let $(c_{i_0}, \ldots, c_{i_p})$ be any p-simplex in $K^{(n-1)}(s_p)$ and let b be its barycenter. Then b is in every $\text{st}(c_{i_j})$ so $\cap_{j=0}^{p} \text{st}(c_{i_j}) \neq \varnothing$ and there exist $p + 1$ elements of $\mathcal{B}$ with nonempty intersection, that is, ord $\mathcal{B} \geqslant p + 1$. In order to prove that no more than $p + 1$ elements of $\mathcal{B}$ can intersect, we first observe the following: Each $\overline{\text{st}(c_i)}$ $= \cup \{s_q : c_i < s_q \in K^{(n)}(s_p)\} = \cup \{\bar{s}_q : c_i < s_q \in K^{(n)}(s_p)\}$ is the polyhedron of a subcomplex of $K^{(n)}(s_p)$. Let b be a vertex of $\overline{\text{st}(c_i)}$ and s the simplex in $K^{(n-1)}(s_p)$ whose barycenter is b. We claim that c_i is a vertex of s. To see this, let t be a simplex in $\overline{\text{st}(c_i)}$ with b and c_i among its vertices. Then t is determined by an ascending sequence of simplexes in $K^{(n-1)}(s_p)$ each of which has c_i as a vertex since t is in $\overline{\text{st}(c_i)}$. But s must occur in this sequence since $b(s) = b$. Thus, c_i is a vertex of s. With this we now suppose that there exist $p + 2$ points $c_{i_0}, \ldots, c_{i_{p+1}}$ such that $\cap_{j=0}^{p+1} \text{st}(c_{i_j}) \neq \varnothing$. This intersection is the polyhedron of a subcomplex of $K^{(n)}(s_p)$ and thus contains a vertex b. By what we have just proved the carrier of b in $K^{(n-1)}(s_p)$ has each c_{i_j}, $j = 0, \ldots, p + 1$, as a vertex and this is impossible since its algebraic dimension is p (see Exercise 2-16). Thus, ord $\mathcal{B} = p + 1$ and the proof of (a) is complete.

To prove (b), we introduce some notation. For each $i = 0, \ldots, p$, we denote by s^i the $(p - 1)$-face of $s_p = (a_0, \ldots, a_p)$ opposite a_i, that is, $s^i = (a_0, \ldots, \hat{a}_i, \ldots, a_p)$. Then $\cap_{i=0}^{p} \bar{s}^i = \varnothing$ since any point in this intersection would have all of its barycentric coordinates zero. Thus, $\cup_{i=0}^{p}(\bar{s}_p - \bar{s}^i) = \bar{s}_p$. Let ϵ be a Lebesgue number for the open cover $\{\bar{s}_p - \bar{s}^i : i = 0, \ldots, p\}$ of $\bar{s}_p$. Suppose $\mathcal{B} = \{B_i : i = 0, \ldots, n\}$ is a closed ϵ-cover of $\bar{s}_p$. We claim that ord $\mathcal{B} \geqslant p + 1$.

Note that diam $B_i < \epsilon$ implies that B_i is contained in some $\bar{s}_p - \bar{s}^j$, that is, each B_i misses some $\bar{s}^j$. Suppose $a_k \in B_i$. Since each $\bar{s}^j$, $j \neq k$, contains the vertex a_k, B_i cannot miss any such $\bar{s}^j$. Therefore, B_i must miss $\bar{s}^k$, that is, $a_k \in B_i$ implies $B_i \cap \bar{s}^k = \varnothing$. In particular, since $\bar{s}^k$ contains all a_j for $j \neq k$, B_i can contain no other vertex of s_p. It follows that $n \geqslant p$, and we may define a map g of $\{0, \ldots, n\}$ onto $\{0, \ldots, p\}$ as follows: For each i in $\{0, \ldots, n\}$ select $g(i)$ in $\{0, \ldots, p\}$ such that $B_i \cap \bar{s}^{g(i)} = \varnothing$. Now, for each $k = 0, \ldots, p$ define $A_k = \cup \{B_i : g(i) = k\}$. Then every B_i is contained in precisely one A_k so, in particular, $\cup_{k=0}^{p} A_k = \bar{s}_p$. Moreover, $a_k \in A_k$ and $A_k \cap \bar{s}^k = \varnothing$ for every $k = 0, \ldots, p$; by Lemma 2-26 there is an x in $\cap_{k=0}^{p} A_k$. Now, for each $k = 0, \ldots, p$ there is an i_k with $x \in B_{i_k}$. Since each B_i is contained in precisely one A_k, $i_k \neq i_j$ if $k \neq j$. Thus, x is contained in $p + 1$ elements $B_{i_0}, \ldots, B_{i_p}$ of $\mathcal{B}$ so ord $\mathcal{B} \geqslant p + 1$ as required.   Q.E.D.

Remark: It follows immediately from Theorem 2-29 that, for any geometric complex K, dim $|K|$ and the algebraic dimension of K coincide. (Verify this!) Therefore, we are justified in using the same symbol "dim" for both, that is, we write dim $|K|$ = dim K. Moreover, if X is a topological polyhedron and (K, h) is a triangulation of X, then dim X = dim $|K|$ = dim K by Theorem 1-54.

Every compact subset A of $\mathbf{R}^n$ is contained in some n-simplex s_n. Since any finite, closed ϵ-cover of $\bar{s}_n$ restricts to a finite, closed ϵ-cover of A whose order does not exceed the order of the original cover we obtain the following corollary:

Corollary 2-30. A compact subspace of $\mathbf{R}^n$ has topological dimension at most n.

With this we can distinguish topologically between the various Euclidean spaces:

Theorem 2-31. $\mathbf{R}^n$ is homeomorphic to $\mathbf{R}^m$ iff $n = m$.

Proof: We need only prove that $n \neq m$ implies $\mathbf{R}^n \not\cong \mathbf{R}^m$. Suppose to the contrary that $m < n$ and h is a homeomorphism of $\mathbf{R}^n$ onto $\mathbf{R}^m$. Let s_n be an n-simplex in $\mathbf{R}^n$. Then dim $\bar{s}_n = n$ and, since $h(\bar{s}_n) \cong \bar{s}_n$, dim $h(\bar{s}_n) = n$ as well. However, $h(\bar{s}_n)$ is a compact subspace of $\mathbf{R}^m$ and $m < n$ so we have contradicted Corollary 2-30. Q.E.D.

Remark: The same proof shows that if $n \neq m$, then an open subset of $\mathbf{R}^n$ cannot be homeomorphic to an open subset of $\mathbf{R}^m$.

Supplementary exercises

2-25 Let $s = \{a_0, \ldots, a_p\}$ be a set of $p + 1$ points in $\mathbf{R}^n$, where $p \leq n$. For each $i = 0, \ldots, p$ let $a_i = (a_i^1, \ldots, a_i^n)$. Show that s is geometrically independent iff the matrix $N(s)$ defined by Equation (15) has rank $p + 1$:

$$(15) \quad N(s) = \begin{bmatrix} a_0^1 & a_0^2 & \cdots & a_0^n & 1 \\ a_1^1 & a_1^2 & \cdots & a_1^n & 1 \\ \cdot & \cdot & & \cdot & \cdot \\ \cdot & \cdot & & \cdot & \cdot \\ \cdot & \cdot & & \cdot & \cdot \\ a_p^1 & a_p^2 & \cdots & a_p^n & 1 \end{bmatrix}$$

2-26 Let $\{b_0, \ldots, b_p\}$ be an arbitrary set of $p + 1$ points in $\mathbf{R}^n$, where $p \leq n$. Show that for any $\epsilon > 0$ we can select points $a_0, \ldots, a_p$ in $\mathbf{R}^n$ with $\|b_i - a_i\| < \epsilon$ for each $i = 0, \ldots, p$ such that $\{a_0, \ldots, a_p\}$ is geometrically independent.

2-27 Let $\{b_0, \ldots, b_p\}$ be a geometrically independent subset of $\mathbf{R}^n$. Show that there exists an $\epsilon > 0$ such that any set $\{a_0, \ldots, a_p\}$ of points in $\mathbf{R}^n$ with $\|b_i - a_i\| < \epsilon$ for each $i = 0, \ldots, p$ is geometrically independent.

2-28 Show that if s_p and t_p are two p-simplexes, then $s_p \cong t_p$ and $\bar{s}_p \cong \bar{t}_p$.

2-29 Let v_1 and v_2 be points in $\mathbf{R}^n$ and B a subset of $\mathbf{R}^n$ with (v_1, B) and (v_2, B) in general position. Show that the cones $v_1 * B$ and $v_2 * B$ are homeomorphic.

2-30 Find a triangulation of the torus similar to that in Figure 2-9 (b) with 7 vertices, 21 edges, and 14 triangles.

2-31 Find a triangulation of the solid cube I^3.

2-32 Show that any triangulation of the n-dimensional cube I^n must contain at least $n!$ n-simplexes.

2-33 Find a triangulation of the annulus $\{(x, y) \in \mathbf{R}^2 : \frac{1}{2} \leq \sqrt{x^2 + y^2} \leq 1\}$.

2-34 Show that if K is a subcomplex of L, then K' is a subcomplex of L'.

2-35 Let s_p and t_q be simplexes. Construct a triangulation of the product $\bar{s}_p \times \bar{t}_q$.

2-36 Prove that a finite product of topological polyhedra is again a topological polyhedron.

Chapter 3
Homotopy theory and the fundamental group

3-1   Introduction

The subject to which we now turn is of central importance in modern topology and is, along with most of the basic concepts of our subject, a contribution of Poincaré. In order to trace the origin of the ideas we shall develop in this chapter, we ask the reader to recall a number of basic facts from real analysis. Specifically, let us consider a connected open subspace X of $\mathbf{R}^2$, two points P_0 and P_1 of X, and two real-valued functions $f(x, y)$ and $g(x, y)$ that are continuous and have continuous partial derivatives in X. Recall that a line integral

(1) $$\int_c f(x, y)\, dx + g(x, y)\, dy$$

of the differential form $f\, dx + g\, dy$ over a piecewise smooth curve c in X from P_0 to P_1 will depend, in general, not only on the endpoints P_0 and P_1 but also on the choice of the path c between them. For certain very important differential forms, however, this is not the case. Let us say that an integral of the form (1) is "independent of path in X" if, for any two points P_0 and P_1 in X, the value of the integral is the same for all paths c in X from P_0 to P_1. From elementary properties of line integrals it follows that the statement "the integral is independent of path in X" is equivalent to the statement "the integral around any closed path in X is zero." The basic result of the subject is that the integral (1) is independent of path in X iff the differential form $f\, dx + g\, dy$ is "exact" in X, that is, iff there exists a function $u(x, y)$ that is continuous and has continuous partial derivatives in X such that $u_x(x, y) = f(x, y)$ and $u_y(x, y) = g(x, y)$ in X. Observe that if $f\, dx + g\, dy$ is exact in X, then $f_y(x, y) = g_x(x, y)$ in X. We might ask if it is true, conversely, that $f_y = g_x$ implies $f\, dx + g\, dy$ is exact. The reader will no doubt recall that this question must be answered in the negative. Consider, for example, the differential form $f\, dx + g\, dy$, where $f(x, y) = -y/(x^2 + y^2)$ and $g(x, y) = x/(x^2 + y^2)$ in the annulus

$$X = \{(x, y) \in \mathbf{R}^2 : \tfrac{1}{2} < (x^2 + y^2)^{1/2} < \tfrac{3}{2}\}$$

shown in Figure 3-1.

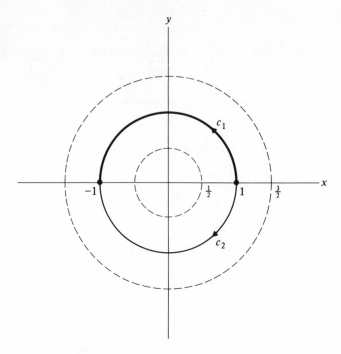

Figure 3-1

A trivial computation shows that $f_y = g_x$ on X. We claim, however, that the integrals of this differential form over the two paths c_1 and c_2 from $P_0 = (1, 0)$ to $P_1 = (-1, 0)$ shown in Figure 3-1 are not equal. Equivalently, we may show that the integral around the closed curve $c = c_1 - c_2$ given parametrically by $x = \cos t$, $y = \sin t$, $0 \le t \le 2\pi$, is not zero. But this is immediate since, on c, $x^2 + y^2 = \cos^2 t + \sin^2 t = 1$, $dx = -\sin t\, dt$ and $dy = \cos t\, dt$ so that

$$\int_c \frac{-y}{(x^2 + y^2)}\, dx + \frac{x}{(x^2 + y^2)}\, dy = \int_0^{2\pi} dt = 2\pi.$$

Thus, the given differential form is not exact on X. The relevant observation from our point of view is that the failure of these two integrals to be equal is entirely because of the existence of the "hole" in X that lies between the two curves, that is, to the fact that c_1 cannot be "continuously deformed" into c_2 without leaving X or, equivalently, that the closed curve c cannot be "shrunk to a point" in X. To see this, recall the following theorem.

Green's Theorem. Let $f(x, y)$ and $g(x, y)$ be continuous and have continuous partial derivatives f_y and g_x in some connected open subspace X of $\mathbf{R}^2$, and let c be a piecewise smooth simple closed curve in X whose interior is also contained in X. Then

$$\int_c f\, dx + g\, dy = \iint_R (g_x - f_y)\, dx\, dy,$$

where R is the closed region bounded by c and c is oriented in such a way that its interior remains to the left as c is traversed.

Now, if c' is any simple closed curve in the annulus X that does not enclose the "hole" $\{(x, y) \in \mathbf{R}^2 : (x^2 + y^2)^{1/2} \leqslant \frac{1}{2}\}$ in X, then the interior of c' is contained in X, Green's Theorem applies, and we conclude from the equality of the partial derivatives f_y and g_x that

$$\int_{c'} \frac{-y}{(x^2 + y^2)}\, dx + \frac{x}{(x^2 + y^2)}\, dy = 0.$$

More generally, we come to the conclusion that if $f\, dx + g\, dy$ is a differential form with $f_y = g_x$ in a connected open subspace X of $\mathbf{R}^2$, then $\int_c f\, dx + g\, dy = 0$ for any closed curve c in X that can be "shrunk to a point in X"; alternatively, for such a differential form $\int_{c_1} f\, dx + g\, dy = \int_{c_2} f\, dx + g\, dy$ whenever c_1 and c_2 are two curves in X with the same initial and terminal points and with the property that either can be "continuously deformed" into the other without leaving X. In particular, if X is "simply connected," that is, if every closed curve in X can be shrunk to a point in X, then a differential form $f\, dx + g\, dy$ is exact in X iff $f_y = g_x$.

The importance of this notion of "simple connectivity" extends far beyond elementary integration theory. Recall how we showed in Chapter 1 that $\mathbf{R}$ and $\mathbf{R}^2$ are not homeomorphic by observing that the removal of a point from $\mathbf{R}$ disconnects $\mathbf{R}$, but removing a point from $\mathbf{R}^2$ does not disconnect $\mathbf{R}^2$ (Section 1-5). This same argument fails to distinguish $\mathbf{R}^2$ and $\mathbf{R}^3$, however, since deleting a point from either $\mathbf{R}^2$ or $\mathbf{R}^3$ fails to disconnect that space. On the other hand, removing a point from $\mathbf{R}^2$ yields a space that is not simply connected, while $\mathbf{R}^3$ minus a point is simply connected. Since it is at least intuitively clear that simple connectivity is a topological property, we have found a simple characteristic of $\mathbf{R}^2$ that distinguishes it topologically from $\mathbf{R}^3$.

Our task in this chapter is to formalize the notion of a "continuous deformation" of one map into another (a path in X is, after all, a map into X) by way of the homotopy concept and thereby make possible precise definitions of simple connectivity and its generalizations in the form of the "fundamental group." Our purpose is to study the local connectivity properties of an arbitrary space X by examining whatever "obstructions" might exist in X to prevent certain closed curves from being shrunk to a point.

3-2   The homotopy relation, nullhomotopic maps, and contractible spaces

Let X and Y be two spaces and $I = [0, 1]$ the closed unit interval. Two maps $f_0, f_1 : X \to Y$ are said to be *homotopic*, written $f_0 \simeq f_1$, if there exists a continuous map $F : X \times I \to Y$ with $F(x, 0) = f_0(x)$ and $F(x, 1) = f_1(x)$ for all $x \in X$. The map F is called a *homotopy* from f_0 to f_1 and we write $F : f_0 \simeq f_1$. Defining $f_t : X \to Y$ for each $t \in I$ by $f_t(x) = F(x, t)$ for all $x \in X$, the homotopy F is seen to represent a family $\{f_t : t \in I\}$ of maps from X into Y, varying continuously with t from f_0 to f_1. Intuitively, if we regard t as a time parameter, F represents a continuous deformation of the map f_0 into the map f_1 with f_t being the stage of the deformation at the instant t.

Theorem 3-1. For any two spaces X and Y the relation $\simeq$ is an equivalence relation on the set $C(X, Y)$ of continuous maps from X to Y.

Proof: The relation is reflexive since, if $f : X \to Y$ is any map, the map $F : X \times I \to Y$ defined by $F(x, t) = f(x)$ for each $x \in X$ and $t \in I$ is a homotopy from f to f. To see that $\simeq$ is symmetric, consider two maps f, $g : X \to Y$ and suppose $F : f \simeq g$. Define $G : X \times I \to Y$ by $G(x, t) = F(x, 1 - t)$. Then G is continuous, $G(x, 0) = F(x, 1) = g(x)$ and $G(x, 1) = F(x, 0) = f(x)$ and thus $G : g \simeq f$. To prove transitivity, let f, g, $h : X \to Y$ and suppose $F : f \simeq g$ and $G : g \simeq h$. Define $H : X \times I \to Y$ by

$$H(x, t) = F(x, 2t), \qquad 0 \leqslant t \leqslant \tfrac{1}{2}$$
$$= G(x, 2t - 1), \qquad \tfrac{1}{2} \leqslant t \leqslant 1.$$

Then $H : f \simeq h$, so the proof is complete.   Q.E.D.

The relation $\simeq$ thus partitions $C(X, Y)$ into equivalence classes that we shall refer to as *homotopy classes*.

Lemma 3-2. Let f_0 and f_1 be homotopic maps of X into Y and g_0 and g_1 homotopic maps of Y into Z. Then $g_0 \circ f_0$ and $g_1 \circ f_1$ are homotopic maps of X into Z.

Proof: Let $F : f_0 \simeq f_1$ and define $G : X \times I \to Z$ by $G = g_0 \circ F$. Then $G : g_0 \circ f_0 \simeq g_0 \circ f_1$. Let $H : g_0 \simeq g_1$ and define $\tilde{f}_1 : X \times I \to Y \times I$ by

$\tilde{f}_1(x, t) = (f_1(x), t)$. Now define $K : X \times I \to Z$ by $K = H \circ \tilde{f}_1$. Then $K : g_0 \circ f_1 \simeq g_1 \circ f_1$. Thus, $g_0 \circ f_0 \simeq g_0 \circ f_1$ and $g_0 \circ f_1 \simeq g_1 \circ f_1$, so by transitivity of the homotopy relation $g_0 \circ f_0 \simeq g_1 \circ f_1$.   Q.E.D.

A map $f : x \to Y$ is called a *homotopy equivalence* with *homotopy inverse* $g : Y \to X$ if $g \circ f \simeq \mathrm{id}_X$ and $f \circ g \simeq \mathrm{id}_Y$. If there exists a homotopy equivalence $f : X \to Y$, we say that X and Y are *homotopically equivalent* or of the same *homotopy type* and we write $X \simeq Y$ or, more explicitly, $f : X \simeq Y$. In the next exercise you will justify the use of the word "equivalent."

Exercise 3-1. Show that, for any spaces X, Y, and Z.
(a) $X \simeq X$.
(b) $X \simeq Y$ implies $Y \simeq X$.
(c) $X \simeq Y$ and $Y \simeq Z$ implies $X \simeq Z$.
Hint: For (c) prove that if $f_1 : X \to Y$ and $f_2 : Y \to Z$ are homotopy equivalences with homotopy inverses $g_1 : Y \to X$ and $g_2 : Z \to Y$, then $f_2 \circ f_1 : X \to Z$ is a homotopy equivalence with homotopy inverse $g_1 \circ g_2 : Z \to X$.

Observe that homeomorphic spaces, that is, spaces of the same "topological type," are of the same homotopy type, but we shall see shortly that the converse is false.

Let Y be a convex subset of $\mathbf{R}^n$ and X an arbitrary space. We claim that any continuous map $f : X \to Y$ is homotopic to a constant map of X into Y. Indeed, if y_0 is any point in Y, then the map $F : X \times I \to Y$ defined by $F(x, t) = ty_0 + (1 - t)f(x)$ is a homotopy from f to the constant map $F(x, 1) = y_0$ (for each $x \in X$). A continuous map that is homotopic to some constant map is said to be *nullhomotopic* or *inessential*. Intuitively, a nullhomotopic map $f : X \to Y$ is one whose range can be "shrunk to a point" in Y. Thus, any map into a convex subset of $\mathbf{R}^n$ is nullhomotopic. In particular, the identity map on such a set is nullhomotopic. A space Y for which id_Y is nullhomotopic is said to be *contractible*. It follows from Lemma 3-2 that contractibility is a topological property (verify this!), so that a contractible space certainly need not be convex. Nevertheless, we claim that any continuous map $f : X \to Y$ with X arbitrary and Y contractible is nullhomotopic: If $F : Y \times I \to Y$ is a homotopy from id_Y to the constant map $F(y, 1) = y_0 \in Y$, then $G : X \times I \to Y$ defined by $G(x, t) = F(f(x), t)$ is a homotopy from f to $G(x, 1) = F(f(x), 1) = y_0$.

Exercise 3-2. Show that, moreover, any continuous map $f : X \to Y$ with X contractible and Y arbitrary is nullhomotopic.

Proposition 3-3. A space X is contractible iff for any space T any two continuous maps $f, g : T \to X$ are homotopic.

Proof: Sufficiency is clear. To prove necessity, suppose X is contractible, let T be an arbitrary space, and $f, g : T \to X$ two continuous maps. Now, $\mathrm{id}_X \simeq c$, where c is some constant map. Thus, $f = \mathrm{id}_X \circ f \simeq c \circ f$ and $g = \mathrm{id}_X \circ g \simeq c \circ g$ by Lemma 3-2. But $c \circ f = c \circ g$, so $f \simeq g$ by transitivity of $\simeq$.   Q.E.D.

Remark: It is not true that if X is contractible and T is arbitrary, then any two maps $f, g : X \to T$ are homotopic, despite the fact that both f and g are nullhomotopic. Nullhomotopic maps need not be homotopic! Indeed, we have the following result.

Exercise 3-3. Show that two constant maps $f, g : X \to Y$, where $f(x) = y_0$ and $g(x) = y_1$ for every $x \in X$, are homotopic iff y_0 and y_1 lie in the same path component of Y.

Proposition 3-4. A space X is contractible iff X has the same homotopy type as a point.

Proof: Suppose first that X is contractible. Then there is a constant map $f : X \to X$, $f(x) = x_0$ for each $x \in X$ such that $\mathrm{id}_X \simeq f$. Regard f as a map of X onto $\{x_0\}$ and define $g : \{x_0\} \to X$ by $g(x_0) = x_0$. Then $g \circ f = f \simeq \mathrm{id}_X$ and $f \circ g = \mathrm{id}_{\{x_0\}}$, so $f : X \simeq \{x_0\}$. Now let $Y = \{y_0\}$ and suppose $f : X \to Y$ is a homotopy equivalence with homotopy inverse $g : Y \to X$. Let $x_0 = g(y_0)$ and $c : X \to X$ the constant map $c(x) = x_0$ for each $x \in X$. Then $c = g \circ f \simeq \mathrm{id}_X$, so X is contractible.   Q.E.D.

In particular, $\mathbf{R}^n$ and each of its convex subsets is homotopically equivalent to a point. It follows that homotopy type is a "coarser" classification of spaces than topological type. Nevertheless, we shall find that in some sense it is the more natural classification since many of our most important constructions, for example, the fundamental group and the simplicial homology groups, are invariants of homotopy type; see Sections 3-5 and 4-12.

3-3   Maps of spheres

Observe that a simplex, being convex, is contractible. A polyhedron, however, need not be contractible, for example, the identity map on $S^0 = \{-1, 1\}$ is not nullhomotopic. Our first major result of this section is that no sphere S^n is contractible. To prove this, we require the next lemma.

Lemma 3-5. Let Y be an arbitrary space and $n \geqslant 0$. Then a map $f : S^n \to Y$ is nullhomotopic iff f has a continuous extension $F : B^{n+1} \to Y$.

Proof: Suppose first that $F : B^{n+1} \to Y$ is a continuous extension of f. Then, for each $x \in S^n$ and $t \in I$, $(1 - t)x \in B^{n+1}$ and the map $H : S^n \times I \to Y$ defined by

(2) $H(x, t) = F((1 - t)x)$

is a homotopy from f to the constant map $H(x, 1) = F(0)$ for each $x \in X$. Conversely, suppose $H : S^n \times I \to Y$ is a homotopy from f to a constant map. Observe that each point of B^{n+1} admits a representation of the form $(1 - t)x$ for some $x \in S^n$ and $t \in I$ and that, moreover, these representations are unique for all points other than the origin (for which $t = 1$ and x is arbitrary). But $H(x, 1)$ is constant, so equation (2) uniquely defines a function $F : B^{n+1} \to Y$ that extends f since $H(x, 0) = f(x)$.   Q.E.D.

Remark: The importance of the homotopy concept in topology is largely due to the relationship between homotopy problems and extension problems, our first indication of which is Lemma 3-5. Observe that the question of whether or not two maps $f, g : X \to Y$ are homotopic is itself essentially an extension problem. Indeed, if we define a map H from the closed subset $(X \times \{0\}) \cup (X \times \{1\})$ of $X \times I$ into Y by $H(x, 0) = f(x)$ and $H(x, 1) = g(x)$ for all $x \in X$, then $f \simeq g$ iff H has a continuous extension to all of $X \times I$. In some cases the existence of such extensions (homotopies) can be inferred from the character of the space Y, for example, if $Y = \mathbf{R}^n$, we can appeal to Corollary 1-46, but, in general, the situation is rarely so simple.

Now consider the map $\mathrm{id}_{S^n} : S^n \to S^n$. Since S^n is not a retract of B^{n+1} by Theorem 2-28, id_{S^n} does not admit a continuous extension to B^{n+1}, so by Lemma 3-5 id_{S^n} is not nullhomotopic, that is, S^n is not contractible. Having shown that the No Retraction Theorem implies that the sphere is not contractible, we leave it to the reader to show that one can also proceed in the opposite direction to obtain the following extension of Theorem 2-28.

Theorem 3-6. The following (valid) statements are equivalent:
(a) S^{n-1} is not a retract of B^n.
(b) B^n has the fixed point property.
(c) S^{n-1} is not contractible.

Exercise 3-4. Complete the proof of Theorem 3-6 by showing that (c) implies (a).

Consider next an arbitrary space X and two maps f, $g : X \to S^n$. Since S^n is not contractible, Proposition 3-3 implies that f and g need not be homotopic. Suppose, however, that f and g have the property that $f(x)$ and $g(x)$ are never antipodal points of S^n, that is, that $f(x) \neq - g(x)$ for each $x \in X$. Then the expression $tg(x) + (1 - t)f(x)$, $t \in I$ and $x \in X$, is never zero so

$$H(x, t) = \frac{tg(x) + (1 - t)f(x)}{\|tg(x) + (1 - t)f(x)\|}$$

is a homotopy between f and g and we have proved the following proposition.

Proposition 3-7. If f and g are two maps of a space X into S^n for which $f(x) \neq - g(x)$ for each $x \in X$, then f and g are homotopic.

Proposition 3-8. If X is an arbitrary space and $f : X \to S^n$ is a nonsurjective map, then f is nullhomotopic.

Exercise 3-5. Prove Proposition 3-8. Hint: Apply Proposition 3-7.

One of the major problems of modern topology is the enumeration of the homotopy classes of maps from S^m to S^n. The case in which $m > n > 1$ is extraordinarily complicated, and even the few results which are known are quite beyond our capabilities here. We shall have somewhat more to say about the case $m = n$ in Section 4-15, but for the present we consider only maps $S^m \to S^n$, where $m < n$. Our object is to use Proposition 3-8 to show that any such map is nullhomotopic and thus, since S^n is pathwise connected (see Exercise 3-15), that there is just one homotopy class. Observe, however, that a continuous map $S^m \to S^n$ may be surjective even if $m < n$. (This follows immediately from the Hahn–Mazurkiewicz Theorem and the fact that S^1 can be mapped onto I by projecting onto $[- 1, 1]$ and composing with a homeomorphism of $[- 1, 1]$ onto I.) Thus, Proposition 3-8 cannot be applied directly. Yet, we will be able to show that any map $S^m \to S^n$ with $m < n$ is homotopic is to a nonsurjective map and is thus nullhomotopic. To do this, however, we must pause to examine more carefully some homotopy properties of maps between polyhedra of geometric complexes. The results we obtain will be crucial in Chapter 4 as well.

Let X be an arbitrary space and L a geometric complex. Two maps f_0, $f_1 : X \to |L|$ are said to be *L-approximate* if for each $x \in X$ there is a simplex $s(x)$ in L such that $f_0(x)$ and $f_1(x)$ both lie in $\overline{s(x)}$. For example, if $X = |K|$ for some geometric complex K, $f : |K| \to |L|$ is a continuous map and $\varphi : |K^{(n)}| \to |L|$ is a simplicial approximation to f, then f and φ are L-approximate. In addition, if $\varphi : |K^{(n)}| \to |L|$ and $\psi : |K^{(m)}| \to |L|$ are contiguous simplicial maps, then φ and ψ are L-approximate.

Lemma 3-9. Let X be an arbitrary space, L a geometric complex and f_0, $f_1 : X \to |L|$ two L-approximate maps. Then f_0 and f_1 are homotopic.

Proof: Suppose $|L| \subseteq \mathbf{R}^n$. Define $F : X \times I \to \mathbf{R}^n$ by $F(x, t) = t f_1(x) + (1 - t)f_0(x)$ for each $t \in I$ and $x \in X$. Then F is continuous, $F(x, 0) = f_0(x)$, and $F(x, 1) = f_1(x)$. For each $x \in X$, $f_0(x)$ and $f_1(x)$ lie in a common closed simplex $s(x)$ of L. Since $\overline{s(x)}$ is convex, $t f_1(x) + (1 - t)f_0(x)$ is in $\overline{s(x)}$ for each $t \in I$. Thus, $F(x, t) \in |L|$ for each $x \in X$ and $t \in I$, so $F : X \times I \to |L|$ is the required homotopy.   Q.E.D.

Proposition 3-10. Let K and L be geometric complexes, m and n nonnegative integers, and $\varphi : |K^{(n)}| \to |L|$ and $\psi : |K^{(m)}| \to |L|$ contiguous equivalent simplicial maps. Then $\varphi \simeq \psi$.

Proof: For contiguous simplicial maps the result follows immediately from Lemma 3-9, so the Proposition itself follows from the transitivity of the homotopy relation.   Q.E.D.

Proposition 3-11. Let K and L be geometric complexes, $f : |K| \to |L|$ a continuous map, n a nonnegative integer and $\varphi : |K^{(n)}| \to |L|$ a simplicial approximation to f. Then $\varphi \simeq f$.

Proposition 3-11, whose proof is immediate from Lemma 3-9, is the key to our investigation of the homotopy classes of maps $S^m \to S^n$, where $m < n$. Before returning to this, however, we record one more result on maps between polyhedra that we shall require in Chapter 4.

Proposition 3-12. Let K and L be geometric complexes and f, $g : |K| \to |L|$ homotopic maps. Then there is an integer $n \geqslant 0$ and simplicial maps φ, $\psi : |K^{(n)}| \to |L|$ in the same contiguity class such that φ is a simplicial approximation to f and ψ is a simplicial approximation to g.

Proof: Let $F : |K| \times I \to |L|$ be a homotopy from f to g. Since $|K| \times I$ is compact, the function F is uniformly continuous (see Exercise 1-39). Thus, if δ is a Lebesgue number for the open cover $\{\mathrm{st}(b_j): b_j \in L_0\}$ of $|L|$, then there exists an $\epsilon > 0$ such that $|t - t'| < \epsilon$ implies $\|F(x, t) - F(x, t')\| < \delta/3$ for all $x \in |K|$. Choose a partition $0 = t_0 < t_1 < \cdots < t_{k-1} < t_k = 1$ of I such that $|t_{i+1} - t_i| < \epsilon$ for each $i = 0, \ldots, k - 1$. Define functions $f_i : |K| \to |L|$ by $f_i(x) = F(x, t_i)$ for $i = 0, \ldots, k - 1$. Then $d(f_{i+1}, f_i) < \delta/3$ for each i. By Proposition 2-20 f_{i+1} and f_i have a common simplicial approximation $\psi_i : |K^{(n_i)}| \to |L|$. Let $n = \max\{n_i : i = 0, \ldots, k - 1\}$. Then each ψ_i induces a simplicial map $\varphi_i : |K^{(n)}| \to |L|$, which is a simplicial approximation to both f_{i+1} and f_i. Since φ_i and φ_{i+1} are both simplicial approximations to f_{i+1}, they are contiguous by

Proposition 2-22. Thus, φ_0 and φ_k are in the same contiguity class. Set $\varphi = \varphi_0$ and $\psi = \varphi_k$. By construction φ is a simplicial approximation to f_0 and ψ is a simplicial approximation to f_k. But $f_0(x) = F(x, 0) = f(x)$ and $f_k(x) = F(x, 1) = g(x)$ for each $x \in |K|$, so the proof is complete. Q.E.D.

We now prove the major result of this section.

Theorem 3-13. Any continuous map of S^m into S^n, where $m < n$, is null-homotopic.

Proof: Let s_{m+1} be an $(m + 1)$-simplex and s_{n+1} an $(n + 1)$-simplex. Since S^m is homeomorphic to $|K_m(s_{m+1})|$ and S^n is homeomorphic to $|K_n(s_{n+1})|$, it suffices to prove that an arbitrary continuous map $f : |K_m(s_{m+1})| \rightarrow |K_n(s_{n+1})|$ is nullhomotopic. Let $\varphi : |K_m^{(l)}(s_{m+1})| \rightarrow |K_n(s_{n+1})|$ be a simplicial approximation to f. By Proposition 3-11 $\varphi \simeq f$ so we need only show that φ is nullhomotopic. But since φ cannot raise the dimension of any simplex in $K_m^{(l)}(s_{m+1})$, $\varphi(|K_m^{(l)}(s_{m+1})|)$ is contained in the union of those simplexes of $K_n(s_{n+1})$ of dimension at most m. Since $m < n$, φ is not surjective, so by Proposition 3-8 it is nullhomotopic. Q.E.D.

3-4   The fundamental group

We now propose to return to the subject of "paths" – which, after all, initially prompted our study of homotopy theory. However, the definitions we have presented thus far are not quite appropriate to the study of paths since a path α has "endpoints," and, in general, we are interested only in deformations of φ that leave these endpoints fixed (see Section 3-1). We therefore begin by carrying out the minor modifications required of our definitions.

A *topological pair* is an ordered pair (X, A) consisting of a space X and a subset A of X. A *(pair) mapping* $f : (X, A) \rightarrow (Y, B)$ of the topological pairs (X, A) and (Y, B) is a continuous map $f : X \rightarrow Y$ for which $f(A) \subseteq B$. Two pair mappings f, $g : (X, A) \rightarrow (Y, B)$ are said to be *homotopic (relative to A)* if there is a continuous map $F : X \times I \rightarrow Y$ with $F(x, 0) = f(x)$, $F(x, 1) = g(x)$ for all $x \in X$ and $F(a, t) = f(a) = g(a)$ for all $a \in A$ and $t \in I$; in this case we write $f \simeq g$ rel A or $f \simeq g[A]$. We call F a *relative homotopy* from f to g. If, as usual, we define $f_t : X \rightarrow Y$ by $f_t(x) = F(x, t)$ for each $t \in I$, we find that, for every t, $f_t|A = f|A = g|A$. Intuitively, then, a relative homotopy is a homotopy in the usual sense, but with the additional property that the set $f(A) = g(A)$ remains fixed throughout the deformation. The relative homotopy relation is an equivalence relation on the set of all pair maps $(X, A) \rightarrow (Y, B)$. A pair map $f : (X, A) \rightarrow (Y, B)$ is a *homotopy equivalence* with *homotopy inverse*

$g:(Y, B) \rightarrow (X, A)$ if $g \circ f \simeq \mathrm{id}_X$ rel A and $f \circ g \simeq \mathrm{id}_Y$ rel B; in this case we say that the pairs (X, A) and (Y, B) are *homotopically equivalent*. Now, if $\alpha, \beta : I \rightarrow X$ are two paths in X with the same initial and terminal points, say, $\alpha(0) = \beta(0) = x_0$ and $\alpha(1) = \beta(1) = x_1$, then α and β may be considered pair maps of $(I, \{0, 1\})$ to $(X, \{x_0, x_1\})$. We say that α and β are *(path) homotopic*, written $\alpha \overset{p}{\simeq} \beta$, if $\alpha \simeq \beta$ rel $\{0, 1\}$.

With these preliminary definitions out of the way we now consider an arbitrary space X and observe that there is a natural "multiplication" defined for certain pairs of paths in X. Specifically, if α is a path from x_0 to x_1 and β is a path from x_1 to x_2, then we may define the *product* $\alpha\beta$ of α and β to be the path from x_0 to x_2 given by

$$\begin{aligned}
\alpha\beta(t) &= \alpha(2t), & 0 \leqslant t \leqslant \tfrac{1}{2} \\
&= \beta(2t - 1), & \tfrac{1}{2} \leqslant t \leqslant 1.
\end{aligned}$$

Intuitively, $\alpha\beta$ traverses first the path α from x_0 to x_1, but in half the time required by α, and then the path β from x_1 to x_2, again in half the time required by β. We also define the *inverse* of any path α from x_0 to x_1 to be the path α^{-1} from x_1 to x_0 given by

$$\alpha^{-1}(t) = \alpha(1 - t)$$

for all $t \in I$. These definitions suggest the possibility of associating with any space X a group whose elements are the paths in X and whose binary operation is the multiplication just defined. Unfortunately, this product is not defined for all pairs of paths since, in order for $\alpha\beta$ to be defined, it is necessary that the terminal point of α and the initial point of β coincide. We can overcome this obstacle rather easily, however, by restricting attention to paths that begin and end at some fixed point x_0 in X, although we then obtain a different group for each "base point" $x_0 \in X$. (This situation is analogous to that in Section 3-1, where we found that all the results from integration theory of interest to us could be phrased in terms of "loops" rather than paths between distinct points.) Even for this restricted class of paths, however, the requirements for a group operation are not satisfied by path multiplication. For example, if α, β, and γ are paths in X all of which begin and end at x_0, then the products $(\alpha\beta)\gamma$ and $\alpha(\beta\gamma)$ are both defined, but, in general, they are not equal. (Write down their definitions explicitly.) We will be able to show, however, that $(\alpha\beta)\gamma$ and $\alpha(\beta\gamma)$ are path homotopic so that our program can be carried out by considering homotopy equivalence classes of paths rather than the paths themselves. We begin with two simple lemmas on path homotopy.

Lemma 3-14. Suppose $\alpha_0 \overset{p}{\simeq} \alpha_1$ and $\beta_0 \overset{p}{\simeq} \beta_1$ are paths in a space X for which $\alpha_0\beta_0$ is defined. Then $\alpha_1\beta_1$ is defined and $\alpha_0\beta_0 \overset{p}{\simeq} \alpha_1\beta_1$.

Exercise 3-6. Prove Lemma 3-14.   Q.E.D.

Lemma 3-15. Let α_0 and α_1 be paths in a space X with $\alpha_0 \overset{\text{p}}{\simeq} \alpha_1$. Then $\alpha_0^{-1} \overset{\text{p}}{\simeq} \alpha_1^{-1}$.

Exercise 3-7. Prove Lemma 3-15.   Q.E.D.

Now, if we denote by $[\alpha]$ the $\overset{\text{p}}{\simeq}$ equivalence class of the path α, then, since path homotopic paths have the same endpoints, the initial and terminal points of $[\alpha]$ are well-defined. Moreover, by Lemmas 3-14 and 3-15 we may define the product and inverses of these equivalence classes in the obvious way, that is, $[\alpha][\beta] = [\alpha\beta]$ whenever $\alpha\beta$ is defined and $[\alpha]^{-1} = [\alpha^{-1}]$ for all α. Now, in the construction of the group we have in mind all that remains is to restrict attention to a class of paths on which the product is defined for every pair.

A path $\alpha : I \to X$ in a space X for which $\alpha(0) = \alpha(1) = x_0 \in X$ is called a *loop* in X *based at* x_0. Observe that the path product of any two loops based at the same point is defined. Denote by $\pi_1(X, x_0)$ the set of all path homotopy equivalence classes of loops in X based at x_0. Then the inverse $[\alpha]^{-1}$ of any $[\alpha]$ in $\pi_1(X, x_0)$ is defined as is the product $[\alpha][\beta]$ of any two elements $[\alpha]$ and $[\beta]$ of $\pi_1(X, x_0)$. Our major result is the next theorem.

Theorem 3-16. Let X be an arbitrary space and x_0 a point in X. Let $\pi_1(X, x_0)$ denote the set of all path homotopy equivalence classes of loops in X based at x_0 and define, for all $[\alpha]$, $[\beta] \in \pi_1(X, x_0)$, the product $[\alpha][\beta]$ by $[\alpha][\beta] = [\alpha\beta]$. With this operation, $\pi_1(X, x_0)$ is a group. If $e : I \to X$ is the constant loop $e(t) = x_0$ for every $t \in I$, then $[e]$ is the identity element of $\pi_1(X, x_0)$. For any $[\alpha] \in \pi_1(X, x_0)$ the inverse $[\alpha]^{-1}$ of $[\alpha]$ in $\pi_1(X, x_0)$ is defined by $[\alpha]^{-1} = [\alpha^{-1}]$.

Proof: We first check associativity. For this it suffices to show that, for all loops α, β, and γ at x_0, $\alpha(\beta\gamma) \overset{\text{p}}{\simeq} (\alpha\beta)\gamma$. Now, in terms of its action on I, $(\alpha\beta)\gamma$ is accomplished by completing the action of α on $[0, \frac{1}{4}]$, the action of β on $[\frac{1}{4}, \frac{1}{2}]$, and the action of γ on $[\frac{1}{2}, 1]$, that is,

$$
\begin{aligned}
(\alpha\beta)\gamma(t_1) &= \alpha(4t_1), & 0 \leq t_1 \leq \tfrac{1}{4} \\
&= \beta(4t_1 - 1), & \tfrac{1}{4} \leq t_1 \leq \tfrac{1}{2} \\
&= \gamma(2t_1 - 1), & \tfrac{1}{2} \leq t_1 \leq 1
\end{aligned}
$$

which we represent by the bottom row in Figure 3-2 (a). Similarly,

$$
\begin{aligned}
\alpha(\beta\gamma)(t_1) &= \alpha(2t_1), & 0 \leq t_1 \leq \tfrac{1}{2} \\
&= \beta(4t_1 - 2), & \tfrac{1}{2} \leq t_1 \leq \tfrac{3}{4} \\
&= \gamma(4t_1 - 3), & \tfrac{3}{4} \leq t_1 \leq 1
\end{aligned}
$$

which is represented by the top row in Figure 3-2 (a).

A homotopy from $(\alpha\beta)\gamma$ to $\alpha(\beta\gamma)$ can then be constructed by allowing

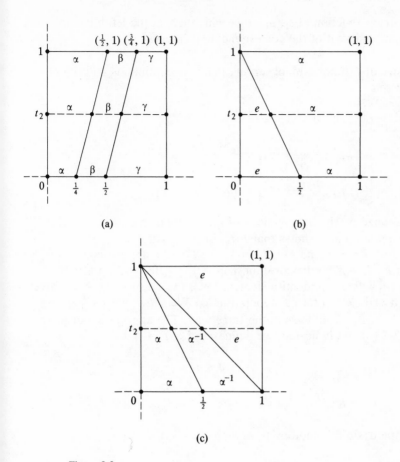

Figure 3-2

the actions of α, β, and γ to be divided at time t_2 as shown on the center row in Figure 3-2 (a). Specifically, we define $F : I \times I \to X$ by

$$F(t_1, t_2) = \alpha \left(\frac{4t_1}{1 + t_2} \right), \qquad 0 \le t_1 \le \frac{t_2 + 1}{4}$$

$$= \beta(4t_1 - 1 - t_2), \qquad \frac{t_2 + 1}{4} \le t_1 \le \frac{t_2 + 2}{4}$$

$$= \gamma \left(1 - \frac{4(1 - t_1)}{2 - t_2} \right), \qquad \frac{t_2 + 2}{4} \le t_1 \le 1.$$

Remark: The t_1-intervals are obtained by simply writing down the equations of the lines in Figure 3-2 (a), that is, $t_2 = 4t_1 - 1$ and $t_2 = 4t_1 - 2$, and solving for t_1. The arguments of α, β, and γ are then obtained by seeking a linear function of t_1, $at_1 + b$, which is 0 at the left endpoint and 1 at the right endpoint of the corresponding t_1-interval.

Continuity of F follows by observing that it is continuous on each of the closed sets

$$A_1 = \left\{ (t_1, t_2) : 0 \leqslant t_1 \leqslant \frac{t_2 + 1}{4} \right\}$$

$$A_2 = \left\{ (t_1, t_2) : \frac{t_2 + 1}{4} \leqslant t_1 \leqslant \frac{t_2 + 2}{4} \right\}$$

$$A_3 = \left\{ (t_1, t_2) : \frac{t_2 + 2}{4} \leqslant t_1 \leqslant 1 \right\}$$

and well-defined on the intersections of these sets (see Lemma 1-22). A trivial computation also shows that $F(t_1, 0) = (\alpha\beta)\gamma(t_1)$, $F(t_1, 1) = \alpha(\beta\gamma)(t_1)$, $F(0, t_2) = x_0$ and $F(1, t_2) = x_0$, so F is the desired homotopy.

Now let $e : I \to X$ be the constant map $e(t) = x_0$ for every $t \in I$. To show that $[e]$ acts as an identity element for $\pi_1(X, x_0)$, it suffices to prove that $e\alpha \overset{\text{p}}{\simeq} \alpha$ and $\alpha e \overset{\text{p}}{\simeq} \alpha$ for each loop α at x_0. We show that $e\alpha \overset{\text{p}}{\simeq} \alpha$ and note that $\alpha e \overset{\text{p}}{\simeq} \alpha$ follows in the same way. The appropriate diagram is Figure 3-2 (b). As indicated, we define $F : I \times I \to X$ by

$$F(t_1, t_2) = x_0, \qquad\qquad 0 \leqslant t_1 \leqslant \frac{1 - t_2}{2}$$

$$= \alpha\left(\frac{2t_1 - 1 + t_2}{1 + t_2} \right), \qquad \frac{1 - t_2}{2} \leqslant t_1 \leqslant 1$$

to obtain the desired homotopy from $e\alpha$ to α.

Exercise 3-8. Complete the proof by showing that $\alpha\alpha^{-1} \overset{\text{p}}{\simeq} e$ for each loop α at x_0. Hint: The appropriate diagram is Figure 3-2 (c).   Q.E.D.

Remark: Although we restricted our attention to loops in the proof of Theorem 3-16, it should be observed that the same arguments show that $(\alpha\beta)\gamma \overset{\text{p}}{\simeq} \alpha(\beta\gamma)$, $e\alpha \overset{\text{p}}{\simeq} \alpha$, and $\alpha\alpha^{-1} \overset{\text{p}}{\simeq} e$ whenever these products are defined.

The group $\pi_1(X, x_0)$ is called the *fundamental group of X at x_0*. From the structure of $\pi_1(X, x_0)$ we can obtain a great deal of information about the local character of the space X at x_0. Suppose, for example, that $\pi_1(X, x_0)$ is the trivial group. Then there is only one homotopy class of loops at x_0, that is, the class containing the trivial loop e. Thus, every loop at x_0 is homotopic to e. Intuitively, this means that near x_0 the space X has no features, for example, "holes," which prevent a loop at x_0 from being "shrunk" to the point x_0. The technique we see emerging here of obtaining topological information by affecting a transition to an algebraic setting characterizes the subject of "algebraic topology" with which we shall become acquainted in this and the next chapter.

If x_0 and x_1 are distinct points in a space X, then, in general, there need not be any relationship between $\pi_1(X, x_0)$ and $\pi_1(X, x_1)$. However, if X is pathwise connected, then all such groups are isomorphic. More generally, we have the following theorem.

Theorem 3-17. Let x_0 and x_1 be points in an arbitrary space X. If $\gamma : I \to X$ is a path in X with $\gamma(0) = x_0$ and $\gamma(1) = x_1$, then the map $\gamma_\# : \pi_1(X, x_0) \to \pi_1(X, x_1)$ defined by $\gamma_\#([\alpha]) = [\gamma^{-1}\alpha\gamma]$ is an isomorphism.

Proof: For each loop α at x_0 in X, the product $\gamma^{-1}\alpha\gamma$ is clearly a loop at x_1. Moreover, by Lemma 3-14 if $\alpha \overset{p}{=} \beta$, then $\gamma^{-1}\alpha\gamma \overset{p}{=} \gamma^{-1}\beta\gamma$, so $\gamma_\#$ is well defined. $\gamma_\#$ is a homomorphism since $\gamma_\#([\alpha])\,\gamma_\#([\beta]) = [\gamma^{-1}\alpha\gamma][\gamma^{-1}\beta\gamma] = [\gamma^{-1}\alpha\gamma\gamma^{-1}\beta\gamma] = [\gamma^{-1}\alpha][\gamma\gamma^{-1}][\beta\gamma] = [\gamma^{-1}\alpha][e_{x_0}][\beta\gamma] = [\gamma^{-1}\alpha][\beta\gamma] = [\gamma^{-1}\alpha\beta\gamma] = \gamma_\#([\alpha\beta])$. Finally, $\gamma_\#$ is an isomorphism since it has an inverse, namely, $\gamma_\#^{-1} = (\gamma^{-1})_\#$.   Q.E.D.

Corollary 3-18. If X is pathwise connected, then $\pi_1(X, x_0) \approx \pi_1(X, x_1)$ for any two points x_0 and x_1 in X.

Remark: If X is pathwise connected, we may by Corollary 3-18 speak of "the fundamental group of X" and write $\pi_1(X)$ without reference to any particular base point. Indeed, we shall often adopt this policy as a matter of convenience, but it is important to observe that, although all of the groups $\pi_1(X, x)$ are isomorphic, they are not, in general, "naturally" isomorphic in the sense that there is no canonical way to identify them. (The isomorphisms defined in Theorem 3-17 depend on the homotopy class of the path γ.) Thus, it is often best to retain reference to the base point even in the pathwise connected case.

The transition from topology to algebra provided by the construction of the fundamental group is rather more complete than we have indicated thus far. We now proceed to show that any continuous map $f : X \to Y$ induces in a natural way a homomorphism $f_* : \pi_1(X, x_0) \to \pi_1(Y, f(x_0))$ for each $x_0 \in X$ and that these induced homomorphisms reflect many of the important properties of the maps from which they arise. It will be convenient to refer to a topological pair (X, x_0) $(= (X, \{x_0\}))$ as a *pointed space* with base point x_0; a pair map $f : (X, x_0) \to (Y, y_0)$ is called simply a *map* of the pointed spaces (X, x_0) and (Y, y_0). For any such map we define

$f_* : \pi_1(X, x_0) \to \pi_1(Y, y_0)$ by $f_*([\alpha]) = [f \circ \alpha]$ for each $[\alpha]$ in $\pi_1(X, x_0)$. Note that if α is a loop at x_0 in X, then $f \circ \alpha$ is indeed a loop at $f(x_0) = y_0$ in Y. Moreover, if α and β are two loops at x_0 with $F : \alpha \overset{p}{\simeq} \beta$, then $f \circ F : f \circ \alpha \overset{p}{\simeq} f \circ \beta$, so $[f \circ \alpha] = [f \circ \beta]$ and f_* is well defined. Since $f_*([\alpha][\beta]) = f_*([\alpha\beta]) = [f \circ (\alpha\beta)] = [(f \circ \alpha)(f \circ \beta)] = [f \circ \alpha][f \circ \beta] = f_*([\alpha])f_*([\beta])$, the map f_* is a homomorphism. Observe that if $Y = X$ and $f = \mathrm{id}_X$, then $f_* = \mathrm{id}_{\pi_1(X, x_0)}$. Now suppose that (Z, z_0) is another pointed space and $g : (Y, y_0) \to (Z, z_0)$ another map. Then $g \circ f : (X, x_0) \to (Z, z_0)$ and $(g \circ f)_*([\alpha]) = [g \circ f \circ \alpha] = g_*([f \circ \alpha]) = g_*(f_*([\alpha])) = g_* \circ f_*([\alpha])$ for each $[\alpha]$ in $\pi_1(X, x_0)$. Thus, $(g \circ f)_* = g_* \circ f_*$. We summarize these results in the next theorem.

Theorem 3-19. Let (X, x_0) and (Y, y_0) be pointed spaces and $f : (X, x_0) \to (Y, y_0)$ a map. Then f induces a homomorphism $f_* : \pi_1(X, x_0) \to \pi_1(Y, y_0)$ defined by $f_*([\alpha]) = [f \circ \alpha]$ for each $[\alpha]$ in $\pi_1(X, x_0)$ that has the following properties:

(a)   If $Y = X$ and $f = \mathrm{id}_X$, then $f_* = (\mathrm{id}_X)_* = \mathrm{id}_{\pi_1(X, x_0)}$.

(b)   If (Z, z_0) is another pointed space and $g : (Y, y_0) \to (Z, z_0)$ another map, then $(g \circ f)_* = g_* \circ f_*$.

Remark: At this point it is important to retrace the steps that led us to Theorem 3-19. Recall that we began by considering the collection of all pointed spaces and observing that there is associated with this collection a natural class of maps. We shall refer to this collection of all pointed spaces together with its distinguished class of maps as the *category of pointed spaces* (*and pair maps*); the pointed spaces themselves are called the *objects* of the category, while the pair maps are its *morphisms*.

We found that there is a procedure for assigning to each object (X, x_0) of this category a group $\pi_1(X, x_0)$ and to each morphism $f : (X, x_0) \to (Y, y_0)$ a homomorphism which we temporarily denote $\pi_1(f) = f_* : \pi_1(X, x_0) \to \pi_1(Y, y_0)$ and, moreover, that this assignment satisfies $\pi_1(\mathrm{id}_X) = \mathrm{id}_{\pi_1(X, x_0)}$ and $\pi_1(g \circ f) = \pi_1(g) \circ \pi_1(f)$. The collection of all groups together with its distinguished class of maps (the homomorphisms) is called the *category of groups* (*and homomorphisms*); again, the groups themselves are the *objects* of the category, while the homomorphisms are its *morphisms*.

Thus, we may view π_1 as a map from the category of pointed spaces to the category of groups that sends objects to objects, morphisms to morphisms, and preserves the identity morphism and compositions; such a map is called a *functor* from the category of pointed spaces to the category of groups. More generally, we shall refer to any collection of mathematically "structured" sets together with some distinguished class of maps (containing the identity map on each set and closed under composition) as a *category* whose *objects* and *morphisms* are the given sets and maps respectively. If $\mathscr{C}$ and $\mathscr{D}$ are two categories, then a *functor F* from $\mathscr{C}$ to $\mathscr{D}$ is a map that assigns to each object X of $\mathscr{C}$ an object $F(X)$ of $\mathscr{D}$ and to each morphism $f : X \to Y$ in $\mathscr{C}$ a morphism $F(f) : F(X) \to F(Y)$ in $\mathscr{D}$ such that

(a)   If $Y = X$ and $f = \mathrm{id}_X$, then $F(f) = \mathrm{id}_{F(X)}$.

(b)   If $g : Y \to Z$ is another morphism of $\mathscr{C}$, then
$$F(g \circ f) = F(g) \circ F(f).$$

Algebraic topology is concerned with the construction of functors from various "topological" categories to "algebraic" categories. The functor itself is the vehicle by which we translate certain topological problems into algebraic problems that are, hopefully, more readily treated. We shall consider a particularly simple, but nonetheless very beautiful application of this technique in Section 3-6, where we translate, via the fundamental group functor π_1, the No Retraction Theorem in dimension two into a trivial algebraic problem and thus obtain a remarkably elegant new proof of this theorem. For the present, however, we must content ourselves with assembling a bit more machinery.

Proposition 3-20. Let (X, x_0) and (Y, y_0) be pointed spaces and f, $g : (X, x_0) \to (Y, y_0)$ two maps with $f \simeq g$ rel $\{x_0\}$. Then the induced homomorphisms f_*, $g_* : \pi_1(X, x_0) \to \pi_1(Y, y_0)$ are identical.

Exercise 3-9. Prove Proposition 3-20.   Q.E.D.

Next we show that two homotopically equivalent pointed spaces have isomorphic fundamental groups; the proof is a typical application of the functorial nature of our construction of π_1.

Theorem 3-21. If (X, x_0) and (Y, y_0) are pointed spaces and $f : (X, x_0) \to (Y, y_0)$ is a homotopy equivalence, then the induced map $f_* : \pi_1(X, x_0) \to \pi_1(Y, y_0)$ is an isomorphism.

Proof: Let $g : (Y, y_0) \to (X, x_0)$ be such that $g \circ f \simeq \mathrm{id}_X$ rel $\{x_0\}$ and $f \circ g \simeq \mathrm{id}_Y$ rel $\{y_0\}$. By Proposition 3-20 $(g \circ f)_* = (\mathrm{id}_X)_*$ and $(f \circ g)_* = (\mathrm{id}_Y)_*$. But by Theorem 3-19 $(g \circ f)_* = g_* \circ f_*$, $(f \circ g)_* = f_* \circ g_*$, $(\mathrm{id}_X)_* = \mathrm{id}_{\pi_1(X, x_0)}$ and $(\mathrm{id}_Y)_* = \mathrm{id}_{\pi_1(Y, y_0)}$. Thus, $g_* \circ f_* = \mathrm{id}_{\pi_1(X, x_0)}$ and $f_* \circ g_* = \mathrm{id}_{\pi_1(Y, y_0)}$, so f_* is an isomorphism with inverse g_*.   Q.E.D.

Proposition 3-20 and Theorem 3-21 both have one rather serious defect in that they refer only to *relative* homotopies. Consider two pathwise connected spaces X and Y and two maps $f_0, f_1 : X \to Y$ that are homotopic. For each $x_0 \in X$, f_0 and f_1 induce homomorphisms $(f_0)_* : \pi_1(X, x_0) \to \pi_1(Y, f_0(x_0))$ and $(f_1)_* : \pi_1(X, x_0) \to \pi_1(Y, f_1(x_0))$. But $\pi_1(Y, f_0(x_0)) \approx \pi_1(Y, f_1(x_0))$, the isomorphisms being constructed, as in Theorem 3-17, from paths in Y from $f_0(x_0)$ to $f_1(x_0)$. In particular, if $F : f_0 \simeq f_1$, then $\gamma : I \to Y$ defined by $\gamma(t) = F(x_0, t)$ is a path in Y from $f_0(x_0)$ to $f_1(x_0)$, so $\gamma_\# : \pi_1(Y, f_0(x_0)) \to \pi_1(Y, f_1(x_0))$ is an isomorphism. We show next that $(f_0)_*$ and $(f_1)_*$ are "essentially" the same in that they differ only by the isomorphism $\gamma_\#$, which compensates for the fact that f_0 and f_1 may send the given base point x_0 in X onto different points in Y.

Proposition 3-22. Let X and Y be pathwise connected spaces and f_0, $f_1 : X \to Y$ homotopic maps with $F : f_0 \simeq f_1$. Let x_0 be an arbitrary point of X and $(f_0)_* : \pi_1(X, x_0) \to \pi_1(Y, f_0(x_0))$ and $(f_1)_* : \pi_1(X, x_0) \to \pi_1(Y, f_1(x_0))$ the induced homomorphisms. If $\gamma : I \to Y$ is the path in Y from $f_0(x_0)$ to $f_1(x_0)$ given by $\gamma(t) = F(x_0, t)$ for each $t \in I$ and $\gamma_\# : \pi_1(Y, f_0(x_0)) \to \pi_1(Y, f_1(x_0))$ the isomorphisms defined by $\gamma_\#([\beta]) = [\gamma^{-1}\beta\gamma]$ for each $[\beta]$ in $\pi_1(Y, f_0(x_0))$, then $(f_1)_* = \gamma_\# \circ (f_0)_*$.

Proof: Let $[\alpha]$ be in $\pi_1(X, x_0)$. We must show that $(f_1)_*([\alpha]) = \gamma_\#((f_0)_*([\alpha]))$, that is, that $[f_1 \circ \alpha] = [\gamma^{-1}(f_0 \circ \alpha)\gamma]$, and for this it suffices to prove that $\gamma^{-1}(f_0 \circ \alpha)\gamma \overset{p}{\simeq} f_1 \circ \alpha$. Consider the map $G : I \times I \to Y$ defined by $G(t_1, t_2) = F(\alpha(t_1), t_2)$. Then G is continuous and

$$G(t_1, 0) = F(\alpha(t_1), 0) = f_0(\alpha(t_1)) = (f_1 \circ \alpha)(t_1)$$
$$G(t_1, 1) = F(\alpha(t_1), 1) = f_1(\alpha(t_1)) = (f_1 \circ \alpha)(t_1)$$
$$G(1, t_2) = F(\alpha(1), t_2) = F(x_0, t_2) = \gamma(t_2)$$
$$G(0, t_2) = F(\alpha(0), t_2) = F(x_0, t_2) = \gamma(t_2) = \gamma^{-1}(1 - t_2).$$

The required homotopy $H : \gamma^{-1}(f_0 \circ \alpha)\gamma \overset{p}{\simeq} f_1 \circ \alpha$ is now obtained as indicated in Figure 3-3. Specifically,

$$H(t_1, t_2) = \gamma^{-1}(2t_1), \qquad\qquad 0 \leq t_1 \leq \frac{1 - t_2}{2}$$

$$= G\left(\frac{4t_1 + 2t_2 - 2}{3t_2 + 1}, t_2\right), \qquad \frac{1 - t_2}{2} \leq t_1 \leq \frac{t_2 + 3}{4}$$

$$= \gamma(4t_1 - 3), \qquad\qquad \frac{t_2 + 3}{4} \leq t_1 \leq 1.$$

Q.E.D.

With the aid of Proposition 3-22 we now show that the fundamental group of a pathwise connected space is a *homotopy invariant*, that is, that pathwise connected spaces of the same homotopy type have isomorphic fundamental groups.

Theorem 3-23. If X and Y are pathwise connected spaces of the same homotopy type, then their fundamental groups are isomorphic.

Proof: Since X and Y are of the same homotopy type, there exist maps $f : X \to Y$ and $g : Y \to X$ with $g \circ f \simeq \mathrm{id}_X$ and $f \circ g \simeq \mathrm{id}_Y$. Choose a

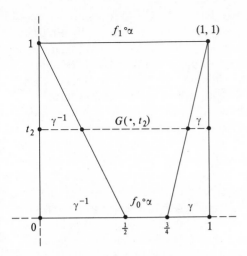

Figure 3-3

point x_0 in the image of g, say, $g(y_0) = x_0$. By Corollary 3-18 we need only show that $f_* : \pi_1(X, x_0) \to \pi_1(Y, f(x_0))$ is an isomorphism. Now, g induces a homomorphism $g_* : \pi_1(Y, f(x_0)) \to \pi_1(X, g \circ f(x_0))$, and by Proposition 3-22 $g_* \circ f_* = (g \circ f)_* = \gamma_\# \circ (\mathrm{id}_X)_* = \gamma_\#$, where $\gamma_\#$ is an isomorphism. Thus, f_* is injective. Let us also denote by g_* the homomorphism of $\pi_1(Y, y_0)$ to $\pi_1(X, g(y_0)) = \pi_1(X, x_0)$ induced by g. Again by Proposition 3-22 $f_* \circ g_* = (f \circ g)_* = \delta_\# \circ (\mathrm{id}_Y)_* = \delta_\#$, where $\delta_\#$ is an isomorphism. Thus, f_* is also surjective and the proof is complete.

Q.E.D.

A space is said to be *simply connected* if it is pathwise connected and has trivial fundamental group.

Corollary 3-24. A contractible space is simply connected.

Exercise 3-10. Prove Corollary 3-24. Hint: Use Proposition 3-3 to show that a contractible space is pathwise connected and then apply Proposition 3-4 and Theorem 3-23.

Remark: We shall see in the next section that a noncontractible space can be simply connected, so the converse of Corollary 3-24 is false.

3-5 Fundamental groups of the spheres

If X is pathwise connected, but not contractible, the computation of $\pi_1(X)$ can be a formidable task indeed. We shall consider a number of relatively simple examples in this section, beginning with the circle S^1.

It will be convenient to think of the unit circle S^1 as the group of complex numbers of modulus 1, that is, $S^1 = \{e^{i\theta} : \theta \in \mathbf{R}\}$, where, by Euler's formula, $e^{i\theta} = \cos\theta + i\sin\theta$. Since S^1 is pathwise connected (verify this!), all of its fundamental groups are isomorphic. We choose to compute $\pi_1(S^1)$ at the base point $e^{i0} = 1$. Thus, we are interested in classifying by homotopy type all maps $\alpha : I \to S^1$ with $\alpha(0) = \alpha(1) = 1 \in S^1$. Intuitively, the situation is quite simple. Consider, for example, the loop α defined by

$$\begin{aligned} \alpha(t) &= e^{\pi i t}, & 0 \leqslant t \leqslant \tfrac{1}{2} \\ &= e^{\pi i(1-t)}, & \tfrac{1}{2} \leqslant t \leqslant 1. \end{aligned}$$

The path α begins at $1 \in S^1$, traces out the circumference of S^1 to the point $e^{\pi i/2} = (0, 1) \in S^1$, and then reverses itself and returns to 1. This path is typical of a class of loops, each of which begins at 1, travels around S^1 without making a complete revolution to a point $e^{i\theta_0}$, and then reverses itself to return to 1. Of course, such paths may proceed at different "speeds" along different arcs, retrace many different subarcs, and so forth, but it is reasonably clear that they are all homotopic to the constant loop at 1.

On the other hand, the loop $\beta : I \to S^1$ defined by $\beta(t) = e^{2\pi i t}$ for all $t \in I$ makes exactly one revolution of S^1, beginning and ending at 1. This map is not homotopic to the constant loop at 1 and is thus in a nontrivial homotopy class. Other examples of loops in this class are those that make one complete revolution of S^1 from 1 and then complete the action of some element of the previous homotopy class, that is, proceed on to a point $e^{i\theta_0}$ without making a second complete revolution and then retrace themselves along this extra bit of arc back to 1. Tracing and retracing this last short arc has the effect of not tracing at all (i.e., is homotopic to the constant loop at 1), so that the entire loop is equivalent to exactly one revolution of S^1.

Finally, we can imagine loops analogous to those in this last homotopy class that make n complete revolutions of S^1 rather than just one; n is positive or negative depending on whether the direction in which these complete revolutions are traversed is counterclockwise or clockwise respectively. Thus, it seems that we obtain one homotopy class for each integer n and we conjecture that $\pi_1(S^1)$ is isomorphic to the group Z of integers.

In order to prove this, we define a homomorphism φ from the additive group of real numbers $\mathbf{R}$ to S^1 by $\varphi(x) = e^{2\pi i x}$ for each $x \in \mathbf{R}$. Note that φ is simply the map that sends $0 \in \mathbf{R}$ onto $e^{2\pi i 0} = 1 \in S^1$ and wraps $\mathbf{R}$ around S^1. Also observe that φ maps the interval $(-\tfrac{1}{2}, \tfrac{1}{2})$ homeomorphically onto $S^1 - \{-1\}$ so we may let $\psi : S^1 - \{-1\} \to (-\tfrac{1}{2}, \tfrac{1}{2})$ be the inverse of $\varphi|(-\tfrac{1}{2}, \tfrac{1}{2})$. We shall require two lemmas; the first says that any loop at 1 in S^1 can be obtained by wrapping a unique path from $0 \in \mathbf{R}$ around S^1 via φ, while the second claims that (path) homotopic loops at $1 \in S^1$ arise in this way from (path) homotopic paths from $0 \in \mathbf{R}$. We prove both lemmas at once.

Lemma 3-25 (Lifting Lemma). Let σ be a loop in S^1 based at 1. Then there is a unique path σ' in $\mathbf{R}$ with initial point 0 such that $\varphi \circ \sigma' = \sigma$.

Lemma 3-26 (Covering Homotopy Lemma). Let σ and τ be loops based at 1 in S^1 and $F : \sigma \overset{p}{\simeq} \tau$. Then there is a unique map $F' : I \times I \to \mathbf{R}$ such that $\varphi \circ F' = F$ and $F' : \sigma' \overset{p}{\simeq} \tau'$.

Proof: Let Y denote either I or $I \times I$, $f : Y \to S^1$ either σ or F and $0 \in Y$ either $0 \in \mathbf{R}$ or $(0, 0) \in I \times I$. Since Y is compact f is uniformly continuous, so there is an $\epsilon > 0$ such that $\|y - y'\| < \epsilon$ implies $\|f(y) - f(y')\| < 1$. In particular, $\|y - y'\| < \epsilon$ implies $f(y) \neq -f(y')$, so $f(y)/f(y') \neq -1$ and $\psi(f(y)/f(y'))$ is defined. Choose an integer N such that $\|y\| < N\epsilon$ for every $y \in Y$. Then each of the quantities

$$\left\| y - \frac{N-1}{N}y \right\|, \left\| \frac{N-1}{N}y - \frac{N-2}{N}y \right\|, \ldots, \left\| \frac{1}{N}y - 0 \right\|$$

is less than ϵ,

so $\quad \psi\left[\dfrac{f(y)}{f((N-1/N)y)} \right], \psi\left[\dfrac{f((N-1/N)y)}{f((N-2/N)y)} \right], \ldots, \psi\left[\dfrac{f((1/N)y)}{f(0)} \right]$

are all defined. Now define $f' : Y \to \mathbf{R}$ by

$$f'(y) = \psi\left[\frac{f(y)}{f((N-1/N)y)} \right] + \cdots + \psi\left[\frac{f((1/N)y)}{f(0)} \right]$$

Then f' is continuous and $f'(0) = \psi(1) + \cdots + \psi(1) = 0$. Moreover, since φ is a homomorphism and $\varphi \circ \psi = \mathrm{id}_{S^1 - \{-1\}}$ we have

$$\varphi \circ f'(y) = \varphi(f'(y))$$

$$= \frac{f(y)}{f((N-1/N)y)} \frac{f((N-1/N)y)}{f((N-2/N)y)} \cdots \frac{f((1/N)y)}{f(0)}$$

$$= \frac{f(y)}{f(0)}$$

$$= f(y),$$

so $\varphi \circ f' = f$ as required. Now, if $f'' : Y \to \mathbf{R}$ also satisfies $f''(0) = 0$ and $\varphi \circ f'' = f$, then $f' - f''$ is a continuous map of Y into the kernel of φ, that is, into the integers Z, because $\varphi(f'(y_0) - f''(y_0)) = \varphi(f'(y_0))/\varphi(f''(y_0)) = f(y_0)/f(y_0) = 1$ for any $y_0 \in Y$. But Y is connected, so $f' - f''$ is a constant map. Since $f'(0) = f''(0) = 0$, $f' - f''$ is identically zero and uniqueness has been proved.

At this point the proof of Lemma 3-25 is complete and all that remains is to show that when $Y = I \times I$, $f = F$ and $f' = F'$, we have $F' : \sigma' \overset{p}{\simeq} \tau'$. We show first that $F' : \sigma' \simeq \tau'$ and then that F' is a homotopy relative to $\{0, 1\}$. First note that $F'(x, 0)$ is a map from I to $\mathbf{R}$ and that, since $\varphi \circ F' = F$, $\varphi(F'(x, 0)) = F(x, 0) = \sigma(x)$ for each $x \in I$. Moreover, $F'(0, 0) = \sigma'(0) = 0$. Thus, by the uniqueness assertion of Lemma 3-25 (already proved), $F'(x, 0) = \sigma'(x)$ for each $x \in I$. Similarly, $F'(x, 1) = \tau'(x)$ for each $x \in I$, so $F' : \sigma' \simeq \tau'$.

Exercise 3-11. Complete the proof by showing that $F'|\{0\} \times I$ and $F'|\{1\} \times I$ are both constant.   Q.E.D.

Now consider an arbitrary loop σ at 1 in S^1. By Lemma 3-25 we may lift σ to a path σ' in $\mathbf{R}$ with initial point 0 such that $\varphi \circ \sigma' = \sigma$. Observe that σ' need not be a loop in $\mathbf{R}$, that is, although $\sigma'(0) = 0$, it need not be the case that $\sigma'(1)$ is 0. However, since $\varphi \circ \sigma' = \sigma$, $\sigma'(1)$ must be in the kernel of φ, that is, $\sigma'(1) \in Z$. Moreover, if τ is a loop at 1 in S^1 with $\tau \overset{p}{\simeq} \sigma$, then by Lemma 3-26 $\tau' \overset{p}{\simeq} \sigma'$ so, in particular, $\tau'(1) = \sigma'(1)$. Thus, the integer $\sigma'(1)$ depends only on the (path) homotopy equivalence class of σ, and we may define a map

$$\chi : \pi_1(S^1, 1) \to Z$$

by $$\chi([\sigma]) = \sigma'(1)$$

for each $[\sigma]$ in $\pi_1(S^1, 1)$.

Remark: Intuitively, the integer $\sigma'(1) = \chi([\sigma])$ counts the number of complete revolutions of S^1 made by any element of $[\sigma]$.

We claim that χ is an isomorphism. We have just seen that χ is well defined. To see that it is a homomorphism, let $[\sigma]$, $[\tau] \in \pi_1(S^1, 1)$ with $\sigma'(1) = m$ and $\tau'(1) = n$. Then $\chi([\sigma]) + \chi([\tau]) = m + n$, and we must show that $\chi([\sigma][\tau]) = m + n$ as well. Now, $\chi([\sigma][\tau]) = \chi([\sigma\tau]) = (\sigma\tau)'(1)$, where $(\sigma\tau)'$ is the path in $\mathbf{R}$ from 0 that lifts $\sigma\tau$. Let τ'' be the path in $\mathbf{R}$ from m to $m + n$ given by $\tau''(t) = \tau'(t) + m$. Then $\sigma'\tau''$ is a path in $\mathbf{R}$ from 0 to $m + n$. Since $\varphi \circ (\sigma'\tau'') = (\varphi \circ \sigma') (\varphi \circ \tau'') = \sigma\tau$, $\sigma'\tau''$ is the lifting of $\sigma\tau$ to $\mathbf{R}$, that is, $(\sigma\tau)' = \sigma'\tau''$. Thus, $(\sigma\tau)'(1) = (\sigma'\tau'')(1) = m + n$, so $\chi([\sigma][\tau]) = m + n$ and χ is a homomorphism.

Exercise 3-12. Show that χ is bijective and thus an isomorphism. Hint: To prove that $\ker \chi = [e]$, use the contractibility of $\mathbf{R}$.

Thus, we have proved the next theorem.

Theorem 3-27. The fundamental group $\pi_1(S^1)$ of the circle is isomorphic to the group Z of integers.

Exercise 3-13. Let X be the "punctured plane" $\mathbf{R}^2 - \{(0, 0)\}$. Show that $\pi_1(X) \approx Z$. Hint: Use Theorem 3-23.

Exercise 3-14. Let X and Y be pathwise connected spaces. Show that $\pi_1(X \times Y) \approx \pi_1(X) \oplus \pi_1(Y)$. Conclude that the fundamental group of the torus is isomorphic to $Z \oplus Z$.

We pause now to obtain from Theorem 3-27 a new proof of the No Retraction Theorem in dimension two. The method of proof is typical of the sort of technique that characterizes algebraic topology, and we advise the reader to linger for some time on the rather beautiful ideas involved.

Our object then is to show that the circle S^1 is not a retract of the disc B^2. Let us suppose to the contrary that $r : B^2 \to S^1$ is a retraction. Thus, if $i : S^1 \hookrightarrow B^2$ is the inclusion map, we have $r \circ i = \mathrm{id}_{S^1}$:

$$
\begin{array}{ccc}
S^1 & \xrightarrow{\;\;i\;\;} & B^2 \\
& \mathrm{id}_{S^1} \searrow & \downarrow r \\
& & S^1
\end{array}
$$

Select an arbitrary base point, say, $1 \in S^1$ and apply the fundamental group functor π_1 to the commutative diagram above to obtain by Theorem 3-19 the following commutative diagram:

$$
\begin{array}{ccc}
\pi_1(S^1, 1) & \xrightarrow{\;\;i_*\;\;} & \pi_1(B^2, 1) \\
\mathrm{id}_{S^1_*} = \mathrm{id}_{\pi_1(S^1,1)} \searrow & & \downarrow r_* \\
& & \pi_1(S^1, 1)
\end{array}
$$

In particular, since $r_* \circ i_* = \mathrm{id}_{\pi_1(S^1,1)}$, i_* is injective. Thus, $\pi_1(S^1, 1)$ must be isomorphic to a subgroup of $\pi_1(B^2, 1)$. But $\pi_1(S^1, 1) \approx Z$ and, since B^2 is contractible, $\pi_1(B^2, 1)$ is trivial (Corollary 3-24), so this is impossible.

Observe that it is the functorial nature of our construction of π_1 that accounts for the success of this method. Indeed, most of the proof is as valid for S^n and B^{n+1} as it is for S^1 and B^2. It is only at the last line, where we note that $\pi_1(S^1)$ cannot be isomorphic to a subgroup of $\pi_1(B^2)$ because the latter is trivial and the former is not, that we required any specific information about S^1 and B^2. Now, it is true that B^{n+1} is contractible and thus has trivial fundamental group for any $n \geq 1$. However, we shall show next that for $n > 1$ the n-sphere S^n, although not contractible (Theorem 3-6), is simply connected so that the proof we have given does not generalize to the case $n > 1$.

Exercise 3-15. Show that S^n is pathwise connected for each $n \geqslant 1$. Hint: Apply Proposition 3-7.

Theorem 3-28. For any $n > 1$ the n-sphere S^n is simply connected.

Proof: By Exercise 3-15 we need only show that $\pi_1(S^n)$ is trivial. Select some base point $p \in S^n$ and let $\alpha : I \to S^n$ be a loop at p. We show that $\alpha \overset{p}{\simeq} e$, where $e : I \to S^n$ is the constant loop $e(t) = p$ for each $t \in I$. Define a map h of the boundary bdy($I \times I$) of $I \times I$ to S^n by

$$h(t_1, 0) = \alpha(t_1)$$
$$h(t_1, 1) = h(0, t_2) = h(1, t_2) = p$$

for all t_1 and t_2 in I.

Exercise 3-16. Construct a homeomorphism φ of B^2 onto $I \times I$ such that $\varphi|S^1$ is a homeomorphism of S^1 onto bdy($I \times I$).

Now, $h \circ (\varphi|S^1) : S^1 \to S^n$. Since $1 < n$, Theorem 3-13 implies that $h \circ (\varphi|S^1)$ is nullhomotopic so that by Lemma 3-5 it extends to a map $F : B^2 \to S^n$. Then $F \circ \varphi^{-1} : I \times I \to S^n$. Let $(t_1, t_2) \in$ bdy($I \times I$). Then $\varphi^{-1}(t_1, t_2) = (\varphi|S^1)^{-1}(t_1, t_2)$ and $F \circ \varphi^{-1}(t_1, t_2) = F((\varphi|S^1)^{-1}(t_1, t_2)) = h \circ (\varphi|S^1)((\varphi|S^1)^{-1}(t_1, t_2)) = h(t_1, t_2)$. Thus, $H = F \circ \varphi^{-1}$ is a continuous extension of h to $I \times I$ and therefore $H : \varphi \overset{p}{\simeq} e$ as required.   Q.E.D.

Despite the failure of our proof to generalize to dimensions greater than two, it is of the utmost importance to observe that this failure is due exclusively to the fact that the functor π_1 is not sufficiently sensitive to distinguish between S^n and B^{n+1} for $n > 1$. In order to obtain a "categorical" proof of the general No Retraction Theorem analogous to that given in dimension two above, we need only construct a functor F from some topological category containing S^n and B^{n+1} to groups for which $F(B^{n+1})$ is trivial and $F(S^n)$ is nontrivial. There are several possibilities for the construction of such functors. At this point we shall very briefly discuss one such construction that we do not intend to pursue.

For any pointed space (X, x_0), $\pi_1(X, x_0)$ measures the number of "essentially distinct" types of loops in X based at x_0. Whether or not there exist nontrivial elements in $\pi_1(X, x_0)$ depends on whether or not there are any "obstructions" in X that prevent certain loops at x_0 from being shrunk to the point x_0. Consider, for example, the space X obtained from B^2 by removing an open disc at its center, that is, X is a circular annulus. Observe that X is homotopically equivalent to S^1. For any x_0 in X there are certain loops at x_0 that cannot be shrunk to x_0, namely, those that surround the "hole" in the center of X. Thus, $\pi_1(X, x_0)$ is nontrivial. Now consider the analogous situation in dimension three, that is, let Y be the space obtained from B^3 by deleting an open ball at its center. This space is homotopically equivalent to the sphere S^2. Observe that, in this case, the "hole" at the center of Y presents no obstacle to the deformation of loops in Y since we can use the extra dimension to push any loop around the hole. From the point of view of the fundamental group this hole is "invisible" and $\pi_1(Y, y_0)$ is trivial for any $y_0 \in Y$. It is precisely this phenomenon that accounts for the inability of the fundamental group functor π_1 to distinguish between S^n and B^{n+1} for $n > 1$. To overcome this difficulty, we observe that the hole in Y would indeed obstruct the deformation of a closed surface in Y that encloses it. This seems to suggest the possibility of using "higher dimensional loops" to detect features of a space that π_1 cannot discern. Specifically, let us consider an arbitrary pointed space (X, x_0). A map $\alpha : I^n \to X$ for which $\alpha(\text{bdy } I^n) = \{x_0\}$ is called an *n-dimensional loop* in X based at x_0. If α and β are two such maps, we define their *product* $\alpha\beta : I^n \to X$ by

$$\alpha\beta(t_1, t_2, \ldots, t_n) = \alpha(2t_1, t_2, \ldots, t_n), \qquad 0 \le t_1 \le \tfrac{1}{2}$$
$$= \beta(2t_1 - 1, t_2, \ldots, t_n), \qquad \tfrac{1}{2} \le t_1 \le 1.$$

The *inverse* α^{-1} of an *n*-dimensional loop α is defined by

$$\alpha^{-1}(t_1, t_2, \ldots, t_n) = \alpha(1 - t_1, t_2, \ldots, t_n).$$

The constant map that carries all of I^n onto x_0 is denoted e. Now, an *n*-dimensional loop is, of course, a pair map of $(I^n, \text{bdy } I^n)$ to (X, X_0) so the appropriate notion of homotopy for such maps is homotopy relative to bdy I^n. This relation partitions the set of all *n*-dimensional loops at x_0 in X into equivalence classes. The equivalence class containing a loop α is denoted $[\alpha]$ and we can show, just as in the 1-dimensional case, that products and inverses of such classes are well defined by $[\alpha][\beta] = [\alpha\beta]$ and $[\alpha]^{-1} = [\alpha^{-1}]$. Next we show that, with these operations, the set of all homotopy equivalence classes of *n*-dimensional loops at x_0 in X forms a group with identity $[e]$. This group is denoted $\pi_n(X, x_0)$ and called the *n*th *homotopy group* of (X, x_0). Finally, we show that each map $f : (X, x_0) \to (Y, y_0)$ induces a homomorphism $\pi_n(f) : \pi_n(X, x_0) \to \pi_n(Y, y_0)$ and that, moreover, $\pi_n(\text{id}_X) = \text{id}_{\pi_n(X, x_0)}$ and $\pi_n(g \circ f) = \pi_n(g) \circ \pi_n(f)$, and this completes the construction of the higher dimensional homotopy functors π_n.

Exercise 3-17. Carry out the construction of the functors π_n, $n > 1$, in detail.

As it turns out, all homotopy groups of the ball B^{n+1} are trivial, but $\pi_n(S^n) \approx Z$ so that we can prove that S^n is not a retract of B^{n+1} just as we did for $n = 1$ with π_n in place of π_1.

We have chosen not to carry out this program in detail here primarily because any serious study of the higher homotopy groups requires a great deal of extremely sophisticated machinery that is quite beyond our reach. Indeed, even the computation of $\pi_k(S^n)$ for $k > n$ has proved to be an enormously difficult task. We elect, therefore, to seek instead alternate methods for defining functors from topological to algebraic categories that accomplish our purpose of distinguishing S^n and B^{n+1} and that are, in addition, somewhat more amenable to treatment by the techniques at our disposal.

Supplementary exercises

3-18   Apply the Simplicial Approximation Theorem to show that there are only countably many homotopy classes of maps from one polyhedron to another.

3-19   Suppose X_1 is homotopically equivalent to Y_1 and X_2 is homotopically equivalent to Y_2. Show that $X_1 \times X_2$ is homotopically equivalent to $Y_1 \times Y_2$.

3-20   Show that the circle S^1 is homotopically equivalent to the cylinder $S^1 \times I$.

3-21   Prove that a retract of a contractible space is contractible.

3-22   A subset A of a space X is a *deformation retract* of X if there is a retraction $r : X \to A$ which, as a map into X, is homotopic to id_X.
(a) Show that a retract need not be a deformation retract.
(b) Prove that if A is a deformation retract of X, then A is homotopically equivalent to X.
(c) Show that every compact convex subset of $\mathbf{R}^n$ is a deformation retract of $\mathbf{R}^n$.

3-23   Let X be a space and A a subspace of X. The pair (X, A) has the *homotopy extension property* with respect to a space Y iff each continuous map $F' : (X \times \{0\}) \cup (A \times I) \to Y$ has an extension to a continuous map $F : X \times I \to Y$. Show that (X, A) has the homotopy extension property with respect to every space Y iff $(X \times \{0\}) \cup (A \times I)$ is a retract of $X \times I$.

3-24   Let C be a closed subset of a space X and f and g two homotopic maps of C into S^n. Show that, if there is an extension F of f to X, then there is also an extension G of g to X with $F \simeq G$.

3-25   Let X and $\tilde{X}$ be pathwise connected and locally pathwise connected and let $p : \tilde{X} \to X$ be a continuous map. The pair $(\tilde{X}, p)$ is a *covering space* for X if p is surjective and if for each $x \in X$ there is an open nbd U of x in X such that $p^{-1}(U)$ is a disjoint union of open sets in $\tilde{X}$, each of which is mapped homeomorphically onto U by p.
(a) Show that if $p : \mathbf{R} \to S^1$ is defined by $p(x) = e^{2\pi i x}$, then $(\mathbf{R}, p)$ is

a covering space for S^1. Interpret $(\mathbf{R}, p)$ as an infinite spiral over S^1, p being the projection.

(b) Construct a covering space for the torus $S^1 \times S^1$.

(c) Suppose $(\tilde{X}, p)$ is a covering space for X. Show that each $f^{-1}(x)$ for x in X is a discrete subset of $\tilde{X}$, that p is a homeomorphism on a nbd of each point of $\tilde{X}$, and that p is an open map.

3-26 A *graph* is a geometric complex of dimension at most one. A *tree* is a pathwise connected graph T such that, for each 1-simplex $s \in T$, $|T| - s$ is not connected.

(a) Construct several graphs that are trees and several that are not.

(b) Show that every tree has a vertex that is the vertex of at most one 1-simplex. Such a vertex is called an *end*.

(c) Prove that if T is a tree, $|T|$ is contractible.

(d) Show that in any tree the number of vertices minus the number of edges is one.

Chapter 4
Simplicial homology theory

4-1   Introduction

In Chapter 3 we found it possible to study the local connectivity properties of an arbitrary space by constructing a functor π_1 from pointed spaces to groups with the property that the image $\pi_1(X,\ x_0)$ of a pointed space $(X,\ x_0)$ is a group which, in some sense, measures the number of "essentially distinct" types of loops in X through x_0. Our categorical proof of the No Retraction Theorem in dimension two provided a rather tantalizing indication of the power and elegance of this sort of algebraic technique. On the other hand, we infer from the simple connectivity of S^n for $n > 1$ that the functor π_1 alone may not provide an entirely satisfactory transition from topology to algebra. We very briefly considered one possible improvement, that is, the construction of analogous higher dimensional homotopy functors π_n, $n > 1$, to measure the number of distinct types of "n-dimensional loops" in X based at x_0. Confronted, however, with the extraordinarily difficult task of computing the higher homotopy groups for even relatively simple spaces, we were persuaded to seek an entirely new line of attack.

Simplicial homology theory studies the global connectivity properties of polyhedra by constructing functors H_p, $p \geqslant 0$, from polyhedra to (finitely generated Abelian) groups. Roughly, the procedure is to determine the number of distinct types of "closed curves" in a polyhedron that do *not* bound a portion of the space, that is, do not disconnect the space. For example, the sphere S^2 and the torus T differ in their connectivity properties because there are no one-dimensional closed curves in S^2 that do not disconnect S^2, whereas there are two distinct types of one-dimensional closed curves on the torus that, taken individually or together, do not disconnect T. It is the task of the one-dimensional simplicial homology functor H_1 to assign to S^2 and T groups $H_1(S^2)$ and $H_1(T)$, which record these differences in structure.

Specifically, we observe that any one-dimensional closed curve on S^2, for example, c in Figure 4-1, bounds a portion of S^2 and thus disconnects it. (If we were to "cut" S^2 along c, the result would be two disjoint con-

 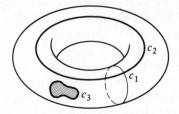

Figure 4-1

nected pieces.) Our construction of the functor H_1 will reflect this property of S^2 by assigning to it the trivial group $H_1(S^2) \approx 0$. (There are no one-dimensional closed curves on S^2 that fail to disconnect it.) Now, although there are closed curves in T that bound portions of T, for example, c_3 in Figure 4-1, these tell us essentially nothing about the global structure of T and we shall ignore them. However, curves such as c_1 or c_2 fail to bound regions of T and thus do not disconnect T. (If we were to cut along either c_1 or c_2, the resulting surface would still be connected.) Even c_1 and c_2 "taken together" do not disconnect T. (Cutting along both c_1 and c_2, we obtain a rectangle; see Figure 2-7.) Allowing for the possibility that a given curve can be "traversed" in two distinct directions and any number of times in either of these directions, we can define, in the usual way, integral multiples of c_1 and c_2 and observe that none of these curves, nor any "sum" of two such curves, bounds a region of T.

Naturally, all of these remarks apply equally well to any curves of the same "type" as c_1 or c_2. Indeed, it seems intuitively clear that we are dealing here with just two "essentially distinct" basic types of nonbounding, closed curves on T (latitudinal and longitudinal curves traversed once in some specified directions) from which the remaining types can be obtained by forming "sums" of "integral multiples." It seems then that the properties of T that interest us here are well represented by the free Abelian group (see the Remark in Section 4-2) generated by these two basic "types" (equivalence classes) of nonbounding closed curves in T (a group that is, of course, isomorphic to $Z \oplus Z$). The functor H_1 as constructed in this chapter does indeed assign to T a group $H_1(T) \approx Z \oplus Z$, the rank of which, that is, 2, is called the first "Betti number" of T and gives mathematical expression to our intuitive feeling that the torus has two "holes" in it.

Higher dimensional homology groups are constructed along similar lines using higher dimensional "closed curves" (closed surfaces, etc.) to detect "holes" in spaces of higher dimension. As our experience with homotopy groups has indicated quite clearly, however, the problem of classifying the closed curves of various dimensions in an arbitrary space is an extraordinarily difficult one. In simplicial homology theory we avoid this difficulty by restricting attention to a class of spaces in which we can isolate a very natural and, more importantly, quite manageable class of closed curves with which to detect holes. Consider once again the torus. Now, T is a topological polyhedron and thus admits a triangulation into curved simplexes. Suppose, for example, that T is triangulated in the manner prescribed by Figure 2-9 (b) (with the edges identified as indicated). Those "chains" of edges in this triangulation that begin and end at the same vertex are called "1-cycles" and form a particularly simple class of one-dimensional closed curves in T. More to the point, however, is the observation that each of the two basic "types" of nonbounding closed curves on T is "represented" by such a cycle: The chain leading from A to D to G and back to A is of the same type as c_1 in Figure 4-1, while that running from A to B to C to A is equivalent to c_2. Stated somewhat differently, it appears that all of the "holes" in T are detected by elements of this class of 1-cycles and that we therefore need not consider more general closed curves. Moreover, it seems intuitively clear that, whatever triangulation of T is chosen, chains of these two basic types must be present, so that this choice may be regarded as a matter of convenience. We are thus led to hope that, by restricting our attention to polyhedra, all of our computations can be carried out using only the "chains" present in any convenient triangulation. Since we shall find that higher dimensional chains and cycles are as easily defined and manipulated as those in dimension one, most of the difficulties encountered in generalizing the construction of π_1 are avoided.

Specifically, we shall begin our investigations by defining homology groups only for geometric complexes (where the chains are even "piecewise linear"). Next we prove that for a given complex K the homology groups of K and its first barycentric subdivision K' are isomorphic. From here it is but a short step to topological invariance, that is, to the fact that if K and L are two complexes with homeomorphic polyhedra, then the homology groups of K and L are isomorphic. We obtain this last result by proving that any continuous map between polyhedra induces, in a natural way, homomorphisms of the corresponding homology groups of the given complexes in each dimension and that this assignment of homomorphisms to continuous maps is functorial. We show, moreover, that homotopic maps between polyhedra induce identical homomorphisms of homology groups and conclude that these groups are, in fact, homotopy invariants, that is, that geometric complexes with polyhedra of the same homotopy type have isomorphic homology groups in each dimension. These invariance results then make it possible to define the homology groups of an arbitrary topological polyhedron.

There are, of course, certain obvious similarities between homotopy and homology theory. For example, the basic operating principle in each is the functorial transition from a topological to an algebraic setting. From the practical point of view, however, there are highly significant differences. In homotopy theory it is clear at the outset that homeomorphic spaces have isomorphic homotopy groups, but the computation of these groups is, in general, inordinately difficult. The situation is reversed in homology theory. Although the topological invariance of the simplicial homology groups is quite a deep theorem, the computation of these groups for a given space is accomplished by an almost mechanical procedure. Clearly, this latter state of affairs possesses distinct advantages over the former in that the difficult part of the theory (the proof of topological invariance) need only be done once. In addition, the simplicial homology groups are all finitely generated and Abelian, so their structure is rather easily discerned. Of course, the price we must pay for these advantages is the rather severely restricted class of spaces to which the theory applies, that is, polyhedra. Although there are more general homology theories that embrace a wider class of spaces (e.g., Čech homology theory and singular homology theory), they will not concern us here.

4-2   Oriented complexes and chains

Our first task then is to define more carefully what is meant by a "chain" in a geometric complex. As motivation we return once again to the one-dimensional case. Now, intuitively, a curve is more than simply a set of points; it is a point-set that has been provided with a definite direction in which it is to be "traversed." This requirement that curves be provided with an "orientation" is often of critical importance. As a particularly simple example consider two real-valued functions $f(x, y)$ and $g(x, y)$ that are continuous and have continuous partial derivatives $f_y(x, y)$ and $g_x(x, y)$ in some domain containing the closed 2-simplex $\bar{s}_2 = [a_0, a_1, a_2]$; see Figure 2-2 (c). Then, according to Green's Theorem (Section 3-1), $\iint_{\bar{s}_2}(g_x - f_y)\, dx\, dy = \int_{\partial s_2} f\, dx + g\, dy$, where ∂s_2 denotes the topological boundary of s_2 and the line integral is taken around ∂s_2 *in the counterclockwise sense.* Now, it is customary to write a formal sum such as

$$\partial s_2 = a_0 a_1 + a_1 a_2 + a_2 a_0$$

to indicate that ∂s_2 may be conveniently thought of as the "sum" of three 1-dimensional segments, each of which has been provided with a definite direction; for example, the symbol $a_0 a_1$ suggests a line segment between a_0 and a_1 and directed from a_0 to a_1. Thus, it appears we might view a "1-dimensional chain" as a "formal sum" of "oriented 1-simplexes" and, more generally, a "*p*-dimensional chain" as a "formal sum" of "oriented *p*-simplexes." It is along these lines that our definition will eventually be

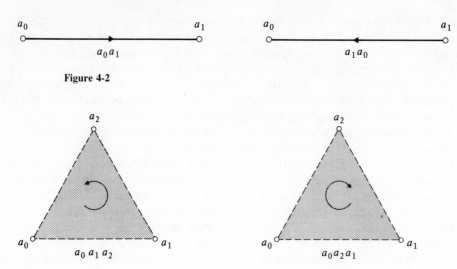

Figure 4-2

(a)   (b)

Figure 4-3

formulated. The first obstacle to overcome, of course, is the definition of
"orientation" for a general p-simplex.

Now, it is intuitively clear that a 1-simplex $s_1 = (a_0, a_1)$ has precisely
two "orientations" (directions) and that each of these is easily specified,
as we have just seen, by selecting one of the two possible orderings of the
vertices a_0 and a_1. Thus, the ordering a_0a_1 indicates the orientation of s_1
from a_0 to a_1, while a_1a_0 determines the "opposite" orientation from a_1 to
a_0 (see Figure 4-2).

For a 2-simplex $s_2 = (a_0, a_1, a_2)$, "orientation" is interpreted as "sense
of rotation" and, again, there are precisely two such orientations (clockwise and counterclockwise). Each of these orientations is easily specified
by ordering the vertices of s_2. For example, if the ordering is $a_0a_1a_2$, then
the indicated sense of rotation is that in which a radius vector turns,
which leads from a fixed interior point of s_2 to a point that varies along
the boundary of s_2 from a_0 to a_1 to a_2 to a_0; the ordering $a_0a_2a_1$ determines the opposite orientation (see Figure 4-3).

In this case, however, there are six possible orderings of the vertices
a_0, a_1, and a_2, and we are led to regard as "equivalent" those orderings
that determine the same sense of rotation in s_2. Thus, $a_0a_1a_2$, $a_1a_2a_0$, and
$a_2a_0a_1$ are equivalent orderings, as are $a_0a_2a_1$, $a_2a_1a_0$, and $a_1a_0a_2$. Observe that two orderings $a_{i_0}a_{i_1}a_{i_2}$ and $a_{j_0}a_{j_1}a_{j_2}$ are equivalent in this sense
iff (j_0, j_1, j_2) is an even permutation of (i_0, i_1, i_2). With this observation
we can formulate our general definitions.

Let $s_p = (a_0, \ldots, a_p)$ be a p-simplex. Two orderings $(a_{i_0}, \ldots, a_{i_p})$ and $(a_{j_0}, \ldots, a_{j_p})$ of the vertices $a_0, \ldots, a_p$ of s_p are considered *equivalent* if $(j_0, \ldots, j_p)$ is an even permutation of $(i_0, \ldots, i_p)$. An equivalence relation is thus defined that, for $p > 0$, partitions the set of all orderings of $a_0, \ldots, a_p$ into two equivalence classes, each of which is called an *orientation* of s_p. An *oriented simplex* is a simplex s_p together with a choice of one of its two possible orientations. In order to specify an orientation for s_p, it suffices to list its vertices in some order since this ordering is a representative of precisely one of the two equivalence classes. If $s_p = (a_0, \ldots, a_p)$, then the symbol

$$\sigma_p = a_0 a_1 \ldots a_p$$

is used to denote the oriented simplex determined by the indicated ordering of the vertices; the other oriented simplex determined by s_p is then denoted $-\sigma_p$. (Which of the two orientations of s_p happens to be denoted σ_p and which $-\sigma_p$ is, of course, entirely arbitrary.) Thus, if $s_2 = (a_0, a_1, a_2)$, then the two oriented simplexes determined by s_2 can be denoted $\sigma_2 = a_0 a_1 a_2 = a_1 a_2 a_0 = a_2 a_0 a_1$ and $-\sigma_2 = a_0 a_2 a_1 = a_2 a_1 a_0 = a_1 a_0 a_2$. Strictly speaking, a 0-simplex, that is, a vertex $s_0 = (a_0)$, has only one orientation. However, it is convenient to consider two orientations for s_0 by formally associating with it the two symbols $\sigma_0 = a_0$ and $-\sigma_0 = -a_0$. This should be regarded as a purely technical device for avoiding "special cases" with no real significance.

A geometric complex in which each simplex has been provided with some fixed orientation is called an *oriented complex*. Note that this definition is quite general: Orientations may be assigned to the simplexes in an entirely arbitrary manner, taking no account whatever of how the simplexes are joined or whether one simplex is a face of another. Indeed, one particularly simple method of orienting an arbitrary complex is to list all of its vertices in some order and require that its simplexes be oriented in the sense specified by the order in which their vertices appear in this list; such an oriented complex is called an *ordered complex*. We shall return to this point somewhat later in our discussion of incidence numbers (Section 4-7). For the present we observe that, since we have determined that a "p-dimensional chain" in a complex should be a "formal sum" of oriented p-simplexes, we need only specify what is meant by a "formal sum" of the elements of an oriented complex in order to complete our definition. As it happens, there is available to us a standard device for the introduction of formal sums into virtually any context in which they are required, that is, free Abelian groups.

Remark: At this point we begin to require certain material that is very often not included in a first course in modern algebra. A brief summary of the relevant concepts with references would therefore seem to be in order. Let us therefore consider an Abelian group G with binary operation $+$ and unit element 0. The inverse of an element g of G is denoted $-g$ and, as usual, we write $g_1 - g_2$ for $g_1 + (-g_2)$ and kg for $g + g + \cdots + g$ (k-summands). If there exists an integer k such that $kg = 0$, then the least such k is called the *order* of g and g is said to be a *torsion element* of G; if no such k exists, then g is said to be of *infinite order*. A subgroup H of G (in particular, H could be G itself) is said to be *generated by* a subset A of G if every $h \in H$ can be written as $h = n_1 a_1 + \cdots + n_k a_k$, where each n_i is an integer and $a_i \in A$ for each $i = 1,$ $\ldots, k$; the elements of A are called *generators* for H. The group G is said to be *finitely generated* if there is a finite subset A of G which generates G; G is *cyclic* if it is generated by a single element. For example, the group Z of integers is cyclic as is the finite group Z_γ of integers modulo an integer γ. Indeed, one can show that any cyclic group is isomorphic either to Z or to some Z_γ (see Birkhoff and MacLane, p. 126, Theorem 9). By forming finite direct sums of these cyclic groups, one obtains examples of finitely generated Abelian groups. The major result in the subject states that, in fact, this is essentially the only way to obtain examples:

Structure Theorem for Finitely Generated Abelian Groups. Every finitely generated Abelian group G is isomorphic to the direct sum of a finite number of cyclic groups.

(For a proof of this fundamental result refer to the excellent treatment in MacDonald, Chapter 5. A more concise, but nonconstructive proof is available Hilton and Wylie, p. 160.)

From the Structure Theorem it follows that every finitely generated Abelian group G is isomorphic to a group of the form

$$Z \oplus \cdots \oplus Z \oplus Z_{\gamma_1} \oplus \cdots \oplus Z_{\gamma_t}.$$

The number of infinite cyclic summands in this direct sum is called the *rank* of G. In general, any group that is isomorphic to a direct sum of infinite cyclic groups is called a *free Abelian group*. Such a group F is characterized by the fact that it has a finite set of generators $\{a_1, \ldots, a_r\}$ with the property that $n_1 a_1 + \cdots + n_r a_r = 0$ implies $n_1 = \cdots = n_r = 0$; such a set of generators is called a *basis* for F. Thus, $F \approx \langle a_1 \rangle \oplus \cdots \oplus \langle a_r \rangle$, where $\langle a_i \rangle$ is the cyclic subgroup of F generated by a_i. Now consider an arbitrary finite set $S = \{s_1, \ldots, s_r\}$. Select a basis $\{a_1, \ldots, a_r\}$ for $Z^r = Z \oplus \cdots \oplus Z$ (r-summands) and define a map $\alpha : S \to Z^r$ by $\alpha(s_i) = a_i$ for each $i = 1, \ldots, r$. Thus, every element of Z^r is uniquely expressible in the form $n_1 \alpha(s_1) + \cdots + n_r \alpha(s_r)$. Gener-

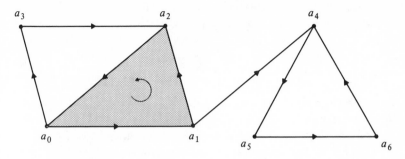

Figure 4-4

ally, it is convenient to adopt the point of view that the one-to-one map α "embeds" the set S into Z^r so that s_i and $\alpha(s_i)$ can be identified. In this way we represent the free Abelian group Z^r as a group of "formal sums" $n_1 s_1 + \cdots + n_r s_r$ of the elements of S. This group of formal sums is called the *free Abelian group generated by the set S*.

Let K be an oriented complex. The free Abelian group generated by the set of all oriented p-simplexes of K is denoted $C_p(K)$ and called the *p-chain group* of K. The elements of $C_p(K)$ are called *p-dimensional chains*, or simply *p-chains*, in K. Every such element is, by definition, a formal sum $c_p = \Sigma n_i \sigma_p^i$, where the n_i are integers and the "sum" is over all oriented p-simplexes σ_p^i in K. Two p-chains $c_p = \Sigma n_i \sigma_p^i$ and $c_p' = \Sigma n_i' \sigma_p^i$ are formally added in $C_p(K)$ to obtain the p-chain $c_p + c_p' = \Sigma(n_i + n_i')\sigma_p^i$. The identity element of $C_p(K)$, which has all coefficients equal to zero, is denoted 0 and the inverse of $c_p = \Sigma n_i \sigma_p^i$ is written $-c_p = \Sigma(-n_i)\sigma_p^i$. Note that each oriented p-simplex in K can be regarded as a p-chain with all coefficients but one equal to zero. Moreover, since $-\sigma_p$ is in $C_p(K)$ for every oriented p-simplex σ_p of K, it is clear that the groups $C_p(K)$ are unaffected by changes in the orientations of the simplexes in K, that is, the p-chain group of K is independent of the choice of orientation for K.

Now, it is important to observe that, although $C_p(K)$ certainly contains all of those objects that, on intuitive grounds, we have chosen to call p-chains, it also has elements we did not anticipate. Consider, for example, the complex K consisting of the 2-simplex (a_0, a_1, a_2), the vertices $a_0, \ldots, a_6$ and the 1-simplexes shown in Figure 4-4, each with the indicated orientation. The 1-chains $a_0 a_1, a_4 a_5 + a_5 a_6, a_0 a_3 + a_3 a_2 + a_2 a_0, a_2 a_0 + a_0 a_1 + a_1 a_4 + a_4 a_5 + a_5 a_6$, and so on all appear in $C_1(K)$ as do all integral multiples of such chains. However, $C_1(K)$ contains, in addition, such chains as $a_0 a_1 + a_1 a_2 + 2 a_6 a_4 - 13 a_3 a_2$, which appear to be rather far removed from our original geometric conception of a 1-chain as a "string" of edges. Nevertheless, from the algebraic point of view it is the group structure of $C_1(K)$ and certain of its subgroups that plays a fundamental role and we shall not be deterred by the "peculiar" appearance of some of its elements.

4-3   Boundary operators

Observe that in the example just given (Figure 4-4) some of the 1-chains in K are "closed" (e.g., $a_0a_1 + a_1a_2 + a_2a_0$), while others are not (e.g., $a_0a_1 + a_1a_4 + a_4a_5$). Among those 1-chains that are "closed" some "bound" regions of the polyhedron (e.g., $a_0a_1 + a_1a_2 + a_2a_0$), while others do not (e.g., $a_4a_5 + a_5a_6 + a_6a_4$). Recalling our discussion of the torus in Section 4-1, it is clear that we are interested primarily in "closed" chains that *do not* "bound" and we now seek a device that will permit us to single out these classes of chains in an arbitrary complex. Consider an oriented 2-simplex $\sigma_2 = a_0a_1a_2$ (see Figure 4-3(a)) and observe that the specified sense of rotation in σ_2 provides a natural sense in which to traverse the topological boundary of σ_2 (from a_0 to a_1 to a_2 and back to a_0). Thus, each 1-face of σ_2 inherits a natural orientation from σ_2 with the property that, with these orientations, the "oriented boundary" of σ_2 as it arose in Green's Theorem (Section 4-2) is just the 1-chain obtained by adding the 1-faces. Specifically,

$$\partial \sigma_2 = a_0a_1 + a_1a_2 + a_2a_0.$$

Indeed, it is just this fact that originally suggested our approach to the definition of a chain. Since we are, after all, interested in distinguishing those ("closed") chains that "bound" regions of the polyhedron (i.e., that are "boundaries") from those that do not, it would behoove us to generalize this process of taking the boundary of an oriented simplex to higher dimensions. To this end we rewrite the given chain as follows (writing $\partial_2\sigma_2$ rather than $\partial\sigma_2$ since we have in mind taking boundaries of simplexes in arbitrary dimensions):

$$\begin{aligned}
\partial_2\sigma_2 &= a_0a_1 + a_1a_2 + a_2a_0 = a_1a_2 + a_2a_0 + a_0a_1 \\
&= a_1a_2 - a_0a_2 + a_0a_1 = \hat{a}_0a_1a_2 - a_0\hat{a}_1a_2 + a_0a_1\hat{a}_2 \\
&= (-1)^0\hat{a}_0a_1a_2 + (-1)^1a_0\hat{a}_1a_2 + (-1)^2a_0a_1\hat{a}_2.
\end{aligned}$$

In this last form the extensions of the concepts of "inherited orientation" and "oriented boundary" to higher dimensions become obvious. If $\sigma_p = a_0 \ldots a_p$, $p > 0$, is an oriented p-simplex and $s_{p-1}^i = (a_0, \ldots, \hat{a}_i, \ldots, a_p)$ is a $(p-1)$-face of σ_p, then the *orientation inherited by s_{p-1}^i from σ_p* is $(-1)^i a_0 \ldots \hat{a}_i \ldots a_p$, where $a_0 \ldots \hat{a}_i \ldots a_p$ is the oriented $(p-1)$-simplex $a_0 \ldots a_{i-1}a_{i+1} \ldots a_p$.

Remark: We caution the reader not to be misled by the notation in this definition. The exponent i of -1 refers to the *position* occupied by the omitted vertex in the original ordering of the vertices of σ_p and not necessarily to its subscript (which is, of course, entirely arbitrary). Thus, although (a_1, a_2) is a face of both $\sigma_2 = a_0 a_1 a_2$ and $\tau_2 = a_1 a_0 a_2$, it inherits different orientations from σ_2 and τ_2: The orientation inherited from σ_2 is $(-1)^0 \hat{a}_0 a_1 a_2 = a_1 a_2$, while from τ_2 (a_1, a_2) inherits the orientation $(-1)^1 a_1 \hat{a}_0 a_2 = -a_1 a_2$.

Exercise 4-1. Show that the orientation inherited by $s^i_{p-1} = (a_0, \ldots, \hat{a}_i, \ldots, a_p)$ from $\sigma_p = a_0 \ldots a_p$ does not depend on the particular vertex ordering $a_0 \ldots a_p$ chosen to represent the orientation of σ_p.

Now, if $\sigma_p = a_0 \ldots a_p$ is an oriented simplex in an oriented complex K, then the *(oriented) boundary* of σ_p, denoted $\partial_p \sigma_p$ or simply $\partial \sigma_p$, is the $(p - 1)$-chain in K consisting of the sum of the $(p - 1)$-faces of σ_p, each with the orientation it inherits from σ_p, that is,

$$\partial_p \sigma_p = \partial \sigma_p = \sum_{i=0}^{p} (-1)^i a_0 \ldots \hat{a}_i \ldots a_p,$$

where it is understood that the coefficient of every $(p - 1)$-simplex in K which is not a face of σ_p is zero. Observe that the orientation inherited by a $(p - 1)$-face s^i_{p-1} of σ_p may or may not be the same as the orientation of s^i_{p-1} in K; this point will be discussed in greater detail in Section 4-7. To cover the case $p = 0$, we take $C_{-1}(K) = 0$ and $\partial_0 \sigma_0 = 0$ for every oriented 0-simplex (vertex) σ_0 of K. (We discuss an alternate approach to the zero-dimensional case in Section 4-6.) As sample computations we consider the oriented simplexes $\sigma_1 = a_0 a_1$, $\sigma_2 = a_0 a_1 a_2$ and $\sigma_3 = a_0 a_1 a_2 a_3$, which we assume lie in some oriented complex K. Then $\partial_1 \sigma_1 = a_1 - a_0$ and, as we already know, $\partial_2 \sigma_2 = a_1 a_2 - a_0 a_2 + a_0 a_1 = a_0 a_1 + a_1 a_2 + a_2 a_0$. In the three-dimensional case we have $\partial_3 \sigma_3 = a_1 a_2 a_3 - a_0 a_2 a_3 + a_0 a_1 a_3 - a_0 a_1 a_2$. Thus, each triangular face of σ_3 is provided with an orientation (sense of rotation) as shown in Figure 4-5, and the boundary of σ_3 is the sum of these oriented faces.

Now, each of these oriented plane faces will induce orientations on its edges. Observe that if two plane faces have a common edge, then this edge inherits opposite orientations (directions) from the two faces. For example, the edge (a_1, a_3) is a face of both $a_0 a_1 a_3$ and $a_1 a_2 a_3$. The orientation (a_1, a_3) inherits from $a_0 a_1 a_3$ is $(-1)^0 \hat{a}_0 a_1 a_3 = a_1 a_3$, whereas it inherits from $a_1 a_2 a_3$ the orientation $(-1)^1 a_1 \hat{a}_2 a_3 = -a_1 a_3$.

Remark: It follows from this last observation that the topological boundary of σ_3 is a "piecewise, smooth, orientable surface" in the sense of advanced calculus and thus $\bar{\sigma}_3$ is just the sort of region to which the three-dimensional analog of Green's Theorem, that is, the Divergence Theorem, can be applied. Each of these theorems affects a transition from an inte-

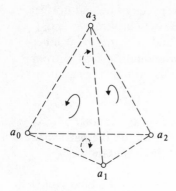

Figure 4-5

gral over a simplex to a lower dimensional integral over the oriented boundary of the simplex. Observe that this interpretation persists even in dimension one via the Fundamental Theorem of the Calculus.

Let K be an arbitrary oriented complex. Thus far we have defined operators ∂_p, which assign to each oriented p-simplex σ_p of K an element $\partial_p\sigma_p$ of $C_{p-1}(K)$. Since the σ_p generate $C_p(K)$, ∂_p extends (by "linearity") to a homomorphism $\partial_p : C_p(K) \to C_{p-1}(K)$. Specifically, if $c_p = \Sigma n_i\sigma_p^i$ is an element of $C_p(K)$, we define $\partial_p c_p = \partial_p(\Sigma n_i\sigma_p^i) = \Sigma n_i\partial_p\sigma_p^i$. Thus we have for any oriented complex K a sequence of homomorphisms of chain groups

$$\cdots \longrightarrow C_{p+1}(K) \xrightarrow{\partial_{p+1}} C_p(K) \xrightarrow{\partial_p} C_{p-1}(K) \xrightarrow{\partial_{p-1}} \cdots \xrightarrow{\partial_0} C_{-1}(K),$$

where by convention $C_i(K) = 0$ and $\partial_i = 0$ for all i greater than the algebraic dimension of K.

4-4   Cycles, boundaries, and homology groups

Recall that our object is to distinguish "closed" p-chains that "bound" from those that do not. The boundary operators ∂ are the basic device for both defining "closed" chains and making this distinction. As motivation for our definitions we return once again to the one-dimensional case. Intuitively, a one-dimensional curve is "closed" if its initial and terminal points coincide, that is, if it has no boundary. Consider, for example, the complex shown in Figure 4-4. Curves (chains) such as $c^1 = a_0a_1 + a_1a_2 + a_2a_0$ or $c^2 = a_4a_5 + a_5a_6 + a_6a_4$ are obviously "closed" and have the property that their boundaries are trivial, for example, $\partial_1c^1 = (a_1 - a_0) + (a_2 - a_1) + (a_0 - a_2) = 0$. However, a chain such as $c^3 = a_0a_1 + a_1a_4 + a_4a_5$ is not "closed" and has boundary $\partial_1c^3 = (a_1 - a_0) + (a_4 - a_1) + (a_5 - a_4) = a_5 - a_0$, which is not zero because $a_0 \neq a_5$, that is, because the initial and terminal points of c^3 do not coincide. In general, we shall refer to a p-chain c_p for which $\partial_p c_p = 0$ as a *p-cycle*. The set of all p-cycles in a complex K, denoted $Z_p(K)$, is precisely the kernel of the homomorphism ∂_p and is thus a subgroup of $C_p(K)$. Of course, $Z_p(K)$ is independent of the particular orientation chosen for K. It is also important to observe that $Z_p(K)$ consists not only of those p-chains that correspond to our intuitive notion of a closed curve such as c^1 and c^2, but also integral multiples and formal sums of such curves; for example, $3c^1 - 8c^2$ is a 1-cycle since $\partial_1(3c^1 - 8c^2) = 3(\partial_1 c^1) - 8(\partial_1 c^2) = 0$. Moreover, the closed curves we consider here need not be "simple" closed curves; for example, $c^4 = a_2 a_0 + a_0 a_1 + a_1 a_4 + a_4 a_5 + a_5 a_6 + a_6 a_4 + a_4 a_1 + a_1 a_2$ is a perfectly reasonable 1-cycle in the complex shown in Figure 4-4.

Intuitively, it seems clear that any p-chain that is the boundary of a $(p + 1)$-simplex (or, more generally, of a $(p + 1)$-chain) is a p-cycle (consider the 1-chain c^1 above). That this is, in fact, true follows directly from the next theorem.

Theorem 4-1. For any oriented complex K and any $p \geq 0$ the composition

$$C_{p+1}(K) \xrightarrow{\partial_{p+1}} C_p(K) \xrightarrow{\partial_p} C_{p-1}(K)$$

is the trivial homomorphism, that is, $\partial_p \circ \partial_{p+1} = 0$.

Proof: Since $\partial_p \circ \partial_{p+1}$ is a homomorphism, it will suffice to prove that $\partial_p \circ \partial_{p+1}(\sigma_{p+1}) = 0$ for each oriented $(p + 1)$-simplex $\sigma_{p+1} = a_0 \ldots a_{p+1}$ in K. For this we simply compute

$$\partial_p \circ \partial_{p+1}(\sigma_{p+1}) = \partial_p \sum_{j=0}^{p+1} (-1)^j a_0 \ldots \hat{a}_j \ldots a_{p+1}$$

$$= \sum_{j=0}^{p+1} (-1)^j \partial_p (a_0 \ldots \hat{a}_j \ldots a_{p+1})$$

$$= \sum_{j=0}^{p+1} (-1)^j \left[\sum_{i=0}^{j-1} (-1)^i a_0 \ldots \hat{a}_i \ldots \hat{a}_j \ldots a_{p+1} \right.$$

$$\left. + \sum_{i=j+1}^{p+1} (-1)^{i-1} a_0 \ldots \hat{a}_j \ldots \hat{a}_i \ldots a_{p+1} \right]$$

$$= \sum_{i<j} (-1)^{i+j} a_0 \ldots \hat{a}_i \ldots \hat{a}_j \ldots a_{p+1}$$

$$+ \sum_{i>j} (-1)^{i+j-1} a_0 \ldots \hat{a}_j \ldots \hat{a}_i \ldots a_{p+1}.$$

Each $a_0 \ldots \hat{a}_k \ldots \hat{a}_l \ldots a_{p+1}$ appears twice in $\partial_p \circ \partial_{p+1}(\sigma_{p+1})$, once with the coefficient $(-1)^{k+l}$ and once with the coefficient $(-1)^{k+l-1}$. Thus, $\partial_p \circ \partial_{p+1}(\sigma_{p+1}) = 0$ as required.   Q.E.D.

A p-chain in an oriented complex K that is the boundary of some $(p + 1)$-chain in K is called a p-*boundary;* the set of all such chains, denoted $B_p(K)$, is precisely the image of the homomorphism ∂_{p+1} and is thus a sub-group of $C_p(K)$. This subgroup is, of course, also independent of the ori-entation of K. According to Theorem 4-1, any p-boundary is a p-cycle, that is, $B_p(K) \subseteq Z_p(K)$ for any p. Since the p-boundaries of K are just those p-chains that "bound" regions of $|K|$, it follows that, in general, not every p-cycle is a p-boundary and it is on this observation that the suc-cess of our method depends. For example, the chains $c^1 = a_0a_1 + a_1a_2 + a_2a_0$ and $c^2 = a_4a_5 + a_5a_6 + a_6a_4$ in Figure 4-4 are both 1-cycles since $\partial_1 c^1 = \partial_1 c^2 = 0$. This fact is purely formal and does not depend at all on whether or not the simplexes $a_0a_1a_2$ and $a_4a_5a_6$ are present in the com-plex. Since $a_0a_1a_2$ is, in fact, in the complex, c^1 bounds a region of the polyhedron, that is, it is a 1-boundary and is thus a closed curve to be ig-nored from the point of view of homology theory. On the other hand, the simplex $a_4a_5a_6$ is not present in the given complex, so c^2 is not a 1-boundary (the simplex that c^2 wants to be the boundary of is missing) and consequently does not disconnect the space (it detects a "hole"). In general, the objects of real interest in homology theory are the p-cycles that are not p-boundaries, that is, $Z_p(K) - B_p(K)$. However, this set has the disadvantage of not being a group and, in addition, of not isolating the basic "types" of nonbounding closed curves in $|K|$. The closest we can come to $Z_p(K) - B_p(K)$ and still retain a group structure is the factor group $Z_p(K)/B_p(K)$ and we are thus led to the following definitions.

Let K be an oriented complex. For each $p \geq 0$, the pth *simplicial ho-mology group* of K, denoted $H_p(K)$, is defined by

$$H_p(K) = Z_p(K)/B_p(K).$$

Observe that $H_p(K)$, which is, of course, independent of the orientation of K, consists of equivalence classes of p-cycles, two p-cycles being iden-tified if they differ by a p-boundary. We say that two p-cycles c_p and c_p' are *homologous*, written $c_p \sim c_p'$, if $c_p - c_p'$ is a p-boundary. Thus, in $H_p(K)$ we do not distinguish between homologous cycles (because they are of the same basic "type"). This identification of homologous cycles is perhaps best understood from the point of view of integration theory. Consider, for example, the "triangular double annulus" R shown in Fig-ure 4-6 (a). Observe that R is the polyhedron of the geometric complex K shown in Figure 4-6 (b). (We have labeled only nine of the nineteen ver-tices in K.) Orient all of the twenty-two 2-simplexes of K in such a way that the indicated sense of rotation in each is counterclockwise; the re-maining simplexes may be oriented arbitrarily. Define 1-chains c^1, c^2, and c^3 as follows: c^1 is the sum of all the 1-simplexes of K around the outside triangle from a_0 to a_1 to a_2 to a_0 (there are twelve terms in the definition

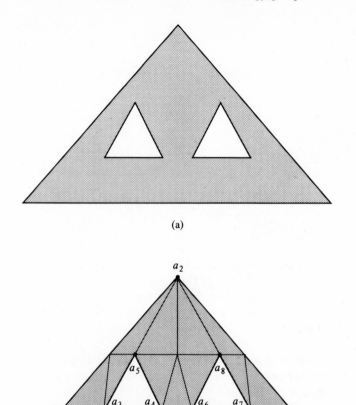

Figure 4-6

of c^1), $c^2 = a_3a_4 + a_4a_5 + a_5a_3$ and $c^3 = a_6a_7 + a_7a_8 + a_8a_6$. Then c^1 and $c^2 + c^3$ are 1-cycles since they have zero boundaries. Moreover, if c is the 2-chain in K consisting of the sum of all 2-simplexes of K with their counterclockwise orientations, then $\partial_2 c = c^1 - (c^2 + c^3)$ so $c^1 \sim c^2 + c^3$. Now, if we consider a differential form $f\,dx + g\,dy$ that is exact on some domain containing R, then the reason for not distinguishing between c^1 and $c^2 + c^3$ becomes clear: As paths of integration for $f\,dx + g\,dy$ they are entirely equivalent, that is,

$$\int_{c^1} f\,dx + g\,dy = \int_{c^2+c^3} f\,dx + g\,dy.$$

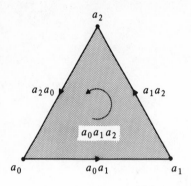

Figure 4-7

Since, for any oriented complex K and any $p \geq 0$, $C_p(K)$ is a free Abelian group over a finite set, the same is true of the subgroups $Z_p(K)$ and $B_p(K)$. It follows that $H_p(K) = Z_p(K)/B_p(K)$ is a finitely generated Abelian group, and we can apply the Structure Theorem for such groups (see Section 4-2) to obtain the next theorem.

Theorem 4-2. Let K be an oriented complex and p a nonnegative integer. Then $H_p(K)$ is isomorphic to the direct sum of a free Abelian group $Z^{\pi_p} = Z \oplus \cdots \oplus Z$ (π_p-summands) and finite cyclic groups Z_{γ_i}, $i = 1$, . . . , t.

The integer π_p is called the pth *Betti number* of K and the γ_i, $i = 1$, . . . , t are the pth *torsion coefficients* of K.

4-5   Elementary examples

We will show later that the homology groups $H_p(K)$ of an oriented complex K depend only on the topology (indeed, only on the homotopy type) of the underlying polyhedron $|K|$ and thus that they are, in particular, independent of the triangulation. Until this is proved, however, we are in a position to construct homology groups only for fixed (oriented) complexes, the choice of the orientation being a matter of convenience. Nevertheless, the forthcoming result on topological invariance should be kept in mind as we consider the following examples.

First consider the standard complex $K(s_2)$, where $s_2 = (a_0, a_1, a_2)$. We orient the simplexes of $K(s_2)$ as follows: $a_0 a_1 a_2$, $a_0 a_1$, $a_1 a_2$, $a_2 a_0$, a_0, a_1, a_2. (See Figure 4-7.) Observe that $C_p(K(s_2)) = 0$ and $\partial_p = 0$ for $p > 2$, so

$H_p(K(s_2)) = 0$ for $p > 2$. We are therefore left with the following sequence of homomorphisms:

$$C_2(K(s_2)) \xrightarrow{\partial_2} C_1(K(s_2)) \xrightarrow{\partial_1} C_0(K(s_2)) \xrightarrow{\partial_0} C_{-1}(K(s_2)).$$

Observe that $C_2(K(s_2))$ is generated by the single element $a_0a_1a_2$ and that a typical element of it is of the form $ma_0a_1a_2$ for $m \in Z$. For any such element of $C_2(K(s_2))$ we have $\partial_2(ma_0a_1a_2) = m\,\partial_2(a_0a_1a_2) = m(a_0a_1 + a_1a_2 + a_2a_0)$. In particular, ker $\partial_2 = Z_2(K(s_2)) = 0$ and Im $\partial_2 = B_1(K(s_2)) = \{m(a_0a_1 + a_1a_2 + a_2a_0) : m \in Z\} \approx Z$. Since $\partial_3 = 0$, Im $\partial_3 = B_2(K(s_2)) = 0$, so $H_2(K(s_2)) = Z_2(K(s_2))/B_2(K(s_2)) \approx 0$.

Now, $C_1(K(s_2))$ is generated by the three oriented 1-simplexes a_0a_1, a_1a_2, and a_2a_0, so a typical element has the form $pa_0a_1 + qa_1a_2 + ra_2a_0$ for integers p, q, and r. For any such element, $\partial_1(pa_0a_1 + qa_1a_2 + ra_2a_0) = (r - p)a_0 + (p - q)a_1 + (q - r)a_2$, which is zero iff $p = q = r$. Thus, ker $\partial_1 = Z_1(K(s_2)) = \{m(a_0a_1 + a_1a_2 + a_2a_0) : m \in Z\} \approx Z$, so $H_1(K(s_2)) = Z_1(K(s_2))/B_1(K(s_2)) \approx Z/Z \approx 0$.

All that remains is to compute $H_0(K(s_2))$. Now, the elements of $H_0(K(s_2))$ are equivalence classes of homologous 0-cycles. By definition each vertex a_i has zero boundary and is thus a 0-cycle. We claim that two such 0-cycles are homologous. Clearly, $a_i \sim a_i$ for any i, so let us suppose that a_i and a_j are distinct vertices of $K(s_2)$. Then a_i and a_j are "connected by a 1-chain" in $K(s_2)$, that is, there are edges in $K(s_2)$ leading from a_i to a_j. For example, a_0 and a_2 are connected by the 1-chain $a_0a_1 + a_1a_2$, while a_1 and a_2 are connected by the 1-chain a_1a_2. Note that the boundary of the 1-chain connecting a_i and a_j is $a_j - a_i$ so $a_i \sim a_j$ as required. Consequently, all of the oriented 0-simplex (vertices) of $K(s_2)$ are in the same equivalence class modulo $B_0(K(s_2))$ so $H_0(K(s_2))$ is a finitely generated Abelian group with one generator and is therefore cyclic. Moreover, $H_0(K(s_2))$ is infinite since, for any i, the 0-cycles ma_i and na_i are homologous iff $m = n$: $ma_i - na_i = (m - n)a_i$ and no 1-chain has boundary of the form ka_i unless $k = 0$. Thus, we have proved $H_0(K(s_2)) \approx Z$, and we summarize our results in the following proposition.

Proposition 4-3. Let s_2 be a 2-simplex and $K(s_2)$ the complex consisting of the faces of s_2. Then $H_0(K(s_2)) \approx Z$ and $H_p(K(s_2)) \approx 0$ for all $p > 0$.

The method by which we proved that $H_0(K(s_2)) \approx Z$ suggests a straightforward generalization. Let K be an oriented complex and suppose that a and b are two vertices of K. We say that a *is connected to* b *by a 1-chain in* K if there is a sequence $a_{i_0}, \ldots, a_{i_k}$ of vertices in K with $a_{i_0} = a$, $a_{i_k} = b$, and $(a_{i_{j-1}}, a_{i_j})$ a 1-simplex in K for each $j = 1, \ldots, k$. This defines an equivalence relation on K_0, the equivalence classes of which are

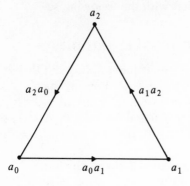

Figure 4-8

all sets of vertices that determine subcomplexes of K. Thus, K is the union of disjoint subcomplexes $K^1, \ldots, K^q$ of K such that two vertices a and b of K can be connected by a 1-chain in K iff a and b belong to the same subcomplex K^r, $1 \le r \le q$, of K. Then $K^1, \ldots, K^q$ are called the *components* of K and K is said to be *connected* if $q = 1$.

Exercise 4-2. Justify this terminology by showing that, for any oriented complex K with components $K^1, \ldots, K^q, |K^1|, \ldots, |K^q|$ are the (connected) components of $|K|$. Hint: A polyhedron is connected iff it is pathwise connected. Show that a complex K is connected iff $|K|$ is connected. For the sufficiency use Theorem 2-19.

Lemma 4-4. Let K be an oriented complex with components $K^1, \ldots, K^q$. Then, for any $p \ge 0$, $H_p(K) \approx H_p(K^1) \oplus \cdots \oplus H_p(K^q)$.

Exercise 4-3. Prove Lemma 4-4.   Q.E.D.

Theorem 4-5. Let K be an oriented complex with components $K^1, \ldots, K^q$. Then $H_0(K) \approx Z^q = Z \oplus \cdots \oplus Z$ (q-summands).

Exercise 4-4. Prove Theorem 4-5.   Q.E.D.

Next we compute the homology groups of the 1-skeleton $K_1(s_2)$ of $K(s_2)$. Let each simplex of $K_1(s_2)$ be oriented as they were in $K(s_2)$ above, that is, the oriented simplexes of $K_1(s_2)$ are a_0a_1, a_1a_2, a_2a_0, a_0, a_1, and a_2. (See Figure 4-8.) Since $C_p(K_1(s_2)) = 0$ for $p > 1$, $H_p(K_1(s_2)) = 0$ for $p > 1$. By Theorem 4-5 $H_0(K_1(s_2)) \approx Z$ and we need only compute $H_1(K_1(s_2))$. Since $C_1(K_1(s_2)) = C_1(K(s_2))$ and $C_0(K_1(s_2)) = C_0(K(s_2))$, we have ker $\partial_1 = Z_1(K_1(s_2)) \approx Z$ as before. However, $C_2(K_1(s_2)) = 0$, so Im $\partial_2 = B_1(K_1(s_2)) = 0$ in this case. Thus, $H_1(K_1(s_2)) \approx Z/0 \approx Z$ and we have proved the proposition:

Proposition 4-6. Let s_2 be a 2-simplex and $K_1(s_2)$ the 1-skeleton of $K(s_2)$. Then $H_0(K_1(s_2)) \approx H_1(K_1(s_2)) \approx Z$ and $H_p(K_1(s_2)) \approx 0$ for each $p > 1$.

Exercise 4-5. Let $s_0 = (a_0)$ and $s_1 = (a_0, a_1)$. By choosing appropriate orientations, compute the homology groups $H_p(K(s_0))$, $H_p(K(s_1))$, and $H_p(K_0(s_1))$ for each $p > 0$.

4-6   Cone complexes, augmented complexes, and the homology groups $H_p(K(s_{n+1}))$

We now begin the task of generalizing these last few examples to arbitrary dimensions. First we compute the homology groups of the standard complex $K(s_{n+1})$, where $s_{n+1} = (a_0, \ldots, a_{n+1})$, $n \geqslant 2$. Now, although they are theoretically possible, it is clear that arguments analogous to those used to compute the groups $H_p(K(s_2))$ will become rather unpleasant in the general case. Fortunately, however, $K(s_{n+1})$ possesses a property that makes the computation of its homology groups relatively easy. We begin our analysis with a general discussion of this property.

Let K be a geometric complex in $\mathbf{R}^m$ and v a point of $\mathbf{R}^m$ that is in general position relative to the polyhedron $|K|$. In particular, v is in general position relative to $|K(s_p)| = \bar{s}_p$ for each simplex $s_p = (a_0, \ldots, a_p)$ in K. By Lemma 2-11 the set $\{v, a_0, \ldots, a_p\}$ is geometrically independent and $v * |K(s_p)| = [v, a_0, \ldots, a_p]$. Thus,

$$(1) \qquad \bigcup_{s \in K} v * |K(s)| = v * |K|.$$

Moreover, $\bigcup_{s \in K} v * |K(s)|$ is a union of simplexes in $\mathbf{R}^m$ since each $v * |K(s_p)|$ is the union of the faces of $(v, a_0, \ldots, a_p)$. Specifically, $\bigcup_{s \in K} v * |K(s)|$ is the union of the collection of simplexes in $\mathbf{R}^m$ consisting of
(a)   The simplexes of K.
(b)   The vertex (v).
(c)   The simplexes of the form $(v, a_0, \ldots, a_p)$, where $(a_0, \ldots, a_p) \in K$.

Exercise 4-6. Show that this collection of simplexes in $\mathbf{R}^m$ forms a complex and thus by equation (1) a triangulation of $v * |K|$.

Thus, for any complex K in $\mathbf{R}^m$ and any point v in $\mathbf{R}^m$ with $(v, |K|)$ in general position, we may define the *cone complex of K over v*, denoted vK, as follows:

(2) $|vK| = v * |K|$

and the simplexes of vK are those defined by (a), (b), and (c) above. More generally, a complex L in $\mathbf{R}^m$ is called a *cone complex* if there is a complex K in $\mathbf{R}^m$ and a $v \in \mathbf{R}^m$ in general position relative to $|K|$ such that $L = vK$.

Example 4-1. Let $s_{n+1} = (a_0, \ldots, a_n, a_{n+1})$ be an $(n + 1)$-simplex and $s_n = (a_0, \ldots, a_n)$ the face of s_{n+1} opposite a_{n+1}. By Lemma 2-11 $K(s_{n+1}) = a_{n+1}K(s_n)$, so $K(s_{n+1})$ is a cone complex. We claim, in addition, that the first barycentric subdivision of $K(s_{n+1})$ is also a cone complex. Indeed, if $K_n(s_{n+1})$ is the n-skeleton of $K(s_{n+1})$, then $b(s_{n+1})$ is in general position relative to $|K'_n(s_{n+1})| = |K_n(s_{n+1})|$ by Lemma 2-10, and by our construction of the barycentric subdivision in Proposition 2-15 and Theorem 2-16 $K'(s_{n+1}) = b(s_{n+1})K'_n(s_{n+1})$.

We shall say that L is an *oriented cone complex* if $L = vK$, where K is an oriented complex and the simplexes of L are oriented as follows: the simplexes of K retain their orientations from K, while if $a_0 \ldots a_p$ is such a simplex, $(v, a_0, \ldots, a_p)$ is provided with the orientation $va_0 \ldots a_p$ ((v) may be oriented arbitrarily). Any cone complex $L = vK$ can be made into an oriented cone complex by orienting K and providing L with this *standard orientation*. If $c_p = \Sigma n_i a_0^i \ldots a_p^i$ is a p-chain in K, we denote by vc_p the $(p + 1)$-chain in $L = vK$ defined by

(3) $vc_p = v \sum n_i a_0^i \ldots a_p^i = \sum n_i va_0^i \ldots a_p^i.$

We shall find that all cone complexes have isomorphic homology groups in each dimension. Indeed, we show that, for any cone complex L, $H_0(L) \approx Z$ and $H_p(L) \approx 0$ for each $p > 0$. The introduction of one more technical device, which essentially amounts to another, sometimes more convenient method of handling the "exceptional" case of 0-dimensional homology groups, will facilitate the discussion. Let K be an oriented complex. The *augmented complex* $\tilde{K}$ of K consists of all the oriented simplexes of K together with one abstract (-1)-dimensional simplex σ_{-1} that acts as a face of every vertex in K and for which $\partial_0\sigma_0 = \sigma_{-1}$ for every σ_0 in K; $\tilde{K}$ is obtained by *augmenting* K. Thus, $C_p(\tilde{K}) = C_p(K)$ for all $p \geqslant 0$, but whereas $C_{-1}(K) = 0$, $C_{-1}(\tilde{K})$ is the free Abelian group on the one generator σ_{-1}. The addition of the simplex σ_{-1} is purely formal and has no effect on the polyhedron of K, that is, $|\tilde{K}| = |K|$. Observe that every subcomplex of an augmented complex is augmented. The cone complex of an augmented complex over some point v is defined as above with $v\sigma_{-1} = v$.

Theorem 4-7. Let K be an oriented complex and $\tilde{K}$ its augmented complex. Then $H_p(\tilde{K}) \approx H_p(K)$ for all $p > 0$ and $H_0(K) \approx H_0(\tilde{K}) \oplus Z$.

Proof: Since the simplicial structures of K and $\tilde{K}$ coincide for $p \geqslant 0$, the first statement is clear. To obtain the second isomorphism, we first observe that $B_0(\tilde{K}) = B_0(K)$ since $\partial_1 : C_1 \to C_0$ is the same for both K and $\tilde{K}$. Thus, it will suffice to show that $Z_0(K) \approx Z_0(\tilde{K}) \oplus Z$, for then $H_0(K) = Z_0(K)/B_0(K) \approx (Z_0(\tilde{K}) \oplus Z)/B_0(K) \approx H_0(\tilde{K}) \oplus Z$.

Exercise 4-7. Complete the proof by showing that $Z_0(K) \approx Z_0(\tilde{K}) \oplus Z$. Hint: Let $\{a_0, \ldots, a_k\}$ be the vertices of K. Then $\{a_0, a_1 - a_0, \ldots, a_k - a_0\}$ generates $Z_0(K)$. Show that $\{a_1 - a_0, \ldots, a_k - a_0\}$ generates $Z_0(\tilde{K})$.   Q.E.D.

Theorem 4-8. Let L be an oriented cone complex. Then $H_p(\tilde{L}) \approx 0$ for each $p \geqslant 0$.

Proof: Since L is an oriented cone complex, there is an oriented complex K and a point v in the ambient Euclidean space with $(v, |K|)$ in general position such that $L = vK$ and L has the standard orientation determined by the orientation of K. First observe that, since a cone complex is connected, Theorem 4-5 implies that $H_0(L) \approx Z$. But by Theorem 4-7 $H_0(L) \approx H_0(\tilde{L}) \oplus Z$, so $H_0(\tilde{L}) \approx 0$. Now let $p > 0$ be fixed. We prove that $H_p(\tilde{L}) \approx 0$ by showing that $Z_p(\tilde{L}) \subseteq B_p(\tilde{L})$ and thus that $Z_p(\tilde{L}) = B_p(\tilde{L})$. Let $c_p \in Z_p(\tilde{L})$. We must show that c_p is a p-boundary. Now, $C_p(\tilde{L})$ is generated by the oriented p-simplexes of K together with simplexes of the form $va_0 \ldots a_{p-1}$, where $a_0 \ldots a_{p-1}$ is an oriented $(p - 1)$-simplex of K. Thus, c_p has a unique representation of the form $c_p = vc_{p-1} + d_p$, where $c_{p-1} \in C_{p-1}(K)$ and $d_p \in C_p(K)$. Apply the boundary operator ∂_p to both sides of this equality to obtain

$$\partial_p c_p = 0 = \partial_p(vc_{p-1}) + \partial_p d_p$$
$$= c_{p-1} - v\,\partial_{p-1}c_{p-1} + \partial_p d_p.$$

Now, the two chains $c_{p-1} + \partial_p d_p$ and $v\,\partial_{p-1}c_{p-1}$ have no generators in common. Since their sum is zero, each must be zero. Thus, computing $\partial_{p+1}(vd_p)$ yields $\partial_{p+1}(vd_p) = d_p - v\,\partial_p d_p = d_p + vc_{p-1} = c_p$, so $c_p \in B_p(\tilde{L})$ as required.   Q.E.D.

Corollary 4-9. Let L be an oriented cone complex. Then $H_0(L) \approx Z$ and $H_p(L) \approx 0$ for all $p > 0$.

By Example 4-1 $K(s_{n+1})$ is a cone complex, so we obtain, in particular, the following theorem.

Theorem 4-10. Let s_{n+1} be an $(n + 1)$-simplex, $n \geqslant -1$, and $K(s_{n+1})$ the complex consisting of the faces of s_{n+1}. Then $H_0(K(s_{n+1})) \approx Z$ and $H_p(K(s_{n+1})) \approx 0$ for all $p > 0$.

4-7 Incidence numbers and the homology groups $H_p(K_n(s_{n+1}))$

Next, we turn our attention to the n-skeleton $K_n(s_{n+1})$ of the standard $(n + 1)$-complex $K(s_{n+1})$, $n \geqslant 1$. Observe that $H_p(K_n(s_{n+1})) \approx 0$ for $p > n$ since $K_n(s_{n+1})$ has no p-simplexes for such p. For $p \leqslant n$ the simplicial structures of $K_n(s_{n+1})$ and $K(s_{n+1})$ agree, so that, for $p < n$, $H_p(K(s_{n+1})) \approx H_p(K(s_{n+1}))$. Thus by Theorem 4-10 $H_0(K_n(s_{n+1})) \approx Z$ and $H_p(K_n(s_{n+1})) \approx 0$ for $0 < p < n$. All that remains is the computation of $H_n(K_n(s_{n+1}))$, and for this we require one more new idea.

Let K be an oriented complex, σ_p an oriented p-simplex in K, and σ_{p-1} an oriented $(p - 1)$-simplex in K. Then σ_{p-1} may or may not be a face of σ_p and, if it is, its orientation may or may not agree with the orientation it would inherit from σ_p. To keep track of these various possibilities, we define the *incidence number* $[\sigma_p, \sigma_{p-1}]$ for σ_p and σ_{p-1} as follows:

$$
\begin{aligned}
[\sigma_p, \sigma_{p-1}] &= 0 && \text{if } \sigma_{p-1} \text{ is not a face of } \sigma_p \\
&= 1 && \text{if } \sigma_{p-1} < \sigma_p \text{ and if the orientation} \\
& && \text{of } \sigma_{p-1} \text{ in } K \text{ agrees with} \\
& && \text{the orientation inherited from } \sigma_p \\
&= -1 && \text{if } \sigma_{p-1} < \sigma_p \text{ and if the orientation} \\
& && \text{of } \sigma_{p-1} \text{ in } K \text{ differs from} \\
& && \text{the orientation inherited from } \sigma_p.
\end{aligned}
$$

Since $\partial_p \sigma_p$ is the $(p - 1)$-chain in K consisting of the sum of the $(p - 1)$-faces of σ_p, each with the orientation it inherits from σ_p, we have

$$
(4) \qquad \partial_p \sigma_p = \sum_i [\sigma_p, \sigma_{p-1}^i] \sigma_{p-1}^i,
$$

where $\{\sigma_{p-1}^i\}$ is the set of all oriented $(p - 1)$-simplexes in K. Thus, if $c_p = \Sigma_j m_j \sigma_p^j$ is a p-chain in K, we find that

$$
(5) \qquad \partial_p c_p = \partial_p \left(\sum_j m_j \sigma_p^j \right)
$$

$$
= \sum_i \left(\sum_j [\sigma_p^j, \sigma_{p-1}^i] m_j \right) \sigma_{p-1}^i.
$$

Now, to compute $H_n(K)$, where $K = K_n(s_{n+1})$, $s_{n+1} = (a_0, a_1, \ldots, a_n, a_{n+1})$, we proceed as follows: Fix an orientation for s_{n+1}, say, $\sigma_{n+1} = a_0 a_1 \ldots a_n a_{n+1}$ and provide the n-simplexes of K with the orientations they inherit from σ_{n+1}. (Note that σ_{n+1} is not in K – we use it simply to provide a convenient orientation for the n-simplexes of K.) Now orient the remaining simplexes of K arbitrarily. Then $B_n(K) = \mathrm{Im}\ \partial_{n+1} = 0$, so $H_n(K) = Z_n(K)$. Let $c_n = \Sigma\ m_j\ \sigma_n^j$ be an n-chain in K, where $\{\sigma_n^j\}$ is the set of all oriented n-simplexes of K. We claim that c_n is an n-cycle iff all of the m_j are equal. To see this, let σ_{n-1}^i be a fixed, but arbitrary $(n-1)$-simplex of K. Then σ_{n-1}^i is a face of precisely two n-simplexes in K, which we denote by σ_n^k and σ_n^l.

Exercise 4-8. Show that σ_{n-1}^i inherits opposite orientations from σ_n^k and σ_n^l.

Thus, $[\sigma_n^k, \sigma_{n-1}^i] = -[\sigma_n^l, \sigma_{n-1}^i]$. Now by equation (5) the coefficient of σ_{n-1}^i in the $(n-1)$-chain $\partial_n c_n$ is

$$\sum_j [\sigma_n^j, \sigma_{n-1}^i] m_j = [\sigma_n^k, \sigma_{n-1}^i] m_k + [\sigma_n^l, \sigma_{n-1}^i] m_l$$
$$= [\sigma_n^k, \sigma_{n-1}^i] m_k - [\sigma_n^k, \sigma_{n-1}^i] m_l$$
$$= [\sigma_n^k, \sigma_{n-1}^i](m_k - m_l) = \pm(m_k - m_l).$$

Since this is true for each $(n-1)$-simplex σ_{n-1}^i in K, it follows that c_n is an n-cycle, that is, $\partial_n c_n = 0$, iff all of the m_j are equal. Thus, $H_n(K) = Z_n(K) = \{\Sigma_j m \sigma_n^j : m \in Z\}$. The map $\varphi : H_n(K) \to Z$ defined by

$$\varphi\left(\sum_j m\ \sigma_n^j\right) = \varphi\left(m \sum_j \sigma_n^j\right) = m$$

is an isomorphism, so $H_n(K) \approx Z$ and we have proved the next theorem.

Theorem 4-11. Let s_{n+1} be an $(n+1)$-simplex and $K_n(s_{n+1})$ the n-skeleton of $K(s_{n+1})$. Then $H_0(K_n(s_{n+1})) \approx H_n(K_n(s_{n+1})) \approx Z$ and $H_p(K_n(s_{n+1})) \approx 0$ for all $0 < p < n$ and all $p > n$.

Remark: If s_{n+1} is an $(n+1)$-simplex, then $|K(s_{n+1})|$ is homeomorphic to the $(n+1)$-ball B^{n+1} and $|K_n(s_{n+1})|$ is homeomorphic to the n-sphere S^n (see Section 2-4). Thus, once we have proved the promised result on topological invariance, Theorems 4-10 and 4-11 will yield the homology groups of B^{n+1} and S^n, respectively.

The examples we have considered thus far will suffice for all of the applications we have in mind and should provide a satisfactory introduction to the methods by which such computations are carried out. Before turning our attention to results of a more general character, we observe (as the reader has no doubt already done) that we have encountered no examples of nonzero torsion coefficients. From the structure of the groups Z_γ it is clear that the existence of nonzero torsion coefficients for a complex K indicates the presence of cycles in K which, although they themselves do not bound, have the property that some integral multiple of them does bound. Now, there are indeed complexes in which this rather peculiar phenomenon occurs (e.g., triangulations of the projective plane or the Klein bottle – see Hilton and Wylie), but we shall have no occasion to consider them.

Exercise 4-9. Let K be an oriented complex of dimension n. For each $0 \leqslant p \leqslant n$, let α_p denote the number of p-simplexes in K and $\pi_p(K)$ the pth Betti number of K. Prove the *Euler–Poincaré Formula:*

$$\sum_{p=0}^{n} (-1)^p \alpha_p = \sum_{p=0}^{n} (-1)^p \pi_p(K).$$

Hint: Show that rank $H_p(K) =$ rank $C_p(K) -$ rank $B_p(K) -$ rank $B_{p-1}(K)$ and take the "alternating sum" of both sides.

Remark: The quantity $\chi(K) = \sum_{p=0}^{n}(-1)^p \alpha_p$ is called the *Euler–Poincaré characteristic* of K.

4-8   Elementary homological algebra

Our next goal is a proof of the topological invariance of the simplicial homology groups. The route we must follow toward this result is rather arduous, but also quite interesting and all of our efforts will be well rewarded. The arguments we give happen to assume a particularly elegant form when stated in the language of "homological algebra," so this section is devoted to a very brief account of the most rudimentary definitions and results of this subject. Most of the new concepts should be sufficiently well motivated by our work in homology theory to require no additional comment.

A sequence $\{C_p\}_{p=-\infty}^{\infty}$ of Abelian groups together with a sequence $\{\partial_p\}_{p=-\infty}^{\infty}$ of homomorphisms $\partial_p : C_p \to C_{p-1}$ such that $\partial_{p-1} \circ \partial_p = 0$ for each p constitute what is called a *chain complex* $C = (C_p, \partial_p)$. The homomorphisms ∂_p are called *boundary maps*. Given two chain complexes $C = (C_p, \partial_p^C)$ and $D = (D_p, \partial_p^D)$, a sequence $\varphi = \{\varphi_p\}_{p=-\infty}^{\infty}$ of homomorphisms $\varphi_p : C_p \to D_p$ for which $\partial_p^D \circ \varphi_p = \varphi_{p-1} \circ \partial_p^C$ for each p is called a *chain map* or *homomorphism* of the chain complexes, and we write $\varphi : C \to D$.

$$\cdots \longrightarrow C_p \xrightarrow{\;\partial_p^C\;} C_{p-1} \longrightarrow \cdots$$

$$\varphi_p \downarrow \qquad\qquad \downarrow \varphi_{p-1}$$

$$\cdots \longrightarrow D_p \xrightarrow[\partial_p^D]{} D_{p-1} \longrightarrow \cdots$$

If $\varphi = \{\varphi_p\}_{p=-\infty}^{\infty} : C \to D$ and $\psi = \{\psi_p\}_{p=-\infty}^{\infty} : D \to E$ are chain maps, then their *composition* $\psi \circ \varphi = \{\psi_p \circ \varphi_p\}_{p=-\infty}^{\infty} : C \to E$ is also a chain map. A chain map φ is an *isomorphism* if each φ_p is an isomorphism of Abelian groups. Denoting the chain map $\{id_{C_p}\}_{p=-\infty}^{\infty}$ of C onto C by id_C for any chain complex C, it follows that $\varphi : C \to D$ is an isomorphism iff there is a chain map $\psi : D \to C$ such that $\psi \circ \varphi = id_C$ and $\varphi \circ \psi = id_D$. If there is an isomorphism of C onto D, we say that C and D are *isomorphic* and write $C \approx D$. If $C = (C_p, \partial_p)$ is a chain complex and C_p' is a subgroup of C_p for each p such that $\partial_{p+1}(C_{p+1}') \subseteq C_p'$, then $C' = (C_p', \partial_p|C_p')$ is a chain complex and is called a *(chain) subcomplex* of C. In this case the sequence $\{i_p\}_{p=-\infty}^{\infty}$ of inclusion maps $i_p : C_p' \hookrightarrow C_p$ is a chain map of C' into C.

Now suppose $C = (C_p, \partial_p^C)$ and $D = (D_p, \partial_p^D)$ are chain complexes and $\varphi : C \to D$ is a chain map. Then ker φ_p is a subgroup of C_p for each p. Moreover, if $c_p \in$ ker φ_p, then $\partial_p^D \circ \varphi_p(c_p) = \partial_p^D(0) = 0$ and thus, since φ is a chain map, $\varphi_{p-1} \circ \partial_p^C(c_p) = 0$ as well, that is, $\partial_p^C c_p \in$ ker φ_{p-1}. Thus, ∂_p^C (ker φ_p) $\subseteq$ ker φ_{p-1}, so (ker φ_p, $\partial_p^C|$ker φ_p) is a subcomplex of C which we call the *kernel* of φ and denote ker φ. Similarly, the *image* of φ, defined by Im $\varphi =$ (Im φ_p, $\partial_p^D|$Im φ_p) is a subcomplex of D. In particular, if $C = (C_p, \partial_p)$, then $\partial = \{\partial_p\}_{p=-\infty}^{\infty}$ may be regarded as a chain map of C into $(C_{p-1}, \partial_{p-1})$:

$$\cdots \longrightarrow C_{p+1} \xrightarrow{\;\partial_{p+1}\;} C_p \xrightarrow{\;\partial_p\;} C_{p-1} \longrightarrow \cdots$$

$$\downarrow \partial_{p+1} \qquad\quad \downarrow \partial_p \qquad\quad \downarrow \partial_{p-1}$$

$$\cdots \longrightarrow C_p \xrightarrow[\partial_p]{} C_{p-1} \xrightarrow[\partial_{p-1}]{} C_{p-2} \longrightarrow \cdots$$

Thus, ker $\partial = $ (ker ∂_p, $\partial_p|$ker ∂_p) and Im $\partial = $ (Im ∂_{p+1}, $\partial_p|$Im ∂_{p+1}) are both subcomplexes of C. In fact, since Im ∂_{p+1} is actually a subgroup of ker ∂_p for each p, Im ∂ is a subcomplex of ker ∂:

$$\cdots \longrightarrow \text{Im } \partial_{p+2} \xrightarrow{\;\partial\;} \text{Im } \partial_{p+1} \xrightarrow{\;\partial\;} \text{Im } \partial_p \longrightarrow \cdots$$

$$\uparrow i_{p+2} \qquad\qquad \uparrow i_{p+1} \qquad\qquad \uparrow i_p$$

$$\cdots \longrightarrow \text{ker } \partial_{p+1} \xrightarrow[\partial]{} \text{ker } \partial_p \xrightarrow[\partial]{} \text{ker } \partial_{p-1} \longrightarrow \cdots$$

The elements of ker ∂_p and Im ∂_{p+1} are called the *p-cycles* and *p-bound-aries* of C, respectively; ker ∂ is generally denoted $Z(C)$, while Im ∂ is written $B(C)$.

Recall that if G and H are Abelian groups, G' and H' are subgroups of G and H, respectively, and $h : G \to H$ is a homomorphism with $h(G') \subseteq H'$; then h induces a homomorphism $\hat{h} : G/G' \to H/H'$ defined by $\hat{h}([g]) = [h(g)]$ for each g in G and the following diagrams commute:

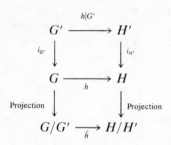

Now, if $C = (C_p, \partial_p)$ is a chain complex and $C' = (C'_p, \partial_p|C'_p)$ is a subcomplex of C, then the groups $\hat{C}_p = C_p/C'_p$ together with the homomorphisms $\hat{\partial}_p : \hat{C}_p \to \hat{C}_{p-1}$ defined by $\hat{\partial}_p([c_p]) = [\partial_p c_p]$ form a chain complex $\hat{C} = (\hat{C}_p, \hat{\partial}_p)$ called the *factor complex* of C by C' and denoted $\hat{C} = C/C'$. In particular, for any $C = (C_p, \partial_p)$, $B(C)$ is a subcomplex of $Z(C)$, and we may form the factor complex $H(C) = Z(C)/B(C)$. The groups $\ker \partial_p/\operatorname{Im} \partial_{p+1}$ are called the *homology groups* of the chain complex C and are denoted $H_p(C)$, while $H(C)$ is called the *homology complex* of C. Observe that in this case all of the connecting homomorphisms $\hat{\partial}_p : H_p(C) \to H_{p-1}(C)$ are identically zero. Indeed, $\hat{\partial}_p$ is defined by $\hat{\partial}_p([c_p]) = [\partial_p c_p]$ for every $c_p \in \ker \partial_p$. But $\partial_p c_p = 0$ for any $c_p \in \ker \partial_p$ so that $\hat{\partial}_p([c_p]) = 0$ and $\hat{\partial}_p = 0$. Thus, $H(C) = (H_p(C), 0)$.

Theorem 4-12. Let C and D be chain complexes. A chain map $\varphi : C \to D$ induces a chain map $\varphi_* : H(C) \to H(D)$ of homology complexes with the following properties:
(a)   If $C = D$ and $\varphi = \mathrm{id}_C$, then $\varphi_* = (\mathrm{id}_C)_* = \mathrm{id}_{H(C)}$.
(b)   If E is another chain complex and $\psi : D \to E$ another chain map, then $(\psi \circ \varphi)_* = \psi_* \circ \varphi_*$.

Exercise 4-10. Prove Theorem 4-12.   Q.E.D.

To complement Theorem 4-12, we require a condition on chain maps sufficient to guarantee that they induce identical chain maps in homology. Two chain maps $\varphi, \psi : C \to D$ are said to be *chain homotopic*, written $\varphi \overset{c}{=} \psi$, if there is a sequence $\gamma = \{\gamma_p\}_{p=-\infty}^{\infty}$ of homomorphisms $\gamma_p : C_p \to D_{p+1}$ such that for each p

$$\partial^D_{p+1} \circ \gamma_p + \gamma_{p-1} \circ \partial^C_p = \varphi_p - \psi_p.$$

The sequence γ is called a *chain homotopy* from φ to ψ, and we write γ : $\varphi \overset{c}{=} \psi$.

Exercise 4-11. Show that chain homotopy is an equivalence relation on the set of all chain maps of C into C.

Theorem 4-13. Let C and D be chain complexes and φ, ψ : $C \to D$ chain homotopic chain maps. Then the induced homomorphisms φ_*, ψ_* : $H(C) \to H(D)$ are identical.

Proof: Let c_p be a p-cycle of C. Then $\varphi(c_p)$ and $\psi(c_p)$ are p-cycles of D, and we must show that they differ by a p-boundary of D and are therefore identified in the factor group. Let $\gamma : \varphi \overset{c}{=} \psi$. Then, by definition, $\partial^D_{p+1} \circ \gamma_p(c_p) + \gamma_{p-1} \circ \partial^C_p(c_p) = \varphi_p(c_p) - \psi_p(c_p)$. But $\partial^C_p(c_p) = 0$ so $\gamma_{p-1} \circ \partial^C_p(c_p) = 0$ and we obtain $\partial^D_{p+1} \circ \gamma_p(c_p) = \varphi_p(c_p) - \psi_p(c_p)$, so that $\varphi_p(c_p)$ and $\psi_p(c_p)$ differ by the boundary of $\gamma_p(c_p)$.   Q.E.D.

Two chain complexes C and D are said to be *chain equivalent* if there exist chain maps $\varphi : C \to D$ and $\psi : D \to C$ such that $\psi \circ \varphi \overset{c}{=} \text{id}_C$ and $\varphi \circ \psi \overset{c}{=} \text{id}_D$.

Corollary 4-14. Chain equivalent complexes have isomorphic homology complexes.

Exercise 4-12. Prove Corollary 4-14.   Q.E.D.

4-9   The homology complex of a geometric complex

With these algebraic preliminaries out of the way we now return to our investigation of homology groups for geometric complexes. Thus, we let K be an oriented geometric complex. By convention $C_p(K) = 0$ for all p greater than the algebraic dimension of K as well as all $p < 0$ in the un-augmented case and all $p < -1$ in the augmented case. Consequently, we obtain from K a chain complex $C(K) = (C_p(K), \partial_p)$ and thereby a chain complex $H(K) = H(C(K))$, which we call the *homology complex* of K.

We show next that an arbitrary simplicial map $\varphi : |K| \to |L|$ induces a chain map of $C(K)$ into $C(L)$ and thus by Theorem 4-12 a chain map of $H(K)$ into $H(L)$ and that, moreover, this assignment of chain maps in homology to simplicial maps is functorial. Now, since φ maps simplexes of K onto simplexes of L, it would seem natural to define φ on the chains of K by linearity and thus induce a chain map. However, φ may lower the dimension of a simplex in K, so that this "natural" image of a p-chain in K may be a mixture of chains of various dimensions in L. To avoid this difficulty, we simply ignore (i.e., map to zero) all of those simplexes in K whose dimension is lowered by φ. More precisely, let K and L be oriented geometric complexes and $\varphi : |K| \to |L|$ a simplicial map. For each p we define a homomorphism $\varphi_p^0 : C_p(K) \to C_p(L)$ as follows: if $a_0 \ldots a_p$ is an oriented simplex of K, then

$$\varphi_p^0(a_0 \ldots a_p) = \varphi(a_0) \ldots \varphi(a_p) \quad \text{if } \varphi(a_i) \neq \varphi(a_j)$$
$$\text{whenever } i \neq j$$
$$= 0 \quad \text{otherwise}$$

and φ_p^0 is defined on the p-chains of K by linearity. We may simplify the definition of φ_p^0 by agreeing that if $b_{i_0}, \ldots, b_{i_p}$ are vertices of an oriented complex that are not all distinct, then $b_{i_0} \ldots b_{i_p}$ is to be interpreted as the zero element of C_p. Thus we may write

$$\varphi_p^0(a_0 \ldots a_p) = \varphi(a_0) \ldots \varphi(a_p)$$

for all oriented simplexes $a_0 \ldots a_p$ in K. Naturally, $\varphi_p^0 = 0$ whenever $C_p(K) = 0$.

Lemma 4-15. Let K and L be oriented geometric complexes and $\varphi : |K| \to |L|$ a simplicial map. Then the sequence $\varphi^0 = \{\varphi_p^0\}_{p=-\infty}^{\infty}$ is a chain map of $C(K)$ to $C(L)$. Moreover,

(a) If $K = L$ and $\varphi = \text{id}_{|K|}$, then $\varphi^0 = (\text{id}_{|K|})^0 = \text{id}_{C(K)}$.

(b) If M is another oriented geometric complex and $\psi : |L| \to |M|$ is another simplicial map, then $(\psi \circ \varphi)^0 = \psi^0 \circ \varphi^0$.

Proof: To show that φ^0 is a chain map, it suffices to prove that $\varphi_{p-1}^0 \circ \partial_p (\sigma_p) = \partial_p \circ \varphi_p^0 (\sigma_p)$ for each p, where $\sigma_p = a_0 \ldots a_p$ is an oriented p-simplex of K. We consider two cases:

Case 1: Suppose the vertices $\varphi(a_0), \ldots, \varphi(a_p)$ are all distinct.

Exercise 4-13. Prove that $\varphi_{p-1}^0 \circ \partial_p (\sigma_p) = \partial_p \circ \varphi_p^0 (\sigma_p)$ under the assumption of Case 1.

Case 2: Suppose that some of the vertices $\varphi(a_0), \ldots, \varphi(a_p)$ coincide. In this case $\partial_p \circ \varphi_p^0 (\sigma_p) = \partial_p(0) = 0$, so we must show that $\varphi_{p-1}^0 \circ \partial_p (\sigma_p) = 0$ as well. Now,

$$(6) \qquad \varphi_{p-1}^0 \circ \partial_p (\sigma_p) = \sum_{i=0}^{p} (-1)^i \varphi_{p-1}^0 (a_0 \ldots \hat{a}_i \ldots a_p).$$

If at least three of the vertices $a_0 , \ldots , a_p$ have the same image under φ, then each of the terms $a_0 \ldots \hat{a}_i \ldots a_p$ contains at least two vertices with the same image under φ so each $\varphi_{p-1}^0 (a_0 \ldots \hat{a}_i \ldots a_p)$ is zero and the result follows. On the other hand, if a_j and a_k are the only vertices with the same image, then by equation (6)

$$\begin{aligned}
\varphi_{p-1}^0 \circ \partial_p(\sigma_p) &= (-1)^j \varphi_{p-1}^0 (a_0 \ldots \hat{a}_j \ldots a_p) \\
&\quad + (-1)^k \varphi_{p-1}^0 (a_0 \ldots \hat{a}_k \ldots a_p) \\
&= (-1)^j \varphi(a_0) \ldots \widehat{\varphi(a_j)} \ldots \varphi(a_p) \\
&\quad + (-1)^k \varphi(a_0) \ldots \widehat{\varphi(a_k)} \ldots \varphi(a_p).
\end{aligned}$$

But $\varphi(a_0) \ldots \widehat{\varphi(a_k)} \ldots \varphi(a_p) = (-1)^{k-j-1} \varphi(a_0) \ldots \widehat{\varphi(a_j)} \ldots \varphi(a_p)$, so this last sum is zero. We have shown that φ^0 is a chain map and the remainder of the lemma is clear.   Q.E.D.

Remark: When there is no possibility of confusion, we will follow the custom of omitting the superscript and using the same symbol for a simplicial map $\varphi:|K| \to |L|$ and the induced chain map $\varphi = \varphi^0 : C(K) \to C(L)$.

Combining Lemma 4-15 and Theorem 4-12 we obtain the following theorem.

Theorem 4-16. Let K and L be oriented geometric complexes and $\varphi:|K| \to |L|$ a simplicial map. Then φ induces a chain map $\varphi_* = (\varphi^0)_* : H(K) \to H(L)$ with the following properties:
(a)   If $K = L$ and $\varphi = \mathrm{id}_{|K|}$, then $\varphi_* = (\mathrm{id}_{|K|})_* = \mathrm{id}_{H(K)}$.
(b)   If M is another oriented geometric complex and $\psi:|L| \to |M|$ another simplicial map, then $(\psi \circ \varphi)_* = \psi_* \circ \varphi_*$.

A major result of this chapter will be the extension of Theorem 4-16 to arbitrary continuous maps between polyhedra. Roughly, our procedure will be as follows: First we show that for any geometric complex K and any $n \geqslant 0$ the homology complexes of K and its nth barycentric subdivision $K^{(n)}$ are isomorphic. (This is the most difficult part of the proof.) Next, given a continuous map $f : |K| \to |L|$, we apply Theorem 2-19 to obtain an $n \geqslant 0$ and a simplicial approximation $\varphi :|K^{(n)}| \to |L|$ to f. Now, φ induces a chain map $\varphi_* : H(K^{(n)}) \to H(L)$. Composing with a natural isomorphism of $H(K)$ onto $H(K^{(n)})$, we obtain a chain map $f_* : H(K) \to H(L)$ that we can show depends only on f. This assignment of f_* to f turns out to be functorial, so we obtain the analog of Theorem 4-16 for arbitrary continuous maps.

4-10 Acyclic carrier functions

Our first task is to show that contiguous equivalent simplicial maps induce chain homotopic chain maps in Lemma 4-15 and thus identical chain maps in homology. The proof is based on the notion of an "acyclic carrier function."

Let K and L be oriented geometric complexes. A function χ, which assigns to each simplex σ_p in K a subcomplex $\chi(\sigma_p)$ of L in such a way that $\chi(\sigma_{p-1}) \subseteq \chi(\sigma_p)$ whenever $\sigma_{p-1} < \sigma_p$, is called a *carrier function*. χ is *acyclic* if $H(\widetilde{\chi(\sigma_p)}) \approx 0$ for each σ_p in K. A chain map $\varphi : C(K) \to C(L)$ is said to be *carried by* χ if, for every σ_p in K, $\varphi(\sigma_p)$ $(= \varphi_p(\sigma_p))$ is a chain in $\chi(\sigma_p)$.

Lemma 4-17. Let K and L be oriented geometric complexes and $\varphi : C(\tilde{K}) \to C(\tilde{L})$ a chain map. If φ is carried by an acyclic carrier function χ, then $\varphi \overset{c}{\simeq} 0$.

Proof: We must define $\gamma_p : C_p(\tilde{K}) \to C_{p+1}(\tilde{L})$ for each p such that $\partial^L_{p+1} \circ \gamma_p + \gamma_{p-1} \circ \partial^K_p = \varphi_p$. The definition is recursive. Let $\gamma_p = 0$ for all $p < -1$. Now suppose $p \geq -1$ and that, for all $q < p$, γ_q has been defined with the following properties:

(a) $\partial^L_{q+1} \circ \gamma_q + \gamma_{q-1} \circ \partial^K_q = \varphi_q$.

(b) $\gamma_q(\sigma_q)$ is a chain in $\chi(\sigma_q)$ for each σ_q in $\tilde{K}$.

We must define γ_p, and for this it suffices to define $\gamma_p(\sigma_p)$ for each σ_p in $\tilde{K}$. First note that $\gamma_{p-1} \circ \partial^K_p (\sigma_p)$ is a chain in $\chi(\sigma_p)$ for each σ_p. Indeed, $\gamma_{p-1} \circ \partial^K_p (\sigma_p)$ is a chain in

$$\bigcup_{\sigma_{p-1} < \sigma_p} \chi(\sigma_{p-1})$$

by the induction hypothesis (b) and, since χ is a carrier function, $\chi(\sigma_{p-1}) \subseteq \chi(\sigma_p)$ whenever $\sigma_{p-1} < \sigma_p$. Moreover, since φ is carried by χ, $\varphi(\sigma_p)$ is a chain in $\chi(\sigma_p)$. Consequently, $\varphi_p(\sigma_p) - \gamma_{p-1} \circ \partial^K_p (\sigma_p) = (\varphi_p - \gamma_{p-1} \circ \partial^K_p)(\sigma_p)$ is a chain in $\chi(\sigma_p)$, that is, $\varphi_p - \gamma_{p-1} \circ \partial^K_p$ is carried by χ.

Exercise 4-14. Show that the p-chain $d_p = \varphi_p(\sigma_p) - \gamma_{p-1} \circ \partial^K_p (\sigma_p)$ is, in fact, a p-cycle in $\chi(\sigma_p)$ for each σ_p in $\tilde{K}$. Hint: Compute $\partial^L_p (d_p)$ using the fact that φ is a chain map, induction hypothesis (b), and Theorem 4-1.

Now, $\chi(\sigma_p)$ is a subcomplex of L and is therefore augmented, that is, $\chi(\sigma_p) = \widetilde{\chi(\sigma_p)}$. Since χ is acyclic, $H_p(\chi(\sigma_p)) \approx 0$, that is, every p-cycle in $\chi(\sigma_p)$ is a p-boundary. Thus, there is a $(p + 1)$-chain d_{p+1} in $\chi(\sigma_p)$

such that $\partial_{p+1}^{L}(d_{p+1}) = d_p$. We define $\gamma_p(\sigma_p)$ by $\gamma_p(\sigma_p) = d_{p+1}$. Thus, γ_p has been defined for every σ_p in $\tilde{K}$ and thus by linearity on all of $C_p(\tilde{K})$. We need only show that properties (a) and (b) are satisfied by γ_p. First observe that $(\varphi_p - \gamma_{p-1} \circ \partial_p^K)(\sigma_p) = \partial_{p+1}^{L}(d_{p+1}) = \partial_{p+1}^{L} \circ \gamma_p(\sigma_p)$ for each σ_p, that is, $\partial_{p+1}^{L} \circ \gamma_p + \gamma_{p-1} \circ \partial_p^K = \varphi_p$, so (a) is satisfied. Moreover, since d_{p+1} is a chain in $\chi(\sigma_p)$, $\gamma_p(\sigma_p)$ is a chain in $\chi(\sigma_p)$ and (b) is also satisfied.
    Q.E.D.

Proposition 4-18. Let K and L be oriented geometric complexes and φ, ψ : $C(\tilde{K}) \to C(\tilde{L})$ chain maps. If φ and ψ are carried by the same acyclic carrier function χ, then $\varphi \stackrel{c}{=} \psi$.

Exercise 4-15. Prove Proposition 4-18. Hint: Use Lemma 4-17.   Q.E.D.

    To obtain an analog of Proposition 4-18 in the unaugmented case, we must impose additional restrictions on the chain maps. Observe that a chain map $\tilde{\varphi}: C(\tilde{K}) \to C(\tilde{L})$ determines by restriction a chain map $\varphi: C(K) \to C(L)$. Conversely, if a chain map $\varphi: C(K) \to C(L)$ is the restriction of some $\tilde{\varphi}: C(\tilde{K}) \to C(\tilde{L})$, then φ is said to be *augmentable* to $\tilde{\varphi}$. (Note that in this case $\tilde{\varphi}$ is uniquely determined by φ since $\tilde{\varphi}(\sigma_{-1}) = \tilde{\varphi}(\partial_0^K(\sigma_0)) = \partial_0^L(\varphi(\sigma_0))$ for any σ_0 in K.) If φ is augmentable to $\tilde{\varphi}$ and if, moreover, $\tilde{\varphi}(\sigma_{-1}) = \sigma_{-1}$, then φ is said to be *proper*. Observe that any chain map induced by a simplicial map as in Theorem 4-16 is proper and that compositions of proper chain maps are proper. Now suppose φ : $C(K) \to C(L)$ is a proper chain map that is carried by the carrier function χ. Define a carrier function $\tilde{\chi}$ of $\tilde{K}$ into the subcomplexes of $\tilde{L}$ by $\tilde{\chi}(\sigma_p) = \widetilde{\chi(\sigma_p)}$ for $p \geqslant 0$ and $\tilde{\chi}(\sigma_{-1}) = \sigma_{-1}$. Then $\tilde{\varphi}$ is carried by $\tilde{\chi}$. Moreover, $\tilde{\chi}$ is acyclic iff χ is acyclic.

Theorem 4-19. Let K and L be oriented geometric complexes and φ, ψ : $C(K) \to C(L)$ proper chain maps. If φ and ψ are carried by the same acyclic carrier function χ, then $\varphi \stackrel{c}{=} \psi$.

Proof: $\tilde{\varphi}$, $\tilde{\psi}$: $C(\tilde{K}) \to C(\tilde{L})$ are both carried by $\tilde{\chi}$ which is acyclic, so by Proposition 4-18 $\tilde{\varphi} \stackrel{c}{=} \tilde{\psi}$. Restricted to the chains in K, $\tilde{\varphi} = \varphi$ and $\tilde{\psi} = \psi$, and by definition of chain homotopy $\varphi \stackrel{c}{=} \psi$.   Q.E.D.

Corollary 4-20. Let K and L be oriented geometric complexes and φ, ψ : $|K| \to |L|$ simplicial maps in the same contiguity class. Then the induced chain maps φ^0, ψ^0 : $C(K) \to C(L)$ are chain homotopic. In particular, the induced chain maps φ_*, ψ_* : $H(K) \to H(L)$ in homology are identical.

Proof: It suffices to consider the case in which φ, ψ : $|K| \to |L|$ are contiguous. Let σ be a simplex of K. Then $\varphi^0(\sigma)$ and $\psi^0(\sigma)$ are simplexes of

L, which are faces of a common simplex of L. Let τ be the simplex of lowest dimension in L that has both $\varphi^0(\sigma)$ and $\psi^0(\sigma)$ as faces. Take $\chi(\sigma)$ to be the subcomplex of L consisting of all the faces of τ, that is, $\chi(\sigma) = K(\tau)$. The function χ thus defined is a carrier function that by construction carries both φ^0 and ψ^0. Moreover, by Theorem 4-10 χ is acyclic. Since the chain maps φ^0 and ψ^0 are induced by simplicial maps, they are proper so $\varphi^0 \stackrel{c}{=} \psi^0$ by Theorem 4-19.   Q.E.D.

4-11 Invariance of homology groups under barycentric subdivision

The entire development of simplicial homology theory will depend on the fact that the homology groups of a geometric complex and its first barycentric subdivision are isomorphic. We prove this statement by showing that the chain complexes $C(K)$ and $C(K')$ are chain equivalent, that is, by exhibiting chain maps $\omega : C(K') \to C(K)$ and $\beta : C(K) \to C(K')$ such that $\omega \circ \beta \stackrel{c}{=} \mathrm{id}_{C(K)}$ and $\beta \circ \omega \stackrel{c}{=} \mathrm{id}_{C(K')}$ and applying Corollary 4-14.

We begin with an arbitrary Sperner map $\omega : K_0' \to K_0$. Then ω induces a simplicial map of $|K'|$ into $|K|$ that in turn induces a chain map of $C(K')$ into $C(K)$, which we also denote by $\omega : C(K') \to C(K)$.

Exercise 4-16. Show that if $\omega_1, \omega_2 : K_0' \to K_0$ are two Sperner maps, then the induced chain maps $\omega_1, \omega_2 : C(K') \to C(K)$ are chain homotopic and thus induce identical chain maps in homology. This accounts for our ability to choose the initial Sperner map arbitrarily. Hint: Apply Corollary 4-20.

The chain map $\beta : C(K) \to C(K')$, called the *barycentric subdivision map,* does not arise from a simplicial map in this way, but is nevertheless intuitively quite natural: β assigns to each σ_p in K the p-chain in K' consisting of the sum of the (suitably oriented) p-simplexes of K' that lie in σ_p. The definition of β is inductive and it is most convenient to consider first the augmented case. Thus, if K is an oriented geometric complex, we define homomorphisms $\beta_p : C_p(\bar{K}) \to C_p(\bar{K}')$ recursively on the generators of $C_p(\bar{K})$ as follows: First let $\beta_{-1}(\sigma_{-1}) = \sigma_{-1}$ and extend β_{-1} to a homomorphism on $C_{-1}(\bar{K})$ by linearity. Now suppose $p > -1$ and that the homomorphisms $\beta_q : C_q(\bar{K}) \to C_q(\bar{K}')$ have been defined for all $q < p$ in such a way that the following two conditions are satisfied:

(a) $\partial_q \circ \beta_q = \beta_{q-1} \circ \partial_q$.

(b) $\beta_q(\sigma_q)$ is a q-chain in $K'(\sigma_q)$ for each $\sigma_q \in \bar{K}$.

Note that (a) and (b) are trivially satisfied for $q = -1$. Now define β_p on each σ_p in $\bar{K}$ as follows: Observe that $\partial_p(\sigma_p)$ is a $(p-1)$-chain in $\bar{K}$ so $\beta_{p-1}(\partial_p(\sigma_p))$ has been defined. Moreover, by (b), $\beta_{p-1}(\partial_p(\sigma_p))$ is a $(p-1)$-chain in $K_{p-1}'(\sigma_p)$. Now, $K'(\sigma_p)$ is the cone complex of $K_{p-1}'(\sigma_p)$ over the barycenter $b = b(\sigma_p)$, so $b\beta_{p-1} \circ \partial_p(\sigma_p)$ is a well-defined p-chain in

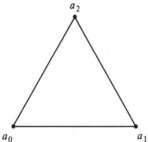

Figure 4-9

$K'(\sigma_p)$ and we may define

$$\beta_p(\sigma_p) = b\beta_{p-1} \circ \partial_p(\sigma_p).$$

Now we extend β_p to a homomorphism by linearity. Property (b) is satisfied by definition. To prove (a) we compute

$$\begin{aligned}
\partial_p\beta_p(\sigma_p) &= \partial_p\big(b\beta_{p-1} \circ \partial_p(\sigma_p)\big) \\
&= \beta_{p-1} \circ \partial_p(\sigma_p) - b\big(\partial_{p-1}\beta_{p-1} \circ \partial_p(\sigma_p)\big) \\
&= \beta_{p-1} \circ \partial_p(\sigma_p) - b\big(\beta_{p-2} \circ \partial_{p-1}\partial_p(\sigma_p)\big) \\
&= \beta_{p-1} \circ \partial_p(\sigma_p) - b\big(\beta_{p-2}(0)\big) \\
&= \beta_{p-1} \circ \partial_p(\sigma_p).
\end{aligned}$$

Letting $\beta_p = 0$ for $p < -1$ and for p greater than the algebraic dimension of K, we find then that $\beta = \{\beta_p\}_{p=-\infty}^{\infty}$ is a chain map of $C(\tilde{K})$ into $C(\tilde{K}')$. Observe also that if $\sigma_0 = a_i$ is one of the vertices of K, then $\beta_0(\sigma_0) = b(\beta_{-1}(\sigma_{-1})) = b\sigma_{-1} = b$, where $b = b(\sigma_0) = b(a_i) = a_i$. Thus,

$$(7) \qquad \beta_0(\sigma_0) = \beta_0(a_i) = a_i.$$

In the unaugmented case we obtain the chain map $\beta : C(K) \to C(K')$ by simply starting the induction at $p = 0$ and using equation (7). In particular, it is thus clear that in the unaugmented case, β is a proper chain map.

As an example we consider the complex $K(s_2)$, where $s_2 = (a_0, a_1, a_2)$. We make $K(s_2)$ into an ordered complex by listing its vertices in the order $a_0a_1a_2$ and by repeating the same ordering on each face. Thus, the oriented simplexes of $K(s_2)$ are $a_0a_1a_2$, a_1a_2, a_0a_1, a_0a_2, a_0, a_1, a_2. We construct $K'(s_2)$ and label its vertices as shown in Figure 4-9.

In the augmented case $\beta_{-1}(\sigma_{-1}) = \sigma_{-1}$. For any 0-simplex a_i of $K(s_2)$ we have $\beta_0(a_i) = a_i = b_i$, $i = 0, 1, 2$. Now consider a 1-simplex of $K(s_2)$, say, $\sigma_1 = a_0a_1$. then

$$\begin{aligned}
\beta_1(\sigma_1) &= \beta_1(a_0a_1) = b_{01}\beta_0 \circ \partial_1(a_0a_1) \\
&= b_{01}\beta_0(a_1 - a_0) = b_{01}(b_1 - b_0) \\
&= b_{01}b_1 - b_{01}b_0.
\end{aligned}$$

Finally, consider the 2-simplex $\sigma_2 = a_0 a_1 a_2$. Then

$$
\begin{aligned}
\beta_2(\sigma_2) &= \beta_2(a_0 a_1 a_2) = b_{012}\beta_1 \circ \partial_2(a_0 a_1 a_2) \\
&= b_{012}\beta_1(a_1 a_2 - a_0 a_2 + a_0 a_1) \\
&= b_{012}[\beta_1(a_1 a_2) - \beta_1(a_0 a_2) + \beta_1(a_0 a_1)] \\
&= b_{012}[(b_{12}b_2 - b_{12}b_1) - (b_{02}b_2 - b_{02}b_0) \\
&\quad + (b_{01}b_1 - b_{01}b_0)] \\
&= b_{012}b_{12}b_2 - b_{012}b_{12}b_1 - b_{012}b_{02}b_2 + b_{012}b_{02}b_0 \\
&\quad + b_{012}b_{01}b_1 - b_{012}b_{01}b_0.
\end{aligned}
$$

Lemma 4-21. For any oriented geometric complex K, $\omega \circ \beta = \mathrm{id}_{C(K)}$.

Proof: We show by induction that $\omega_p \circ \beta_p = \mathrm{id}_{C_p(K)}$ for each p. Since this is clear for $p = -1$ and $p = 0$, we suppose that $\omega_{k-1} \circ \beta_{k-1} = \mathrm{id}_{C_{k-1}(K)}$. Then, for any σ_k in K

$$
\begin{aligned}
\omega_k \circ \beta_k(\sigma_k) &= \omega_k(\beta_k(\sigma_k)) = \omega_k(b\beta_{k-1} \circ \partial_k(\sigma_k)) \\
&= \omega(b)\omega_{k-1} \circ \beta_{k-1} \circ \partial_k(\sigma_k) \\
&= \omega(b)\partial_k(\sigma_k).
\end{aligned}
$$

But $\omega(b)$ is by definition a vertex of σ_k so $\omega(b)\partial_k(\sigma_k) = \sigma_k$, and we have $\omega_k \circ \beta_k(\sigma_k) = \sigma_k$ as required.   Q.E.D.

Remark: $\beta \circ \omega$ is *not* equal to $\mathrm{id}_{C(K')}$ so β and ω are not inverses. However, we show next that $\beta \circ \omega$ and $\mathrm{id}_{C(K')}$ are chain homotopic so that, in homology, $\beta_* \circ \omega_* = \mathrm{id}_{H(K')}$. Thus, β and ω are "close enough" to being inverses to induce inverse isomorphisms in homology.

Lemma 4-22. For any oriented geometric complex K, $\beta \circ \omega \overset{c}{=} \mathrm{id}_{C(K')}$.

Proof: To prove that $\beta \circ \omega \overset{c}{=} \mathrm{id}_{C(K')}$, we construct an acyclic carrier function χ that carries both $\beta \circ \omega$ and $\mathrm{id}_{C(K')}$, note that both of these chain maps are proper, and apply Theorem 4-19. Let σ_p' be an arbitrary simplex in K'. We define $\chi(\sigma_p')$ as follows: If b is the leading vertex of σ_p', then $b = b(\sigma_p)$ for some σ_p in K, and we let $\chi(\sigma_p')$ be the barycentric subdivision of the complex consisting of the faces of σ_p, that is, $\chi(\sigma_p') = K'(\sigma_p)$.

Exercise 4-17. Show that the function χ thus defined is an acyclic carrier function that carries both $\beta \circ \omega$ and $\mathrm{id}_{C(K')}$.   Q.E.D.

Combining Lemmas 4-21 and 4-22 we find that $C(K)$ and $C(K')$ are chain equivalent. Applying Corollary 4-14 we obtain the next theorem.

Theorem 4-23. Let K be an oriented geometric complex. Then the homology complexes $H(K)$ and $H(K')$ are isomorphic.

Corollary 4-24. Let K be an oriented geometric complex and $n \geqslant 0$ an integer. Then the homology complexes $H(K)$ and $H(K^{(n)})$ are isomorphic.

Although Corollary 4-24 is a trivial consequence of Theorem 4-23, we will require explicit representations for the isomorphisms we have in mind. Thus, consider an oriented geometric complex K and an integer $n > 0$. For each $i = 1, \ldots, n$ we have a chain map $\omega : C(K^{(i)}) \rightarrow C(K^{(i-1)})$ and a chain map $\beta : C(K^{(i-1)}) \rightarrow C(K^{(i)})$; we inflict a minor abuse on our notation by using the same symbols ω and β for each i. We denote by $\omega^{(n)} : C(K^{(n)}) \rightarrow C(K)$ and $\beta^{(n)} : C(K) \rightarrow C(K^{(n)})$ the compositions

$$C(K^{(n)}) \underset{\beta}{\overset{\omega}{\rightleftarrows}} C(K^{(n-1)}) \underset{\beta}{\overset{\omega}{\rightleftarrows}} \cdots \underset{\beta}{\overset{\omega}{\rightleftarrows}} C(K') \underset{\beta}{\overset{\omega}{\rightleftarrows}} C(K).$$

Then $\omega^{(n)}$ and $\beta^{(n)}$ are chain maps. Moreover, $\omega^{(n)} \circ \beta^{(n)} = \mathrm{id}_{C(K)}$ follows from Lemma 4-21 and a minor modification of the proof of Lemma 4-22 yields $\beta^{(n)} \circ \omega^{(n)} \overset{c}{=} \mathrm{id}_{C(K^{(n)})}$. (Verify this!) Thus, in homology

$$\omega_*^{(n)} \circ \beta_*^{(n)} = \mathrm{id}_{H(K)}$$
and $$\beta_*^{(n)} \circ \omega_*^{(n)} = \mathrm{id}_{H(K^{(n)})}$$

so that $\omega_*^{(n)}$ and $\beta_*^{(n)}$ are inverse isomorphisms.

Remark: By convention $\omega^{(0)} = \beta^{(0)} = \mathrm{id}_{C(K)}$.

4-12 Homomorphisms induced by continuous maps

With the proof of Corollary 4-24 behind us we are finally in a position to demonstrate the categorical nature of simplicial homology theory. Let K and L be oriented geometric complexes and $f : |K| \rightarrow |L|$ an arbitrary continuous map. By Theorem 2-19 there is an integer $n \geqslant 0$ and a simplicial map $\varphi : |K^{(n)}| \rightarrow |L|$, which is a simplicial approximation to f. φ induces a chain map $\varphi^0 : C(K^{(n)}) \rightarrow C(L)$ as in Lemma 4-15 that in turn induces a chain map $\varphi_* : H(K^{(n)}) \rightarrow H(L)$ in homology by Theorem 4-16. Now, if $\beta_*^{(n)} : H(K) \rightarrow H(K^{(n)})$ is the isomorphism induced by the iterated barycentric subdivision map $\beta^{(n)} : C(K) \rightarrow C(K^{(n)})$, then

$$\varphi_* \circ \beta_*^{(n)} = (\varphi \circ \beta^{(n)})_* : H(K) \rightarrow H(L)$$

is a homomorphism. We claim that this homomorphism depends only on the map f. For the proof we require a simple lemma that in turn depends on the observation in the following exercise.

Exercise 4-18. Let $C, D, E,$ and F be chain complexes and $\varphi_1 \overset{c}{=} \varphi_2 : C \rightarrow D$, $\theta : D \rightarrow E$ and $\eta : F \rightarrow C$ chain maps. Show that $\theta \circ \varphi_1 \overset{c}{=} \theta \circ \varphi_2$ and $\varphi_1 \circ \eta \overset{c}{=} \varphi_2 \circ \eta$.

Lemma 4-25. Let K and L be oriented geometric complexes and suppose $\varphi : |K| \to |L|$ and $\psi : |K^{(n)}| \to |L|$ are contiguous, simplicial maps. Then $\varphi^0 \circ \omega^{(n)} \overset{c}{=} \psi^0$ and $\psi^0 \circ \beta^{(n)} \overset{c}{=} \varphi^0$.

Proof: Let $u^{(n)}:|K^{(n)}| \to |K|$ be an arbitrary, standard map and let $u^{(n)}$ induce $\tilde{\omega}^{(n)}:C(K^{(n)}) \to C(K)$. Since φ and ψ are contiguous, $\varphi \circ u^{(n)} : |K^{(n)}| \to |L|$ and $\psi:|K^{(n)}| \to |L|$ are also contiguous. By functoriality $\varphi \circ u^{(n)}$ induces $\varphi^0 \circ \tilde{\omega}^{(n)}$, which must therefore be chain homotopic to ψ^0 by Corollary 4-20. But $\tilde{\omega}^{(n)} \overset{c}{=} \omega^{(n)}$ since both arise from standard maps. By Exercise 4-18 $\varphi^0 \circ \tilde{\omega}^{(n)} \overset{c}{=} \varphi^0 \circ \omega^{(n)}$, and thus $\varphi^0 \circ \omega^{(n)} \overset{c}{=} \psi^0$ by Exercise 4-11. Another application of Exercise 4-18 yields $\varphi^0 \circ \omega^{(n)} \circ \beta^{(n)} \overset{c}{=} \psi^0 \circ \beta^{(n)}$. But $\omega^{(n)} \circ \beta^{(n)} = \text{id}_{C(K)}$, so we obtain $\varphi^0 \overset{c}{=} \psi^0 \circ \beta^{(n)}$.   Q.E.D.

Now suppose $\varphi:|K^{(n)}| \to |L|$ and $\psi:|K^{(m)}| \to |L|$ are both simplicial approximations to $f:|K| \to |L|$. Then φ and ψ induce $\varphi^0 : C(K^{(n)}) \to C(L)$ and $\psi^0 : C(K^{(m)}) \to C(L)$. We must show that $(\varphi^0 \circ \beta^{(n)})_* = (\psi^0 \circ \beta^{(m)})_*$, and for this it suffices to show that $\varphi^0 \circ \beta^{(n)} \overset{c}{=} \psi^0 \circ \beta^{(m)}$. Assume $m \geq n$ and let $\beta^{(m-n)} : C(K^{(n)}) \to C(K^{(m)})$ be the iterated barycentric subdivision map. Then $\beta^{(m)} = \beta^{(m-n)} \circ \beta^{(n)}$. By Proposition 2-22 φ and ψ are contiguous, and we conclude from Lemma 4-25 that $\varphi^0 \overset{c}{=} \psi^0 \circ \beta^{(m-n)}$. But then by Exercise 4-18 $\varphi^0 \circ \beta^{(n)} \overset{c}{=} \psi^0 \circ \beta^{(m-n)} \circ \beta^{(n)} = \psi^0 \circ \beta^{(m)}$ as required. We have thus shown that a continuous map $f : |K| \to |L|$ induces a well-defined chain map in homology that is denoted

$$f_* : H(K) \to H(L)$$

and defined as follows: Choose an arbitrary simplicial approximation $\varphi : |K^{(n)}| \to |L|$ to f and let

$$(8) \qquad f_* = (\varphi^0 \circ \beta^{(n)})_* = \varphi_* \circ \beta_*^{(n)}.$$

Remark: Observe that if $f : |K| \to |L|$ is a simplicial map, then the induced homomorphisms defined by equation (8) and by Theorem 4-16 are identical. In particular, by Theorem 4-16(a), $(\text{id}_{|K|})_* = \text{id}_{H(K)}$.

Theorem 4-26. Let K and L be oriented geometric complexes and f, $g : |K| \to |L|$ homotopic continuous maps. Then the induced chain maps f_*, $g_* : H(K) \to H(L)$ are identical.

Proof: Since f and g are homotopic, Proposition 3-12 yields an $n \geq 0$ and simplicial maps φ, $\psi : |K^{(n)}| \to |L|$ in the same contiguity class such that φ is a simplicial approximation to f and ψ is a simplicial approximation to g. Then $f_* = \varphi_* \circ \beta_*^{(n)}$ and $g_* = \psi_* \circ \beta_*^{(n)}$. But by Corollary 4-20 $\varphi_* = \psi_*$, so $f_* = g_*$.   Q.E.D.

To complete our generalization of Theorem 4-16, we must show that if $f : |K| \to |L|$ and $g : |L| \to |M|$ are continuous, then $(g \circ f)_* = g_* \circ f_*$ and for this we require several preliminary observations. Let K and L be geometric complexes and $\varphi : |K| \to |L|$ a simplicial map. Define φ' : $|K'| \to |L'|$ as follows: Let $b = b(s)$ be an arbitrary vertex of K', set $\varphi'(b) = b(\varphi(s))$ and now extend φ' to K' by linearity.

Exercise 4-19. Show that $\varphi' : |K'| \to |L'|$ is a simplicial map.

Thus, by iteration, the simplicial map $\varphi : |K| \to |L|$ induces, for any $m > 0$, a simplicial map

$$\varphi^{(m)} : |K^{(m)}| \to |L^{(m)}|.$$

Observe that it follows trivially from Lemma 3-9 that φ and $\varphi^{(m)}$ are homotopic.

Lemma 4-27. Let K and L be oriented geometric complexes and φ : $|K| \to |L|$ a simplicial map. Then $(\varphi')^0 \circ \beta_K = \beta_L \circ \varphi^0$, where $\beta_K : C(K) \to C(K')$ and $\beta_L : C(L) \to C(L')$ are the barycentric subdivision maps.

Exercise 4-20. Prove Lemma 4-27. Hint: Proceed by induction.   Q.E.D.

Now suppose K, L, and M are geometric complexes and $f : |K| \to |L|$ and $g : |L| \to |M|$ are continuous maps. Then there exist integers n and m and simplicial maps $\varphi:|K^{(n)}| \to |L|$ and $\psi:|L^{(m)}| \to |M|$ such that φ is a simplicial approximation to f and ψ is a simplicial approximation to g. Now, φ induces a simplicial map $\varphi^{(m)}:|K^{(n+m)}| \to |L^{(m)}|$, and by interating Lemma 4-27 we obtain

$$(9) \qquad (\varphi^{(m)})^0 \circ \beta_K^{(m)} = \beta_L^{(m)} \circ \varphi^0,$$

where $\beta_K^{(m)}:C(K^{(n)}) \to C(K^{(n+m)})$ and $\beta_L^{(m)}:C(L) \to C(L^{(m)})$ are the iterated barycentric subdivision maps. Observe also that $\varphi^{(m)} \simeq \varphi \simeq f$ so that $g \circ f \simeq g \circ \varphi^{(m)} \simeq \psi \circ \varphi^{(m)}$. By Theorem 4-26 $(g \circ f)_* = (\psi \circ \varphi^{(m)})_*$. But then

$$
\begin{aligned}
(g \circ f)_* &= (\psi \circ \varphi^{(m)})_* = ((\psi \circ \varphi^{(m)})^0 \circ \beta_K^{(n+m)})_* & \\
&= ((\psi \circ \varphi^{(m)})^0)_* \circ (\beta_K^{(n+m)})_* & \text{by Theorem 4-12} \\
&= (\psi^0 \circ (\varphi^{(m)})^0)_* (\beta_K^{(m)} \circ \beta_K^{(n)})_* & \text{by Lemma 4-15} \\
&= \psi_* \circ (\varphi^{(m)})_* \circ (\beta_K^{(m)})_* \circ (\beta_K^{(n)})_* & \text{by Theorem 4-12} \\
&= \psi_* \circ ((\varphi^{(m)})^0 \circ \beta_K^{(m)})_* \circ (\beta_K^{(n)})_* & \text{by Theorem 4-12} \\
&= \psi_* \circ (\beta_L^{(m)} \circ \varphi^0)_* \circ (\beta_K^{(n)})_* & \text{by (9)} \\
&= \psi_* \circ (\beta_L^{(m)})_* \circ \varphi_* \circ (\beta_K^{(n)})_* & \text{by Theorem 4-12} \\
&= (\psi \circ \beta_L^{(m)})_* \circ (\varphi \circ \beta_K^{(n)})_* & \text{by Theorem 4-12} \\
&= g_* \circ f_*.
\end{aligned}
$$

We have thus completed the proof of the next theorem.

Theorem 4-28. Let K and L be oriented geometric complexes and f : $|K| \to |L|$ a continuous map. Then f induces a chain map $f_* : H(K) \to H(L)$ with the following properties:
(a)   If $K = L$ and $f = \mathrm{id}_{|K|}$, then $f_* = (\mathrm{id}_{|K|})_* = \mathrm{id}_{H(K)}$.
(b)   If M is another oriented geometric complex and $g : |L| \to |M|$ another continuous map, then $(g \circ f)_* = g_* \circ f_*$.

Combining Theorems 4-26 and 4-28 we obtain the following corollary.

Corollary 4-29. Let K and L be oriented geometric complexes for which the polyhedra $|K|$ and $|L|$ are of the same homotopy type. Then $H(K) \approx H(L)$.

Exercise 4-21. Prove Corollary 4-29. Hint: Let $f : |K| \to |L|$ be a homotopy equivalence with homotopy inverse $g : |L| \to |K|$, and show that f_* and g_* are inverse isomorphisms.

Remark: In particular, geometric complexes with homeomorphic poly-hedra have isomorphic homology complexes. More precisely, if f : $|K| \to |L|$ is a homeomorphism, then $f_* : H(K) \to H(L)$ is an isomorphism with inverse $f_*^{-1} = (f^{-1})_*$. However, it is important to observe that the added strength of Corollary 4-29 would seem to indicate that, at least from the point of view of algebraic topology, classification of spaces by homotopy type rather than homeomorphism type is more natural.

For convenience, we shall summarize the major results of this section in terms of homology groups rather than homology complexes.

Theorem 4-30. Let K and L be oriented geometric complexes and f : $|K| \to |L|$ a continuous map. Then, for each integer p, f induces a homo-morphism $f_{*p} = (f_*)_p : H_p(K) \to H_p(L)$ with the following properties:
(a)   If $K = L$ and $f = \mathrm{id}_{|K|}$, then $f_{*p} = (\mathrm{id}_{|K|})_{*p} = \mathrm{id}_{H_p(K)}$ for each p.
(b)   If M is another oriented geometric complex and $g : |L| \to |M|$ another continuous map, then $(g \circ f)_{*p} = g_{*p} \circ f_{*p}$ for each p.
Moreover, homotopic continuous maps induce identical homomorphisms in homology in each dimension.

4-13 Homology groups of topological polyhedra

Now let X be a topological polyhedron. Our goal is to define the simpli-cial homology groups $H_p(X)$ of X. By definition X admits a triangulation, that is, there exists a pair (K, h) consisting of an (oriented) geometric complex K and a homeomorphism h of $|K|$ onto X. Now, it is tempting to define $H_p(X)$ to be $H_p(K)$. However, if (K_1, h_1) and (K_2, h_2) are two triangulations of X, then although $H_p(K_1)$ and $H_p(K_2)$ are isomorphic for each p (Corollary 4-29), they are, in general, not identical (as sets). Since we clearly have no reason for preferring one triangulation over another, this procedure would determine $H_p(X)$ only up to isomorphism. We remedy this defect with the following construction.

Let X denote an arbitrary topological polyhedron and fix an integer p. For each triangulation (K, h) of X denote by $H_p(K, h)$ the pth simplicial homology group of the complex K and consider the set $\cup H_p(K, h)$, where the union is over all triangulations (K, h) of X. Define a relation $\sim$ on $\cup H_p(K, h)$ as follows: If $x_1 \in H_p(K_1, h_1)$ and $x_2 \in H_p(K_2, h_2)$, then $x_1 \sim x_2$ iff the isomorphism $(h_2^{-1} \circ h_1)_{*p} : H_p(K_1, h_1) \to H_p(K_2, h_2)$ carries x_1 onto x_2, that is, iff $(h_2^{-1} \circ h_1)_{*p}(x_1) = x_2$.

Exercise 4-22. Prove that the relation $\sim$ thus defined on $\cup H_p(K, h)$ is an equivalence relation.

We denote by $H_p(X)$ the set of all equivalence classes of $\cup H_p(K, h)$ modulo the relation $\sim$. If $\xi \in H_p(X)$, $x \in \xi$ and $x \in H_p(K, h)$, then x is a *representative* of ξ in the triangulation (K, h). Thus, each $\xi \in H_p(X)$ has a representative in each (K, h). Now, if $\xi, \eta \in H_p(X)$, we define $\xi + \eta$ as follows: Find representatives x_1 and x_2 of ξ and η, respectively, in some triangulation (K, h) of X and let $\xi + \eta$ be the element of $H_p(X)$ containing $x_1 + x_2$. It follows from the fact that $(h_2^{-1} \circ h_1)_{*p}$ is an isomorphism for any two triangulations (K_1, h_1) and (K_2, h_2) of X that this definition does not depend on the choice of (K, h). Similarly, if $\xi \in H_p(X)$ and x is a representative of ξ in (K, h) we may define $- \xi$ to be the element of $H_p(X)$ which contains $-x$. If 0 is the identity element of $H_p(K, h)$ for any triangulation (K, h) of X, then the equivalence class containing 0 is also denoted 0 and acts as an identity element for the addition just defined in $H_p(X)$. $H_p(X)$ has thus been endowed with an Abelian group structure. Indeed, $H_p(X)$ is isomorphic to each $H_p(K, h)$. We call $H_p(X)$ the pth *simplicial homology group of X.*

Now suppose X and Y are topological polyhedra and $f : X \to Y$ is a continuous map. Let (K, h) and (L, g) be triangulations of X and Y, respectively. Then $g^{-1} \circ f \circ h : |K| \to |L|$ is continuous, so $(g^{-1} \circ f \circ h)_{*p} : H_p(K, h) \to H_p(L, g)$ is a well-defined homomorphism. Let ξ be an arbitrary element of $H_p(X)$ and choose a representative x of ξ in (K, h). Let η be the element of $H_p(Y)$ containing $(g^{-1} \circ f \circ h)_{*p}(x) \in H_p(L, g)$. We define $f_{*p} : H_p(X) \to H_p(Y)$ by $f_{*p}(\xi) = \eta$. To show that this definition does not depend on the choice of (K, h) and (L, g), we let (K_1, h_1) and (K_2, h_2) be triangulations of X and (L_1, g_1) and (L_2, g_2) triangulations of Y.

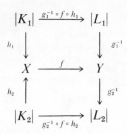

Now let $\xi \in H_p(X)$ and x_1 and x_2 be representatives of ξ in (K_1, h_1) and (K_2, h_2), respectively. Then $x_1 \sim x_2$. Now let $y_1 = (g_1^{-1} \circ f \circ h_1)_{*p}(x_1)$ and $y_2 = (g_2^{-1} \circ f \circ h_2)_{*p}(x_2)$. We must show that $y_1 \sim y_2$. But

$$
\begin{aligned}
(g_2^{-1} \circ g_1)_{*p}(y_1) &= (g_2^{-1} \circ g_1)_{*p} \circ (g_1^{-1} \circ f \circ h_1)_{*p}(x_1) \\
&= (g_2^{-1} \circ g_1 \circ g_1^{-1} \circ f \circ h_1)_{*p}(x_1) && \text{by Theorem 4-30} \\
&= (g_2^{-1} \circ f \circ h_1)_{*p}(x_1) \\
&= (g_2^{-1} \circ f \circ h_1)_{*p} \circ (h_1^{-1} \circ h_2)_{*p}(x_2) && \text{since } x_1 \sim x_2 \\
&= (g_2^{-1} \circ f \circ h_1 \circ h_1^{-1} \circ h_2)_{*p}(x_2) && \text{by Theorem 4-30} \\
&= (g_2^{-1} \circ f \circ h_2)_{*p}(x_2) \\
&= y_2.
\end{aligned}
$$

Finally, observe that if f, $F : X \to Y$ are homotopic, continuous maps and (K, h) and (L, g) are triangulations of X and Y, respectively, then $g^{-1} \circ f \circ h$ and $g^{-1} \circ F \circ h$ are also homotopic so that $(g^{-1} \circ f \circ h)_{*p} = (g^{-1} \circ F \circ h)_{*p}$. Consequently, $f_{*p} = F_{*p}$. A trivial application of Theorem 4-30 shows that the assignment of homomorphisms $f_{*p} : H_p(X) \to H_p(Y)$ to continuous maps $f : X \to Y$ is functorial and we arrive at the next theorem.

Theorem 4-31. Let X and Y be topological polyhedra and $f : X \to Y$ a continuous map. Then, for each integer p, f induces a homomorphism $f_{*p} :$ $H_p(X) \to H_p(Y)$ with the following properties:
(a)   If $X = Y$ and $f = \mathrm{id}_X$, then $f_{*p} = (\mathrm{id}_X)_{*p} = \mathrm{id}_{H_p(X)}$ for each p.
(b)   If Z is another topological polyhedron and $g : Y \to Z$ another continuous map, then $(g \circ f)_{*p} = g_{*p} \circ f_{*p}$ for each p.
Moreover, homotopic, continuous maps induce identical homomorphisms in homology in each dimension.

Theorem 4-31 represents the culmination of all of our efforts in simplicial homology theory in that it completes the construction of the functors H_p from topological polyhedra to finitely generated Abelian groups. We ask the reader to carry out our first major application in this exercise:

Exercise 4-23. Apply Theorems 4-10, 4-11, and 4-31 to prove directly that the sphere S^n is not a retract of the ball B^{n+1}. Hint: Imitate the proof for $n = 1$ in Section 3-6 with H_n in place of π_1.

Now, despite the simplicity and elegance of this proof of the No Retraction Theorem (and thus the Brouwer Fixed Point Theorem), it in itself would scarcely be sufficient reward for the prodigous efforts required to obtain the results of the last few sections. We turn now to our major application of homology theory, a deep and powerful generalization of the Brouwer Fixed Point Theorem due to Lefschetz.

4-14 The Hopf Trace Theorem

The Lefschetz Fixed Point Theorem depends on a beautiful generalization of the Euler–Poincaré Formula (Exercise 4-9) known as the Hopf Trace Theorem. For a proof of this result we shall require a number of algebraic preliminaries, most notably an extension of the notion of "rank" for finitely generated Abelian groups.

Let G be a finitely generated Abelian group. A set $g_1, \ldots, g_k$ of elements of G is said to be *linearly independent* if, whenever $n_1, \ldots, n_k$ are integers for which $\Sigma_{i=1}^{k} n_i g_i = 0$, then $n_1 = \cdots = n_k = 0$. Since G is finitely generated, it contains maximal linearly independent sets and, moreover, all such sets contain the same number of elements. If $e_1, \ldots, e_k$ is a maximal linearly independent set in G, then for each $g \in G$ there exist integers $n \neq 0$ and $n_1, \ldots, n_k$ such that

$$(10) \qquad ng = \sum_{i=1}^{k} n_i e_i.$$

Exercise 4-24. Show that in equation (10) the quotients n_i/n are unique, that is, that if $m \neq 0$ and $m_1, \ldots, m_k$ are integers for which $mg = \Sigma_{i=1}^{k} m_i e_i$, then $m_i/m = n_i/n$ for each $i = 1, \ldots, k$.

By virtue of Exercise 4-24 it is customary to *formally* rewrite equation (10) as

$$(11) \qquad g = \sum_{i=1}^{k} (n_i/n) e_i,$$

even though multiplication by rational numbers is generally not defined in G. We regard (11) as simply a convenient notation, entirely equivalent to equation (10).

Now let φ be an endomorphism of the finitely generated Abelian group G, that is, a homomorphism of G into itself, and let $e_1, \ldots, e_k$ be a maximal linearly independent set in G. For each $i = 1, \ldots, k$ there exist integers n_i and $n_{i1}, \ldots, n_{ik}$ such that $n_i(\varphi(e_i)) = \Sigma_{j=1}^{k} n_{ij} e_j$, where the quotients $m_{ij} = n_{ij}/n_i$ are all unique. We define the *trace* of φ, denoted tr φ, by

$$\text{tr } \varphi = \sum_{i=1}^{k} m_{ii}.$$

Exercise 4-25. Show that tr φ is well-defined by proving that it is independent of the maximal, linearly independent set in G with respect to which it is computed.

Now we claim that for any finitely generated Abelian group G

(12)   rank $G = \text{tr}(\text{id}_G)$

so that the trace is indeed a generalization of the notion of rank. To prove claim (12), we note that since G is a direct sum of cyclic groups (see the Structure Theorem in Section 4-2), we can produce a maximal, linearly independent set in G of the form $e_1, \ldots, e_r, t_1, \ldots, t_\ell$, where $r = $ rank G, each e_i has infinite order and each t_i is a torsion element, that is, has finite order. We compute $\text{tr}(\text{id}_G)$ relative to this set as follows: First note that

$$(\text{id}_G)(e_1) = e_1 = 1 \cdot e_1 + 0 \cdot e_2 + \cdots +$$
$$0 \cdot e_r + 0 \cdot t_1 + \cdots + 0 \cdot t_l$$

$$\cdot$$
$$\cdot$$
$$\cdot$$

$$(\text{id}_G)(e_r) = e_r = 0 \cdot e_1 + 0 \cdot e_2 + \cdots +$$
$$1 \cdot e_r + 0 \cdot t_1 + \cdots + 0 \cdot t_l$$

Now, since each t_i is torsion, there exist nonzero integers n_i, $i = 1, \ldots, l$ such that $n_i t_i = 0 = 0 \cdot e_1 + 0 \cdot e_2 + \cdots + 0 \cdot e_r + 0 \cdot t_1 + \cdots + 0 \cdot t_l$. Since the quotients $0/n_i$ are all zero, the corresponding m_{ii} are zero and thus do not contribute to $\text{tr}(\text{id}_G)$. Thus, $\text{tr}(\text{id}_G) = 1 + 1 + \cdots + 1$ (r-summands) $= r = $ rank G as required.

Now let H be a subgroup of G and recall that the basic tool used in the proof of the Euler–Poincaré Formula is the equality rank $G = $ rank $G/H + $ rank H. We obtain an analog of this formula for the trace as follows: Let $\varphi : G \to G$ be an endomorphism of G and H a subgroup of G with $\varphi(H) \subseteq H$. Then φ induces two endomorphisms $\varphi|H : H \to H$ and $\hat{\varphi} : G/H \to G/H$, where $\hat{\varphi}$ is defined, as usual, by $\hat{\varphi}([g]) = [\varphi(g)]$ for each $g \in G$.

Lemma 4-32. If $\varphi : G \to G$ is an endomorphism of the finitely generated Abelian group G and H is a subgroup of G with $\varphi(H) \subseteq H$, then

$$\text{tr } \varphi = \text{tr } \hat{\varphi} + \text{tr } (\varphi|H).$$

Proof: Let $e_1, \ldots, e_k$ and $[e_1'], \ldots, [e_m']$ be maximal, linearly independent sets in H and G/H, respectively. Then $e_1, \ldots, e_k, e_1', \ldots, e_m'$ is a maximal independent set in G. Thus, there exist rational numbers m_{ij}, n_{ij}, and $\hat{n}_{ij}$ such that the following formal relationships hold:

(13) $\varphi(e_i) = \sum_{j=1}^{k} m_{ij}e_j, \; i = 1, \ldots, k$

(14) $\varphi(e_i') = \sum_{j=1}^{k} n_{ij}e_j + \sum_{j=1}^{m} \hat{n}_{ij}e_j', \; i = 1, \ldots, m.$

From (13) it is clear that $\text{tr}(\varphi|H) = \Sigma_{i=1}^{k} m_{ii}$, while from (14) it follows that $\hat{\varphi}([e_i']) = \Sigma_{j=1}^{m} \hat{n}_{ij}[e_j']$ and thus $\text{tr } \hat{\varphi} = \Sigma_{i=1}^{m} \hat{n}_{ii}$. Consequently.

(15) $\text{tr } \hat{\varphi} + \text{tr}(\varphi|H) = \sum_{i=1}^{m} \hat{n}_{ii} + \sum_{i=1}^{m} m_{ii}.$

Now, from equation (13) we see that for each $i = 1, \ldots, k$ the coefficient of e_i in the expansion of $\varphi(e_i)$ is m_{ii}, while it follows from equation (14) that the coefficient of e_i' in the expansion of $\varphi(e_i')$ is n_{ii} for each $i = 1, \ldots, m$. But $\text{tr } \varphi$ is the sum of these coefficients, so the result follows from equation (15).   Q.E.D.

Now consider an oriented, geometric complex K and a chain map $\varphi : C(K) \to C(K)$. Then φ induces a chain map $\varphi_* : H(K) \to H(K)$. For each p, $\varphi_p : C_p(K) \to C_p(K)$ and $\varphi_{*p} : H_p(K) \to H_p(K)$ are endomorphisms of the finitely generated Abelian groups $C_p(K)$ and $H_p(K)$, respectively. Our major result is

Theorem 4-33 (Hopf Trace Theorem). Let K be an oriented, geometric complex, $\varphi : C(K) \to C(K)$ a chain map and $\varphi_* : H(K) \to H(K)$ the induced chain map in homology. Then

$$\sum_{p} (-1)^p \text{tr } \varphi_p = \sum_{p} (-1)^p \text{tr } \varphi_{*p}.$$

Proof: We omit parenthetical reference to K since no other complex is considered. Observe that, since φ is a chain map, both $\varphi|Z_p$ and $\varphi|B_p$ are endomorphisms for any p. By Lemma 4-32 $\text{tr } \varphi_p = \text{tr } \hat{\varphi}_p + \text{tr}(\varphi|Z_p)$, where $\hat{\varphi}_p : C_p/Z_p \to C_p/Z_p$ is the induced endomorphism. But $B_{p-1} = \text{Im } \partial_p \approx C_p/Z_p$ and, if we identify B_{p-1} and C_p/Z_p and again use the fact that φ is a chain map, we find that $\text{tr } \varphi_p = \text{tr}(\varphi_{p-1}|B_{p-1})$ so

(16) $\text{tr } \varphi_p = \text{tr}(\varphi_{p-1}|B_{p-1}) + \text{tr}(\varphi_p|Z_p).$

Similarly, since B_p is a subgroup of Z_p which is invariant under $\varphi_p|Z_p$, we obtain

(17) $\text{tr}(\varphi_p|Z_p) = \text{tr } \varphi_{*p} + \text{tr}(\varphi_p|B_p).$

Combining (16) and (17), we find that

(18) $\text{tr } \varphi_p = \text{tr}(\varphi_{p-1}|B_{p-1}) + \text{tr}(\varphi_p|B_p) + \text{tr } \varphi_{*p}.$

Taking the alternating sum of both sides of (18) and noting that each term tr $(\varphi_i|B_i)$ occurs twice with opposite signs, we obtain the desired equality.   Q.E.D.

4-15   The Lefschetz Fixed Point Theorem

Now let X be a topological polyhedron and $f : X \to X$ a continuous map. For each p, f induces an endomorphism $f_{*p} : H_p(X) \to H_p(X)$. We define the *Lefschetz number* $L(f)$ of f by

$$L(f) = \sum_p (-1)^p \, \text{tr} \, f_{*p}.$$

Observe that by Theorem 4-31 if f and g are homotopic maps of X into itself, then $L(f) = L(g)$.

Theorem 4-34 (Lefschetz Fixed Point Theorem). Let X be a topological polyhedron and $f : X \to X$ a continuous map. If $L(f) \neq 0$, then f has a fixed point.

Proof: Exercise 4-26. Show that it suffices to prove Theorem 4-34 in the case in which X is the polyhedron of a geometric complex.

Thus, we let K be a geometric complex and $f : |K| \to |K|$ a continuous map. We will assume that f has no fixed points and show that $L(f) = 0$.
Since f is assumed to have no fixed points, $d(x, f(x)) > 0$ for each $x \in |K|$. Since $|K|$ is compact, $\delta = \inf\{d(x, f(x)) : x \in |K|\}$ is greater than zero. Choose $n \geq 0$ so that $\mu(K^{(n)}) < \delta/3$. Since $|K^{(n)}| = |K|$, we may regard f as a map of $|K^{(n)}|$ into itself. Select $m \geq n$ and a simplicial map $\varphi : |K^{(m)}| \to |K^{(n)}|$, which is a simplicial approximation to f. Then φ induces a chain map $\varphi^0 : C(K^{(m)}) \to C(K^{(n)})$. If $\beta^{(m-n)} : C(K^{(n)}) \to C(K^{(m)})$ is the iterated, barycentric, subdivision map, then $\psi = \varphi^0 \circ \beta^{(m-n)}: C(K^{(n)}) \to C(K^{(n)})$ is a chain map. We show next that, for each p, tr $\psi_p = 0$. By expression (11) of Chapter 2 and $\mu(K^{(n)}) < \delta/3$ it follows that $d(\psi(x), f(x)) < \delta/3$ for each $x \in |K|$. It then follows that $d(\varphi(x), x) > \frac{2}{3}\delta$ for each x. Now let $s \in K^{(m)}$ and let $t \in K^{(n)}$ be the carrier of s. For any $x \in \bar{s}$ and $y \in \bar{t}$, $d(x, y) < \delta/3$, so that $d(\varphi(x), y) > \delta/3$. Thus, $\varphi(s) \neq t$ and it follows that, for each $\sigma_p \in K^{(n)}$, $\psi(\sigma_p)$ is a chain in which σ_p appears with coefficient zero. Thus, tr $\psi_p = 0$ for each p. Consequently, $\Sigma_p (-1)^p$ tr $\psi_p = 0$. By Theorem 4-33 we therefore have $\Sigma_p (-1)^p$ tr $\psi_{*p} = 0$. By the definition (8) of f_{*p}, $\psi_{*p} = f_{*p}$, so $L(f) = \Sigma_p (-1)^p$ tr $f_{*p} = 0$ as required.   Q.E.D.

Corollary 4-35. Let X be a topological polyhedron and $f : X \to X$ a continuous map. If f is homotopic to a map without fixed points, then $L(f) = 0$.

Exercise 4-27. Apply Theorem 4-34 to obtain another proof of the Brouwer Fixed Point Theorem. Hint: B^n is contractible, so every continuous map of B^n into itself is homotopic to a constant map. Let $f : B^n \to B^n$ be a constant map and show that $L(f) = 1$ by applying Corollary 4-20 and Theorem 4-10.

Finally, let us consider a map $f : S^n \to S^n$, where $n \geq 1$. $H_n(S^n)$ is infinite cyclic and therefore generated by a single element, say, ξ. Moreover, $f_{*n}(\xi) = d\xi$ for some integer d. We call d the *degree* of f and denote it deg f. Thus,

$$f_{*n}(\xi) = (\deg f)\, \xi.$$

Observe that deg $f = \operatorname{tr} f_{*n}$ so, in particular, degree is a homotopy invariant, that is, if f and g are homotopic maps of S^n into itself, then deg f = deg g. That the converse is also true, that is, two maps $f, g : S^n \to S^n$ of the same degree are homotopic, is a deep result of Hopf which we shall neither prove nor use, but which we observe provides a complete enumeration of the homotopy classes of maps $S^n \to S^n$ (see Section 3-3).

Exercise 4-28. Show that deg $(\operatorname{id}_{S^n}) = 1$ and that deg $f = 0$ for every constant map $f : S^n \to S^n$.

Remark: Observe that Exercise 4-28 provides another proof of the noncontractibility of S^n. In addition, Proposition 3-8 implies that a nonsurjective map of S^n into itself has degree zero.

Exercise 4-29. Show that for any continuous map $f : S^n \to S^n$, $L(f) = 1 + (-1)^n$ deg f.

Exercise 4-30. Show that if $f : S^n \to S^n$ is homotopic to a map without fixed points, then deg $f = (-1)^{n+1}$. Conclude that, if n is even, id_{S^n} is not homotopic to a map without fixed points.

Remark: In Section 5-1 we apply Exercise 4-30 to the problem of determining which spheres admit "nonvanishing, continuous tangent vector fields."

Supplementary exercises

4-31 Find all homology groups of the annulus $\{(x, y) \in \mathbf{R}^2 : \frac{1}{2} \leq (x^2 + y^2)^{1/2} \leq 1\}$ and the cylinder $S^1 \times I$.

4-32 Find all homology groups of the Möbius strip.

4-33 Two polyhedra with isomorphic homology groups in each dimen-

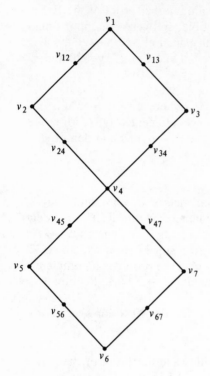

Figure 4-10

sion clearly need not be homeomorphic; show that they need not even be of the same homotopy type.

4-34   Derive the Euler–Poincaré Formula (Exercise 4-9) from the Hopf Trace Theorem.

4-35   Let K be the complex whose polyhedron is the "figure eight" shown in Figure 4-10 with vertices v_1, v_2, v_3, v_4, v_5, v_6, v_7, v_{12}, v_{13}, v_{24}, v_{34}, v_{45}, v_{56}, and v_{47} and 1-simplexes (v_1, v_{12}), (v_{12}, v_2), (v_2, v_{24}), (v_{24}, v_4), (v_4, v_{34}), (v_{34}, v_3), (v_3, v_{13}), (v_{13}, v_1), (v_4, v_{45}), (v_{45}, v_5), (v_5, v_{56}), (v_6, v_{67}), (v_{67}, v_7), (v_7, v_{47}), and (v_{47}, v_4).
Let $f : |K| \to |K|$ be the simplicial map induced by the following vertex assignment:

v_1	v_2	v_3	v_4	v_5	v_6	v_7	v_{12}	v_{13}	v_{24}	v_{34}	v_{45}	v_{56}	v_{67}	v_{47}
v_1	v_3	v_2	v_4	v_6	v_4	v_6	v_3	v_2	v_3	v_2	v_5	v_7	v_5	v_7

Show that $L(f) = 0$ even though f has fixed points. Thus, the converse of the Lefschetz Fixed Point Theorem is false.

4-36 Let U be a bounded open subset of $\mathbf{R}^n$. Show that bdy U is not a retract of $\overline{U}$.

4-37 Let $f, g : S^n \to S^n$ be continuous maps. Show that $\deg(fg) = \deg f \cdot \deg g$.

4-38 Show that for each integer $n \geq 1$ and each integer $m \geq 1$ there is a continuous map $f : S^n \to S^n$ with $\deg f = m$.

Chapter 5
Differential techniques

5-1 Introduction

The topological structure of subspaces of $\mathbf{R}^n$ is so extremely rich and variegated that a systematic treatment of all such spaces is quite out of the question. Indeed, our investigations thus far have proved most fruitful in just those situations in which the scope of our inquiry was most limited, the most striking example being our study of polyhedra. Motivated by our interest in dimension-theoretic questions and our suspicion that the linear structure of $\mathbf{R}^n$ should be decisive in resolving them, we defined the class of polyhedra in Chapter 2, restricted our attention to this very limited class of spaces, and were rewarded with results whose depth and beauty we could scarcely have anticipated. We now propose to exploit yet another aspect of the structure of $\mathbf{R}^n$ in order to obtain topological information about another restricted, but extremely important class of its subspaces.

A typical element of the class of spaces with which we will be concerned here is the sphere $S^2 = \{(x, y, z) \in \mathbf{R}^3 : x^2 + y^2 + z^2 = 1\}$. Consider the map h defined on the hemisphere $z > 0$, which projects a point (x, y, z) in S^2 onto the point $(x, y, 0)$ in the xy-plane. The hemisphere $z > 0$ is open in S^2 and the map h is continuous, one-to-one, and onto the open unit disc in the copy $\{(x, y, z) \in \mathbf{R}^3 : z = 0\}$ of $\mathbf{R}^2$. Moreover, on this disc the inverse of h is given by $(x, y, 0) \rightarrow (x, y, (1 - x^2 - y^2)^{1/2})$ and is therefore also continuous. We find then that the map h is a homeomorphism of the open hemisphere $z > 0$ in S^2 onto an open disc in $\mathbf{R}^2$. Similar arguments show that each of the open hemispheres $z < 0$, $y > 0$, $y < 0$, $x > 0$, and $x < 0$ is likewise homeomorphic to an open disc in $\mathbf{R}^2$, for example $y < 0$ is homeomorphic to the open unit disc in the copy $\{(x, y, z) \in \mathbf{R}^3: y = 0\}$ of $\mathbf{R}^2$ via the map $(x, y, z) \rightarrow (x, 0, z)$ which, on this disc, has inverse $(x, 0, z) \rightarrow (x, -(1 - x^2 - z^2)^{1/2}, z)$. Since these hemispheres cover S^2, we conclude that each point of S^2 has a nbd homeomorphic to an open subset of $\mathbf{R}^2$. A space X in which every point has a nbd homeomorphic to an open subset of $\mathbf{R}^k$ is called a *topological manifold* of *dimension k*.

Exercise 5-1. Show that the dimension of a topological manifold is well-defined. Hint: See the Remark following Theorem 2-31.

We have just shown that S^2 is a topological manifold of dimension two, and entirely analogous arguments prove that S^k is a k-dimensional topological manifold for any positive integer k. Although other examples of topological manifolds abound (find some!), it is important to observe that certain very simple spaces fail to have the required property. For example, the closed unit interval [0, 1] is not a topological manifold; although any point in (0, 1) has a nbd in [0, 1] homeomorphic to an open subset of **R**, this is not true for either of the endpoints 0 or 1. (Why not?) Equally important is the observation that no "new" structure of R^n is required for the definition of a topological manifold. If, however, we return to our example S^2, we find that a great deal more can be said. Consider again the homeomorphism h of the open hemisphere $z > 0$ onto the open unit disc in R^2 that, after eliminating the superfluous coordinate, is given by $h(x, y, z) = (x, y)$. Observe that in addition to being continuous h is "smooth" in the sense that each of its coordinate functions $h_1(x, y, z) = x$ and $h_2(x, y, z) = y$ has continuous partial derivatives of all orders and that, moreover, the same is true of the inverse map $h^{-1}(x, y) = (x, y, (1 - x^2 - y^2)^{1/2})$ on the open unit disc. (All of the terms we use rather loosely in this introduction will be defined precisely later in the chapter.) A homeomorphism that is smooth and has a smooth inverse will be referred to as a "diffeomorphism"; a space X in which every point has a nbd diffeomorphic to an open subset of R^k is called a "differentiable manifold" of dimension k. A differentiable manifold is, of course, also a topological manifold, but the converse is false since smoothness of the local homeomorphisms prohibits the existence of sharp "peaks" and other such "singular points" in a differentiable manifold. Thus, for example, the subspace $\{(x, |x|) \in R^2 : x \in R\}$ of R^2 is a topological – but not a differentiable – manifold since a homeomorphism can "straighten out" the the corner at (0, 0), whereas a diffeomorphism cannot. (Check to see that the projection of this subspace onto the first coordinate space is a homeomorphism, but not a diffeomorphism. Then construct for yourself a surface in R^3 that is a topological, but not a differentiable 2-manifold.)

The class of differentiable manifolds is important largely because it arises naturally in the construction of mathematical models for certain physical phenomena in the theory of differential equations. Since we shall have occasion somewhat later to make use of one result from this subject, it will be to our advantage to obtain some idea of just how this comes about. Let us then consider quite generally a physical situation whose state changes with time and that possesses three properties we shall call "determinacy," "finite dimensionality," and "smoothness."

Loosely speaking, a process is "deterministic" if its entire future course and its entire past are uniquely determined by its state at any given instant of time. This strong form of determinism is, of course, a basic assumption of much of classical (Newtonian) mechanics, but should be recognized as an assumption with, at best, a limited range of applicability. The set of all possible states of a process is its "phase space." For example, in classical mechanics the motion of a system of n particles under the influence of various forces is a deterministic process, each state of which is described by a set of $6n$ numbers (three components each for the position and momentum of each of the n particles). Thus, a typical element of the phase space is a $6n$-tuple of real numbers specifying a state of the system at some particular instant. In general, of course, this phase space will be a proper subset of $\mathbf{R}^{6n}$ since not all $6n$-tuples of real numbers can correspond to possible states of the system. (For example, if $n \geq 2$, no point in the phase space can have all of its components equal since this would imply, in particular, that all of the particles occupied the same position at some instant.) For a given initial state of the system, the set of all its subsequent and previous states will be described by a curve in the phase space – one state for each instant t of time. In general, we say that a process is "finite dimensional" if its phase space is finite dimensional, that is, if a finite number of real parameters suffice to describe any state of the system. Even in classical physics examples of processes that cannot be considered finite dimensional abound, for example, fluid motion and the vibration of strings and membranes. Finally, an evolutionary process is said to be "smooth" if the state of the system varies smoothly with time and if, moreover, the state at any fixed time t_0 depends smoothly on the initial state. Observe that a finite dimensional smooth process must have a phase space that is also "smooth", that is, a differentiable manifold, since a singular point of the phase space would correspond to a state of the system at which abrupt (nondifferentiable) changes of state occur. Again, smoothness is an assumption, appropriate to some situations (e.g., the motion of the planets in our solar system about the sun), but not to others (e.g., the motion of colliding billiard balls on a table).

Having seen how differentiable manifolds arise in the analysis of physical phenomena satisfying certain conditions, we now examine somewhat more closely the mathematical implications of these hypotheses. Let us suppose then that the process we have under consideration possesses all three of the properties we have discussed and let us denote by X the phase space of the process. For each $x \in X$ and $t \in \mathbf{R}$ we let $g^t(x)$ denote the state of the process at time t, given that its initial state was x. ($g^t(x)$ is uniquely determined since the process is assumed deterministic.)

For each $t \in \mathbf{R}$ this defines a map $g^t : X \to X$ that sends $x \in X$, regarded as the initial state of a process, to the state $g^t(x)$ of that process at time t. Observe that $g^0 = \mathrm{id}_X$ and $g^{t+s} = g^t \circ g^s$ for all t and s in $\mathbf{R}$ (again by determinacy). A family $\{g^t : t \in \mathbf{R}\}$ of maps of X into itself with these two properties is called a "one-parameter group of transformations" of X. Note, in particular, that for any $t \in \mathbf{R}$, $g^t \circ g^{-t} = g^0 = \mathrm{id}_X$, so each g^t has an inverse defined on all of X and is thus bijective. Reversing our point of view, we define for each $x \in X$ a map $g_x : \mathbf{R} \to X$ by $g_x(t) = g^t(x)$ for each $t \in \mathbf{R}$. Thus, g_x maps t in $\mathbf{R}$ onto the state at time t of the process whose initial state is x. g_x is thus a curve in X whose image is the set of all states of our process, assuming that the initial state was x. Generally, it is more convenient to combine both of these points of view by defining a map $g : \mathbf{R} \times X \to X$ by $g(t, x) = g^t(x) = g_x(t)$ for each $(t, x) \in \mathbf{R} \times X$. Having assumed that our process is finite dimensional and smooth, we conclude that X is a differentiable manifold and that this map g is smooth. In particular, all of the maps g^t, $t \in \mathbf{R}$, and g_x, $x \in X$, are smooth. Moreover, being bijective with smooth inverse, each g^t is a diffeomorphism of X onto itself. We shall refer to $\{g^t : t \in \mathbf{R}\}$ as a "one-parameter group of diffeomorphisms" of X; it is this group that describes the essential nature of the system. The question then is, "For each given phenomenon, how do we find this group?" The laws of physics provide us with no direct information about the map g. Rather, these laws generally describe only the rate at which the state of the system varies at each point of the phase space. Consider, for example, a pendulum with arm of length l and bob of mass m and let $x_1 = x_1(t)$ denote the angle between the pendulum arm and the vertical at time t. An elementary application of Newton's laws of motion shows that, under suitable conditions, $x_1(t)$ satisfies the differential equation

(1) $$\frac{d^2 x_1}{dt^2} + \frac{G}{l} \sin x_1 = 0,$$

where G is the gravitational constant. Now, a state of this system is described by two real parameters, that is, the position x_1 and angular velocity dx_1/dt of the pendulum bob. Introducing the variable $x_2 = x_2(t) = dx_1/dt$, we find that $dx_2/dt = d^2 x_1/dt^2$ so that by equation (1) we must have

(2)
$$\frac{dx_1}{dt} = x_2$$
$$\frac{dx_2}{dt} = -\frac{G}{l} \sin x_1.$$

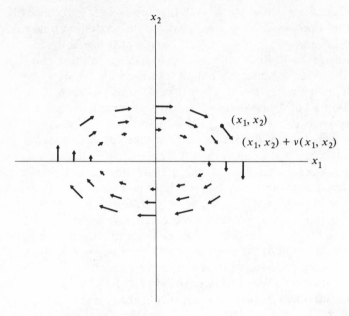

x_2

(x_1, x_2)

$(x_1, x_2) + v(x_1, x_2)$

x_1

Figure 5-1

If the oscillations of the pendulum are "small," $\sin x_1 \approx x_1$ and we may replace (2) by the "linearized" system

(3)
$$\frac{dx_1}{dt} = x_2$$
$$\frac{dx_2}{dt} = -\frac{G}{l} x_1.$$

The phase space of system (3) is then a nbd of the origin in the $x_1 x_2$-plane. A natural way to interpret (3) is to view it as assigning to each point (x_1, x_2) in the phase space a vector $v(x_1, x_2) = (dx_1/dt, dx_2/dt) = (x_2, -(G/l) x_1)$ that describes the rate at which the state of the system is varying at the point (x_1, x_2); see Figure 5-1.

This then is essentially all the information physics provides. The problem of constructing from this the corresponding function g (and thus the one-parameter group of diffeomorphisms) is one of mathematics and, more particularly, of differential equations. For each initial state (x_1^0, x_2^0) of the system, what is required is a curve in the phase space through (x_1^0, x_2^0) with the property that the "velocity vector" at each point (x_1, x_2) of the curve is just $v(x_1, x_2)$. We need not concern ourselves with the specific procedures whereby such constructions are carried out. Suffice it to say that in certain (rare) cases, for example, the linearized pendulum

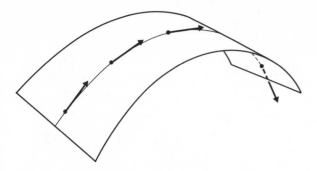

Figure 5-2

problem (3), it is possible to solve the equations directly and thus explicitly write down the function g. In most cases, for example, the nonlinear system (2), this is either impossible or extremely complicated and we must turn instead to qualitative methods. The relevant observation from our point of view is that the laws of science, when applied to a particular phenomenon, generally take the form of an assignment to each point of the phase space of a vector tangent to the manifold, which indicates the rate at which the state of the process varies at that point (see Figure 5-2). Such an assignment will be referred to as a "tangent vector field" on the manifold. Generally, certain regularity conditions are imposed on vector fields; for the time being we will require only that the magnitude and direction of the assigned vectors vary continuously from point to point in the manifold.

Locally, all of the interesting behavior of a vector field v on a manifold X occurs at the "zeros" of v, that is, at those points $x \in X$ where $v(x)$ is the zero vector. Essentially, when $v(x) \neq 0$, then (by continuity) v is nearly constant in magnitude and direction near x. If, however, x is a zero of v, then the direction of v may change radically in any nbd of x. For example, the origin is a zero of the vector field associated with the system (3), and from Figure 5-1 it is clear that near (0, 0) the field "circulates" and therefore assumes every direction. A great many other possibilities exist, some of which are shown in Figure 5-3.

By this time you have no doubt asked yourself (several times) just what all of this material on one-parameter groups of transformations, vector fields, and differential equations has to do with topology. Although an entirely satisfactory answer would require quite a deep analysis, which we do not propose to carry out here, we will have occasion in Section 5-13 to investigate the matter somewhat more closely. On an intuitive level, however, it should be clear that there must exist important relationships between the topology of a manifold and the type of vector field that can

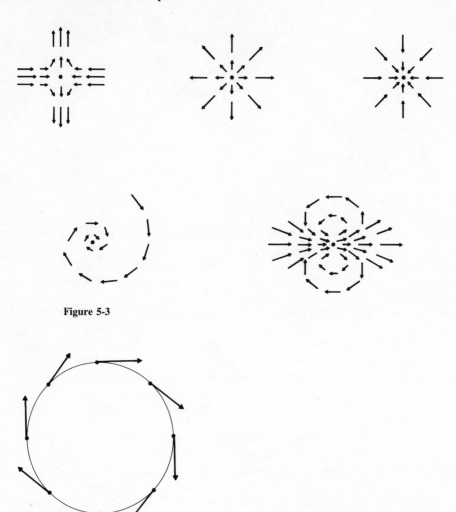

Figure 5-3

Figure 5-4

be defined on that manifold. Suppose we ask, for example, whether or not a given manifold admits a continuous vector field with no zeros. The circle S^1 surely does admit such a field (see Figure 5-4). From this we can easily construct a continuous, nonzero vector field on the torus $S^1 \times S^1$. If you now try to construct such a vector field on the 2-sphere S^2, you will find that all of your efforts come to naught. We shall next use the very special geometrical properties of spheres to give an *ad hoc* definition of "continuous tangent vector field" on S^k and prove a classical theorem to the effect that S^k admits a nonvanishing field of tangent vectors iff k is odd.

A *continuous tangent vector field on the sphere* S^k is a continuous map $v = (v_1, \ldots, v_{k+1})$ of S^k into $\mathbf{R}^{k+1}$ such that, for each $x = (x_1, \ldots, x_{k+1}) \in S^k \subseteq \mathbf{R}^{k+1}$, $x \cdot v(x) = \Sigma_{i=1}^{k+1} x_i v_i(x) = 0$. Such a vector field is *nonvanishing* if $v(x)$ is not the zero vector of $\mathbf{R}^{k+1}$ for any $x \in S^k$.

Theorem 5-1. There exists a nonvanishing continuous tangent vector field on the sphere S^k iff k is odd.

Proof: First assume that k is odd, that is, that $k = 2m - 1$ for some positive integer m. For each $x = (x_1, \ldots, x_{2m}) \in S^{2m-1} \subseteq \mathbf{R}^{2m}$, let $v(x) = (-x_{m+1}, \ldots, -x_{2m}, x_1, \ldots, x_m)$. Then v is continuous, $\Sigma_{i=1}^{2m} x_i v_i(x) = 0$, and $v(x) \neq 0$ for each $x \in S^{2m-1}$.

Now suppose k is even and let v be a continuous tangent vector field on S^k. We assume that $v(x) \neq 0$ for each $x \in S^k$ and obtain a contradiction. Since $v(x)$ is nonzero for each x, we may normalize (divide by $\|v(x)\|$) and assume that $v(x) \cdot v(x) = 1$ for each x. Define a map F of $S^k \times I$ into $\mathbf{R}^{k+1}$ by $F(x, t) = x \cos \pi t + v(x) \sin \pi t$. Then $F(x, t) \cdot F(x, t) = x \cdot x \cos^2 \pi t + 2x \cdot v(x) \cos \pi t \sin \pi t + v(x) \cdot v(x) \sin^2 \pi t = \cos^2 \pi t + \sin^2 \pi t = 1$ for all x and t. Thus, $F : S^k \times I \to S^k$. Moreover, $F(x, 0) = x$ and $F(x, 1) = -x$ so, being continuous, F is a homotopy from id_{S^k} to the antipodal map on S^k. Since the antipodal map has no fixed points on S^k, this contradicts Exercise 4-30.   Q.E.D.

Although they are not our major concern here, you will certainly want to pursue further the deep and beautiful relationships between vector fields and the topology of manifolds (see Guillemin and Pollack; Hirsch; and Milnor, 1965). Our rather prolonged introduction to this subject was intended to serve several purposes. Hopefully, you have gained some insight into the origin of our interest in manifolds and, more particularly, of some of the basic concepts with which we will be dealing. We felt it particularly important that some contact be made with science which, it should be clear, is still a powerful stimulus to mathematical research, even in a subject as "abstract" as topology.

5-2   Smooth maps

Let U be an open subset of $\mathbf{R}^n$ and $f : U \to \mathbf{R}^m$ a continuous map with coordinate functions $f_1, \ldots, f_m$. Then f is said to be *smooth* if all of its partial derivatives $\partial^k f_i / \partial x_{i_1} \ldots \partial x_{i_k}$ exist and are continuous on U. More generally, if X and Y are arbitrary subsets of $\mathbf{R}^n$ and $\mathbf{R}^m$, respectively, then a continuous map $f : X \to Y$ is said to be *smooth* if there is an

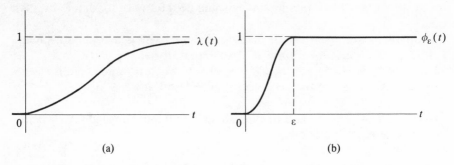

(a)   (b)

Figure 5-5

open subset U of $\mathbf{R}^n$ containing X and a smooth map $F : U \to Y$ that agrees with f on X. Observe that if $f : X \to Y$ and $g : Y \to Z$ are both smooth, then $g \circ f : X \to Z$ is also smooth. The identity map on any set is smooth as is any constant map. We pause now to construct several non-trivial examples of smooth maps that will be useful later as "separating functions."

Example 5-1. Define a real-valued function of a real variable $\lambda : \mathbf{R} \to \mathbf{R}$ by

$$\begin{aligned} \lambda(t) &= 0 && \text{for } t \leq 0 \\ &= e^{-1/t} && \text{for } t > 0. \end{aligned}$$

Now, λ is smooth on each of the intervals $t < 0$ and $t > 0$ so that, to show that it is smooth on all of $\mathbf{R}$, we need only prove that each derivative of $e^{-1/t}$ on $t > 0$ approaches 0 as t approaches 0 through positive values. But for any $n > 0$ the nth derivative of $e^{-1/t}$ on $t > 0$ has the form $q(1/t)e^{-1/t}$, where q is a polynomial of degree $2n$. Thus,

$$\lim_{t \to 0^+} \frac{d^n}{dt^n}(e^{-1/t}) = \lim_{t \to 0^+} q(1/t)e^{-1/t} = \lim_{x \to \infty} \frac{q(x)}{e^x} = 0$$

by l'Hospital's rule. Observe that $0 \leq \lambda(t) < 1$ for all $t \in \mathbf{R}$ (see Figure 5-5 (a)). Now let $\epsilon > 0$ be given and define $\varphi_\epsilon : \mathbf{R} \to \mathbf{R}$ by

$$\varphi_\epsilon(t) = \frac{\lambda(t)}{\lambda(t) + \lambda(\epsilon - t)}$$

for each $t \in \mathbf{R}$. Then φ_ϵ is smooth on all of $\mathbf{R}$, $0 \leq \varphi_\epsilon(t) \leq 1$, $\varphi_\epsilon(t) = 0$ for $t \leq 0$, and $\varphi_\epsilon(t) = 1$ for $t \geq \epsilon$ (see Figure 5-5 (b)). Finally, consider a fixed $x^0 \in \mathbf{R}^n$ and let $B_r(x_0)$ be the closed r-ball about x_0 in $\mathbf{R}^n$, where $r > 0$. Let $\epsilon > 0$ be given and define $\psi_\epsilon : \mathbf{R}^n \to \mathbf{R}$ by

$$\psi_\epsilon(t) = 1 - \varphi_\epsilon(\|y - x_0\| - r)$$

for each $y \in \mathbf{R}^n$. Now, $\|y - x_0\|$ is not smooth at $y = x_0$, but on a nbd of x_0, $\|y - x_0\| - r$ is less than zero so $\varphi_\epsilon(\|y - x_0\| - r) = 0$ and $\psi_\epsilon(y) = 1$. Thus, ψ_ϵ is locally constant at $y = x_0$ and consequently is smooth there. It follows that ψ_ϵ is smooth on all of $\mathbf{R}^n$. Moreover, $0 \leqslant \psi_\epsilon(y) \leqslant 1$ for all $y \in \mathbf{R}^n$, $\psi_\epsilon(y) = 1$ for $y \in B_r(x_0)$, and $\psi_\epsilon(y) = 0$ for $y \notin \text{int } B_{r+\epsilon}(x_0)$.

Smoothness is, of course, a much stronger condition than continuity, but the actual gulf that exists between these two concepts cannot be fully appreciated until we have seen an example of a continuous function that fails to have even a first derivative at every point in its domain. We outline the construction of one such function in the next exercise.

Exercise 5-2. Let $f_1 : \mathbf{R} \to \mathbf{R}$ be the function of period 1 that on $[-\frac{1}{2}, \frac{1}{2}]$ is defined by $f_1(x) = |x|$. For each $n > 1$ define $f_n : \mathbf{R} \to \mathbf{R}$ by $f_n(x) = 4^{-n+1}f_1(4^{n-1}x)$. Then f_n has period 4^{-n+1} and assumes a maximum value of $\frac{1}{2} \cdot 4^{-n+1}$. Finally, define $f : \mathbf{R} \to \mathbf{R}$ at each $x \in \mathbf{R}$ by $f(x) = \Sigma_{n=1}^\infty f_n(x) = \Sigma_{n=1}^\infty f_1(4^{n-1}x)/4^{n-1}$. Then f is continuous on all of $\mathbf{R}$. Now let x_0 be an arbitrary real number. For each $n > 0$ choose h_n to be either 4^{-n-1} or -4^{-n-1} so that $|f_n(x_0 + h_n) - f_n(x_0)| = |h_n|$. Then $|f_m(x_0 + h_n) - f_m(x_0)|$ is $|h_n|$ for $m \leqslant n$ and vanishes for $m > n$. Thus, $(f(x_0 + h_n) - f(x_0))/h_n$ is an integer that is even if n is even and odd if n is odd. Therefore, f must fail to be differentiable at x_0.

Nevertheless, it is our intention to use the properties of smooth maps to obtain topological information and, in particular, to investigate the behavior of arbitrary continuous maps. Our situation then is analogous to that in Chapters 2, 3, and 4, where we studied continuous maps on polyhedra by exploiting properties of the much more restricted class of simplicial maps. Recall that our ability to do this rested in large measure on the fact that continuous maps of polyhedra admit arbitrarily good simplicial approximations. Our first major goal in this chapter is to prove a similar approximation theorem for smooth maps.

5-3   The Stone–Weierstrass Theorem

The theorem we prove in this section is among the most important in all of modern mathematics. Although we could content ourselves with the classical form of the result (due to Weierstrass), we shall prove the more general, algebraic version of Stone since little extra effort is required and the rewards are considerable. In order to do so, however, we must introduce some new terminology.

Let X be a compact subspace of $\mathbf{R}^n$ and $C(X)$ the set of all continuous, real-valued functions on X. If addition and scalar multiplication are defined as usual on $C(X)$ by $(f + g)(x) = f(x) + g(x)$ and $(\alpha f)(x) = \alpha f(x)$

for all f, $g \in C(X)$ and $\alpha \in \mathbf{R}$, then $C(X)$ has the structure of a real vector space. If we also define a product on $C(X)$ by $(fg)(x) = f(x)g(x)$ for all f and g in $C(X)$, then $C(X)$ becomes an algebra with identity, the identity being the function, denoted 1, that takes the value $1 \in \mathbf{R}$ at each $x \in X$. Now let $\mathscr{A}$ be a subalgebra of $C(X)$, that is, $\mathscr{A}$ is a linear subspace of $C(X)$ that contains the product of each pair of its elements. We say that $\mathscr{A}$ *separates points* of X if, whenever x and y are distinct points of X, there is an element h of $\mathscr{A}$ with $h(x) \neq h(y)$. $\mathscr{A}$ is said to be *uniformly closed* if whenever $\{f_n\}_{n=1}^{\infty}$ is a sequence of elements of $\mathscr{A}$ that converges uniformly on X to some function f (necessarily in $C(X)$ by Apostol, Theorem 13-3, or Buck, Chapter 4, Theorem 17), then f is also in $\mathscr{A}$. If $\mathscr{A}$ is an arbitrary subalgebra of $C(X)$, the *uniform closure* of $\mathscr{A}$, denoted $\overline{\mathscr{A}}$, is the set of all elements of $C(X)$ that are uniform limits of sequences of elements of $\mathscr{A}$; equivalently, $\overline{\mathscr{A}}$ is the set of all elements of $C(X)$ that can be uniformly approximated by elements of $\mathscr{A}$ to any desired degree of accuracy. $\overline{\mathscr{A}}$ is also a subalgebra of $C(X)$.

Example 5-2. The Subalgebra of Smooth Functions. Let $\mathscr{S}(X)$ be the set of all smooth, real-valued functions on the compact subspace X of $\mathbf{R}^n$. Since sums, scalar multiples, and products of smooth, real-valued functions are smooth, $\mathscr{S}(X)$ is a subalgebra of $C(X)$; see Exercise 1-38. Moreover, $\mathscr{S}(X)$ separates points of X: If x and y are distinct points of X, then there is an $r > 0$ and $\epsilon > 0$ such that $x \in B_r(x)$ and $y \in \mathbf{R}^n - \text{int } B_{r+\epsilon}(x)$, so that the restriction to X of the function ψ_ϵ constructed in Example 5-1 takes the value 1 at x and 0 at y. However, $\mathscr{S}(X)$ is *not* uniformly closed, in general, since a uniform limit of smooth functions need not be smooth. (This will follow from Theorem 5-3, but you should construct a specific example, e.g., by Fourier series.) Our object is to prove that the uniform closure of $\mathscr{S}(X)$ in $C(X)$ is all of $C(X)$, that is, that every continuous, real-valued function on X can be uniformly approximated by smooth functions to any desired degree of accuracy. The form of the Stone–Weierstrass Theorem we prove here states, much more generally, that *any* uniformly closed subalgebra of $C(X)$ that separates points and contains a nonzero constant function must be all of $C(X)$. For the proof we require the following lemma.

Lemma 5-2. Let λ be a real variable. For each $\epsilon > 0$, there exists a polynomial $p_\epsilon(\lambda)$ with real coefficients such that $\big| |\lambda| - p_\epsilon(\lambda) \big| < \epsilon$ for all $\lambda \in [-1, 1]$.

Exercise 5-3. Prove Lemma 5-2. Hint: According to the Binomial Theorem from real analysis (Apostol, Section 13-20), the expansion

$$1 - \sqrt{1 - t} = \sum_{n=1}^{\infty} \left| \binom{\frac{1}{2}}{n} \right| t^n$$

is valid on $(-1, 1)$ and thus on $[0, 1)$. Show that the series converges at $t = 1$ and then that the convergence is uniform on $[0, 1]$. Now make the change of variable $t = 1 - \lambda^2$ for $\lambda \in [-1, 1]$.   Q.E.D.

Theorem 5-3 (Stone–Weierstrass Theorem). Let X be a compact subspace of $\mathbf{R}^n$ and $\mathscr{A}$ a uniformly closed subalgebra of $C(X)$ that separates points of X and contains a nonzero constant function. Then $\mathscr{A} = C(X)$.

Proof: Since $\mathscr{A}$ contains a nonzero constant function and is a subalgebra of $C(X)$, it contains all constant functions, including the multiplicative identity 1. Note that if X consists of a single point, then every element of $C(X)$ is a constant function and the result is trivial. Thus, we may assume that X contains more than one point.

First we prove that if $f \in \mathscr{A}$, then the function $|f|$, defined by $|f|(x) = |f(x)|$ is also in $\mathscr{A}$. (Why is $|f|$ continuous?) Observe that, since X is compact, $|f|$ assumes a maximum value on X. Let us first assume that $\max|f| \leq 1$. Let $\epsilon > 0$ be given and let $p_\epsilon(\lambda)$ be the polynomial whose existence is guaranteed by Lemma 5-2. The function $p_\epsilon(f)$, defined by $p_\epsilon(f)(x) = p_\epsilon(f(x))$ for each $x \in X$, is in $\mathscr{A}$ and, for each x, $\big| |f(x)| - p_\epsilon(f(x)) \big| < \epsilon$. Thus, $|f|$ can be uniformly approximated by elements of $\mathscr{A}$. Since $\mathscr{A}$ is uniformly closed, $|f| \in \mathscr{A}$. Now, if $\max |f| > 1$, then $\mu|f| \in \mathscr{A}$, where $\mu = 1/\max|f|$ so, since $\mathscr{A}$ is a subalgebra of $C(X)$, $f = (1/\mu)(\mu|f|)$ is in $\mathscr{A}$ also.

Next we show that if $f_1, \ldots, f_k$ are in $\mathscr{A}$, then the functions $\max(f_1, \ldots, f_k)$ and $\min(f_1, \ldots, f_k)$ defined by $\max(f_1, \ldots, f_k)(x) = \max(f_1(x), \ldots, f_k(x))$ and $\min(f_1, \ldots, f_k)(x) = \min(f_1(x), \ldots, f_k(x))$ are also in $\mathscr{A}$. By induction it suffices to show that $\max(f_1, f_2)$ and $\min(f_1, f_2)$ are in $\mathscr{A}$, and this follows immediately from what we have just proved since we may write $\max(f_1, f_2) = \frac{1}{2}(f_1 + f_2) + \frac{1}{2}|f_1 - f_2|$ and $\min(f_1, f_2) = \frac{1}{2}(f_1 + f_2) - \frac{1}{2}|f_1 - f_2|$.

Now let f be an arbitrary element of $C(X)$ and let $\epsilon > 0$ be given. In order to show that $f \in \mathscr{A}$, it will suffice to find a $g \in \mathscr{A}$ with $|g(z) - f(z)| < \epsilon$ for all $z \in X$. Fix a point $x \in X$. Now let y be any point of X distinct from x. Since $\mathscr{A}$ separates points of X, there is an $h \in \mathscr{A}$ with $h(x) \neq h(y)$. But then we can find real numbers λ and μ such that the function $f_{xy} : X \to \mathbf{R}$ defined by $f_{xy} = \lambda h + \mu 1$ satisfies $f_{xy}(x) = f(x)$ and $f_{xy}(y) = f(y)$. Observe that f_{xy} is in $\mathscr{A}$. Consider the open set $G_y = \{z \in X : f_{xy}(z) < f(z) + \epsilon\}$. Then both x and y are in G_y. Still regarding x as fixed, obtain one such open set G_y for each $y \neq x$ in X, and by compactness select finitely many such sets $G_{y_1}, \ldots, G_{y_m}$ that cover X. Define $f_x : X \to \mathbf{R}$ by $f_x = \min(f_{xy_1}, \ldots, f_{xy_m})$. Then $f_x \in \mathscr{A}$, $f_x(x) = f(x)$ and $f_x(z) < f(z) + \epsilon$ for each $z \in X$. Now define an open set H_x by $H_x = \{z \in X : f_x(z) > f(z) - \epsilon\}$. Then $x \in H_x$.

Now repeat this whole process for each $x \in X$, thus producing one function $f_x \in \mathscr{A}$ and one open set H_x in X for each $x \in X$. Select finitely many of the H_x, say, $H_{x_1}, \ldots, H_{x_k}$, that cover X and define $g : X \to \mathbf{R}$ by $g = \max(f_{x_1}, \ldots, f_{x_k})$. Then $g \in \mathscr{A}$ and $g(z) > f(z) - \epsilon$ for each $z \in X$. But each f_{x_i} satisfies $f_{x_i}(z) < f(z) + \epsilon$ for each $z \in X$, so the same is true of g. Thus, $f(z) - \epsilon < g(z) < f(z) + \epsilon$ or $|g(z) - f(z)| < \epsilon$ for each $z \in X$ as required.   Q.E.D.

In particular, if $\mathscr{S}(X)$ is the subalgebra of $C(X)$ consisting of all smooth real-valued functions on the compact space X, then, by Example 5-2, $\overline{\mathscr{S}(X)} = C(X)$. Now, let $f : X \to \mathbf{R}^m$ be a continuous map with coordinate functions $f_1, \ldots, f_m$, and let $\epsilon > 0$ be given. For each $i = 1, \ldots, m$, let $g_i : X \to \mathbf{R}$ be a smooth function on X with $|g_i(z) - f_i(z)| < \epsilon/\sqrt{m}$ for all $z \in X$. Define $g : x \to \mathbf{R}^m$ by $g = (g_1, \ldots, g_m)$. Then g is smooth and, for any $z \in X$,

$$\|g(z) - f(z)\| = ((g_1(z) - f_1(z))^2 + \cdots$$
$$+ (g_m(z) - f_m(z))^2)^{1/2} < (m(\epsilon/\sqrt{m})^2)^{1/2} = \epsilon.$$

We have thus proved the following corollary.

Corollary 5-4. Let X be a compact subspace of $\mathbf{R}^n$ and $f : X \to \mathbf{R}^m$ a continuous map. For any $\epsilon > 0$ there exists a smooth map $g : X \to \mathbf{R}^m$ such that $\|g(z) - f(z)\| < \epsilon$ for all $z \in X$.

Just as the Simplicial Approximation Theorem links the topological and simplicial structures of polyhedra, so Corollary 5-4 will act as our bridge between the topological and differentiable points of view in $\mathbf{R}^n$.

5-4   Derivatives as linear transformations

Further progress in the study of smooth maps depends on a careful reexamination of the concept of differentiability in the simplest case of real-valued functions of a real variable. By definition a function $f : \mathbf{R} \to \mathbf{R}$ is differentiable at $x_0 \in \mathbf{R}$ if there exists a real number, denoted $f'(x_0)$, such that

$$(4) \qquad \lim_{h \to 0} \frac{f(x_0 + h) - f(x_0)}{h} = f'(x_0).$$

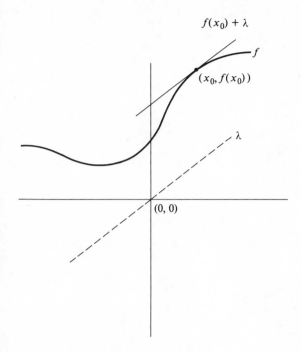

Figure 5-6

If we define a linear transformation $\lambda : \mathbf{R} \to \mathbf{R}$ by $\lambda(h) = f'(x_0) \cdot h$ for each $h \in \mathbf{R}$, then equation (4) can be written

(5) $$\lim_{h \to 0} \frac{f(x_0 + h) - (f(x_0) + \lambda(h))}{h} = 0,$$

which we interpret as saying that the affine function $f(x_0) + \lambda$ is a good approximation to f near x_0. This is, of course, just another way of stating the elementary fact that the graph of $f(x_0) + \lambda$ is the line tangent to the graph of f at $(x_0, f(x_0))$; see Figure 5-6.

Observe that it is precisely this property of having a good affine approximation at each point that distinguishes differentiable functions. For example, the function whose graph is shown in Figure 5-7 fails to be differentiable at the origin because no linear function can be regarded as a good approximation to it on any nbd of zero.

With the essential nature of differentiability for real-valued functions of a real variable exposed, it becomes clear how these ideas generalize to higher dimensional situations: A map $f : \mathbf{R}^n \to \mathbf{R}^m$ will be said to be "differentiable" at a point $x_0 \in \mathbf{R}^n$ if there is a linear transformation of $\mathbf{R}^n$ into $\mathbf{R}^m$ whose translation to $f(x_0)$ is a "good" approximation to f near x_0.

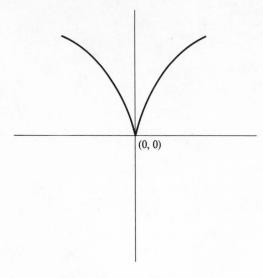

(0, 0)

Figure 5-7

However, since we are, after all, dealing with a local property of the maps under consideration, we prefer to formulate our precise definitions for maps defined on open subsets of Euclidean spaces. Thus, suppose $f: U \to V$ is a map, where U is open in $\mathbf{R}^n$ and V is open in $\mathbf{R}^m$. (We could take V to be all of $\mathbf{R}^m$, but we choose not to do so for reasons that will become clear later.) We say that f is *differentiable at* $x_0 \in U$ if there exists a linear transformation $\lambda : \mathbf{R}^n \to \mathbf{R}^m$ such that

$$(6) \qquad \lim_{h \to 0} \frac{\|f(x_0 + h) - (f(x_0) + \lambda(h))\|}{\|h\|} = 0.$$

f is said to be *differentiable* on U if it is differentiable at each $x_0 \in U$.

Exercise 5-4. Show that if $f: U \to V$ is differentiable at $x_0 \in U$, then the linear transformation $\lambda : \mathbf{R}^n \to \mathbf{R}^m$ satisfying (6) is unique.

By virtue of Exercise 5-4 we may define the *derivative of* f *at* x_0, denoted df_{x_0}, to be the unique linear transformation $df_{x_0} : \mathbf{R}^n \to \mathbf{R}^m$ which satisfies

$$\lim_{h \to 0} \frac{\|f(x_0 + h) - (f(x_0) + df_{x_0}(h))\|}{\|h\|} = 0$$

Remark: In the one-dimensional case this definition alters the traditional terminology of elementary calculus, shifting emphasis from the number $f'(x_0)$ to the function $\lambda(h) = f'(x_0) \cdot h$.

Exercise 5-5. (a) Let $T : \mathbf{R}^n \to \mathbf{R}^m$ be a linear transformation. Show that there exists a constant M such that $\|T(h)\| \leq M\|h\|$ for each $h \in \mathbf{R}^n$.

(b)   Show that if $f : U \to V$ is differentiable at $x_0 \in U$, then f is continuous at x_0.

Here, as in the calculus, the derivative df_{x_0} of a differentiable map $f : U \to V$ is our single most important source of information about the local behavior of f near x_0. In order to exploit this tool most effectively, however, we must gather some basic information about its operation. The material with which we conclude this section consists of a collection of standard results from advanced calculus, although the form in which we present them may be new to the reader. We suggest that this material be approached as an exercise in translation from the language of advanced calculus to that adopted here. (Assistance is available in Guillemin and Pollack, Chapter 2, where all of these results are discussed in detail.)

Theorem 5-5 (Chain Rule). Let $U \subseteq \mathbf{R}^n$, $V \subseteq \mathbf{R}^m$, and $W \subseteq \mathbf{R}^p$ be open sets. If $f : U \to V$ is differentiable at $x_0 \in U$ and $g : V \to W$ is differentiable at $f(x_0) \in V$, then $g \circ f : U \to W$ is differentiable at x_0 and $d(g \circ f)_{x_0} : \mathbf{R}^n \to \mathbf{R}^p$ is given by

(7) $d(g \circ f)_{x_0} = dg_{f(x_0)} \circ df_{x_0}.$

Our next two results clarify the relationship between the concepts of smoothness and differentiability for a map $f : U \to V$. First suppose that f is differentiable at $x_0 \in U$. Then $df_{x_0} : \mathbf{R}^n \to \mathbf{R}^m$ is a linear transformation and, as such, has a matrix relative to the standard bases in $\mathbf{R}^n$ and $\mathbf{R}^m$. This $m \times n$ matrix is called the *Jacobian* matrix of f at x_0 and is denoted $f'(x_0)$.

Remark: The reason for this notation should be clear: If $f : \mathbf{R} \to \mathbf{R}$ is differentiable at $x_0 \in \mathbf{R}$, then the Jacobian $f'(x_0)$ is the 1×1 matrix whose single entry is the number that is denoted $f'(x_0)$ in the calculus.

Theorem 5-6. Let $U \subseteq \mathbf{R}^n$ and $V \subseteq \mathbf{R}^m$ be open sets and $f = (f_1, \ldots, f_m) : U \to V$ a map that is differentiable at $x_0 \in U$. Then each of the partial derivatives $\partial f_i / \partial x_j$ exists at x_0 and the Jacobian $f'(x_0)$ of f at x_0 is given by

$$
(8) \quad f'(x_0) = \begin{bmatrix} \dfrac{\partial f_1}{\partial x_1}(x_0) & \cdots & \dfrac{\partial f_1}{\partial x_n}(x_0) \\ \cdot & & \cdot \\ \cdot & & \cdot \\ \cdot & & \cdot \\ \dfrac{\partial f_m}{\partial x_1}(x_0) & \cdots & \dfrac{\partial f_m}{\partial x_n}(x_0) \end{bmatrix}
$$

From the Jacobian $f'(x_0)$ we can evaluate the derivative df_{x_0} at any point in $\mathbf{R}^n$ by a simple matrix multiplication: if $h = (h_1, \ldots, h_n) \in \mathbf{R}^n$, then by (8)

$$
(9) \quad df_{x_0}(h) = \begin{bmatrix} \dfrac{\partial f_1}{\partial x_1}(x_0) & \cdots & \dfrac{\partial f_1}{\partial x_n}(x_0) \\ \cdot & & \cdot \\ \cdot & & \cdot \\ \cdot & & \cdot \\ \dfrac{\partial f_m}{\partial x_1}(x_0) & \cdots & \dfrac{\partial f_m}{\partial x_n}(x_0) \end{bmatrix} \begin{bmatrix} h_1 \\ \cdot \\ \cdot \\ \cdot \\ h_n \end{bmatrix}
$$

Theorem 5-6 states that a differentiable map has first order partial derivatives, but, of course, these partials need not be continuous. (Find an example in the one-dimensional case.) In particular, a differentiable map need not be smooth. On the other hand, we have the following theorem.

Theorem 5-7. Let $U \subseteq \mathbf{R}^n$ and $V \subseteq \mathbf{R}^m$ be open sets and $f = (f_1, \ldots, f_m) : U \to V$ a map. If the first order partial derivatives $\partial f_i / \partial x_j$ exist and are continuous on a nbd of $x_0 \in U$, then f is differentiable at x_0. In particular, a smooth map on an open subset of $\mathbf{R}^n$ is differentiable.

Remark: Suppose $f : U \to V$ has continuous first partial derivatives at $x_0 \in U$. Define for each $h \in \mathbf{R}^n$ the *directional derivative* of f at x_0 in the direction h, denoted $D_h f(x_0)$, by

$$
(10) \quad D_h f(x_0) = \lim_{t \to 0} \frac{f(x_0 + th) - f(x_0)}{t}
$$

$$
= \lim_{t \to 0} \left(\frac{f_1(x_0 + th) - f_1(x_0)}{t}, \ldots, \frac{f_m(x_0 + th) - f_m(x_0)}{t} \right).
$$

Now, it is a standard result of basic vector calculus that $D_h f(x_0)$ exists and is a linear function of h (indeed, each $\lim_{t \to 0}(1/t)(f_i(x_0 + th) - f_i(x_0))$ is just the dot product of $h/\|h\|$ with the gradient of f_i at x_0). Moreover, if e_i is the ith standard basis vector in $\mathbf{R}^n$, then

$$
D_{e_i} f(x_0) = \left(\frac{\partial f_1}{\partial x_i}(x_0), \ldots, \frac{\partial f_m}{\partial x_i}(x_0) \right)
$$

so that, by (9), $D_{e_i}f(x_0) = df_{x_0}(e_i)$. Regarded as functions of h, $D_h f(x_0)$ and $df_{x_0}(h)$ are therefore linear transformations of $\mathbf{R}^n$ to $\mathbf{R}^m$ that agree on a basis for $\mathbf{R}^n$. Thus,

(11) $D_h f(x_0) = df_{x_0}(h)$

and we find that (10) gives an explicit representation for the derivative of f at x_0 as a limit.

Several of the most important properties of the ordinary derivative have immediate generalizations to higher dimensions. In each of the following, U is an open subset of $\mathbf{R}^n$.

(a)   If $f : U \to \mathbf{R}^m$ is a constant map, then $df_{x_0} = 0$ for each $x_0 \in U$.

(b)   If $f : \mathbf{R}^n \to \mathbf{R}^m$ is a linear map, then $df_{x_0} = f$ for each $x_0 \in U$.

(c)   If $f, g : U \to \mathbf{R}$ are differentiable at $x_0 \in U$, then fg is differentiable at x_0 and $d(fg)_{x_0} = g(x_0)\, df_{x_0} + f(x_0)\, dg_{x_0}$. Moreover, if $g(x_0) \neq 0$, then f/g is differentiable at x_0 and

$$d\left(\frac{f}{g}\right)_{x_0} = \frac{g(x_o)\, df_{x_0} - f(x_0)\, dg_{x_0}}{(g(x_0))^2}.$$

(d)   If $f = (f_1, \ldots, f_m) : U \to \mathbf{R}^m$, then f is differentiable at $x_0 \in U$ iff each $f_i : U \to \mathbf{R}$ is differentiable at x_0 and, in this case, $df_{x_0} = (d(f_1)_{x_0}, \ldots, d(f_m)_{x_0})$.

From the Chain Rule (Theorem 5-5) and property (b) we can prove a result that should indicate quite clearly the usefulness of the derivative in analyzing smooth maps. A smooth map $f : X \to Y$ of subsets X and Y of two Euclidean spaces is called a *diffeomorphism* if it is bijective and if the inverse map $f^{-1} : Y \to X$ is also smooth; if such a map exists, we say that X and Y are *diffeomorphic*. A diffeomorphism is also a homeomorphism, but the converse is false, for example, the map $f : \mathbf{R} \to \mathbf{R}$ defined by $f(x) = x^3$ is a homeomorphism (indeed, even a smooth homeomorphism), but its inverse $f^{-1}(x) = x^{1/3}$ is not differentiable at $x = 0$.

Theorem 5-8. If $U \subseteq \mathbf{R}^n$ and $V \subseteq \mathbf{R}^m$ are open sets and $f : U \to V$ is a diffeomorphism, then, for each $x_0 \in U$, df_{x_0} is an isomorphism with inverse $d(f^{-1})_{f(x_0)}$. In particular, $m = n$.

Proof: Since $f^{-1} \circ f = \mathrm{id}_U$, we have by the chain rule $d(f^{-1})_{f(x_0)} \circ df_{x_0} = d(\mathrm{id}_U)_{x_0}$. But it follows from property (b) that $d(\mathrm{id}_U)_{x_0} = \mathrm{id}_{\mathbf{R}^n}$ so $d(f^{-1})_{f(x_0)} \circ df_{x_0} = \mathrm{id}_{\mathbf{R}^n}$. Similarly, $df_{x_0} \circ d(f^{-1})_{f(x_0)} = \mathrm{id}_{\mathbf{R}^m}$, so $d(f^{-1})_{f(x_0)}$ acts as a two-sided inverse for df_{x_0}, that is, $d(f^{-1})_{f(x_0)} = (df_{x_0})^{-1}$ and the result follows.   Q.E.D.

Consequently, open subsets of $\mathbf{R}^n$ and $\mathbf{R}^m$ cannot be diffeomorphic unless $m = n$. The reader should compare the ease with which we obtained this fact with the prodigious efforts required to prove the corresponding topological result (see the Remark following Theorem 2-31).

The converse of Theorem 5-8 is false, in general. Indeed, a map f: $\mathbf{R}^n \to \mathbf{R}^n$ can be such that df_{x_0} is an isomorphism for each $x_0 \in \mathbf{R}^n$ and yet nevertheless fail to be a diffeomorphism onto its image.

Exercise 5-6. Define $f: \mathbf{R}^2 \to \mathbf{R}^2$ by $f(x, y) = (e^x \cos y, e^x \sin y)$. Show that f is differentiable at all points of $\mathbf{R}^2$ and that, for any $(x_0, y_0) \in \mathbf{R}^2$, $df_{(x_0, y_0)}$ is an isomorphism, but that f is not one-to-one on $\mathbf{R}^2$. Hint: To show that $df_{(x_0, y_0)}$ is an isomorphism, it suffices to show that the matrix $f'(x_0, y_0)$ is nonsingular, that is, that $\det f'(x_0, y_0) \neq 0$.

That this particular phenomenon can occur is hardly surprising since the derivative df_{x_0} depends only on the local behavior of f near x_0 and can therefore not be expected to provide any global information about f. There is, however, an important partial converse to Theorem 5-8. Let us say that a smooth map $f: X \to Y$ between two arbitrary spaces is a *local diffeomorphism at* $x_0 \in X$ if it maps some nbd of x_0 in X diffeomorphically onto a nbd of $f(x_0)$ in Y; f is a *local diffeomorphism* if it is a local diffeomorphism at each x_0 in X. Now observe that if $U \subseteq \mathbf{R}^n$ and $V \subseteq \mathbf{R}^m$ are open sets and $f: U \to V$ is a local diffeomorphism at $x_0 \in U$, then it follows immediately from Theorem 5-8 that df_{x_0} is an isomorphism and, in particular, that $m = n$. The next result we borrow from advanced calculus is the Inverse Function Theorem, which states that the converse is also true.

Theorem 5-9 (Inverse Function Theorem). Let U and V be open subsets of $\mathbf{R}^n$ and $f: U \to V$ a smooth map whose derivative df_{x_0} at $x_0 \in U$ is an isomorphism. Then f is a local diffeomorphism at x_0.

Corollary 5-10. Let U and V be open subsets of $\mathbf{R}^n$ and $f: U \to V$ a smooth map. Then f is a local diffeomorphism at $x_0 \in U$ iff df_{x_0} is an isomorphism.

Corollary 5-10 is an extremely powerful result whose importance in the study of smooth maps can scarcely be overestimated. Observe, in particular, the ease with which we can now determine whether or not a given smooth map $f: U \to V$ is a diffeomorphism on some nbd of a point $x_0 \in U$: f is a local diffeomorphism at x_0 iff $\det f'(x_0) \neq 0$.

Exercise 5-7. Let $U \subseteq \mathbf{R}^n$ be open and $f: U \to \mathbf{R}^n$ a smooth map that is one-to-one and for which df_{x_0} is an isomorphism for each $x_0 \in U$. Show that $V = f(U)$ is open in $\mathbf{R}^n$ and that $f: U \to V$ is a diffeomorphism.

5-5   Differentiable manifolds

Although we have defined smoothness for maps between arbitrary subsets of Euclidean spaces, our discussion of derivatives has thus far been limited to maps on open sets. It is our task now to isolate a class of spaces to which the notion of the derivative of a smooth map extends naturally. The reasoning by which we arrive at the appropriate definition is simplicity itself: Derivatives are defined for smooth maps on open subsets of Euclidean spaces and depend only on the local behavior of the map at the point in question, so it is only natural to restrict our attention to spaces with the property that every point has a nbd that looks just like an open set in some $\mathbf{R}^k$.

A subset X of $\mathbf{R}^n$ is said to be a *differentiable manifold* (or *smooth manifold*) of *dimension* k if each $x \in X$ has a nbd V in X that is diffeomorphic to some open subset U of $\mathbf{R}^k$. Any particular diffeomorphism φ of U onto V is called a *parametrization* of V, while the inverse diffeomorphism $\varphi^{-1} : V \to U$ is called a *coordinate system* on V. The set V itself is called a *coordinate neighborhood* in X. When we write the diffeomorphism φ^{-1} in coordinates, $\varphi^{-1} = (x_1, \ldots, x_k)$, we obtain k smooth real-valued functions $x_1, \ldots, x_k$ on V, which we call *coordinate functions* on V; for any $v \in V$ the numbers $x_1(v), \ldots, x_k(v)$ are the *coordinates* of v relative to the coordinate system φ^{-1}. If both X and Z are smooth manifolds in $\mathbf{R}^n$ and $Z \subseteq X$, then Z is said to be a *(smooth) submanifold* of X.

Remark: Henceforth, the word *manifold* unmodified will mean "smooth manifold." By convention a manifold of dimension zero is a discrete space.

Note that it follows immediately from Theorem 5-8 (or Exercise 5-1) that the dimension of a smooth manifold is well defined. Observe also that, since a manifold is locally connected, its components are open as well as closed (Theorem 1-35), and it therefore consists of a disjoint union of connected, open submanifolds. As a result, there is no real loss of generality in considering only connected manifolds.

Example 5-3. (a) $\mathbf{R}^k$ is a k-dimensional smooth manifold, as is each of its open subsets.

(b) In Section 5-1 we showed that the 2-sphere $S^2 = \{(x, y, z) \in \mathbf{R}^3 : x^2 + y^2 + z^2 = 1\}$ can be covered by six open sets (the hemispheres $z > 0$, $z < 0$, $y > 0$, $y < 0$, $x > 0$, and $x < 0$) each of which is homeomorphic to an open disc in $\mathbf{R}^2$. For example, the map φ from the open unit disc in $\mathbf{R}^2$ onto $z > 0$ defined by $\varphi(x, y) = (x, y, (1 - x^2 - y^2)^{1/2})$ is a homeomorphism with inverse $\varphi^{-1}(x, y, z) = (x, y)$. Now, φ is clearly smooth on the open unit disc and φ^{-1} is smooth on $z > 0$ since it extends to a smooth map on all of $\mathbf{R}^3$ (the projection map). The same arguments apply to all of the hemispheres, and we conclude that S^2 is a smooth manifold of dimension two. The map φ is a parametrization of the coordinate nbd $z > 0$, and φ^{-1} is a coordinate system on this coordinate nbd with coordinate functions $(x, y, z) \to x$ and $(x, y, z) \to y$.

(c) Arguments entirely analogous to those in (b) show that, for any positive integer k, the sphere S^k is a k-dimensional differentiable manifold in $\mathbf{R}^{k+1}$.

(d) In particular, we conclude from (c) that the circle S^1 is a smooth 1-manifold. Aside from $\mathbf{R}$ itself, the most obvious examples of 1-manifolds distinct from S^1 are the open intervals in $\mathbf{R}$. In the next exercise you will show that all such intervals are diffeomorphic to $\mathbf{R}$.

Exercise 5-8. Show that any open interval in $\mathbf{R}$ is diffeomorphic to $\mathbf{R}$.

In Section 5-11 we show that any compact, connected 1-manifold is homeomorphic to S^1. It also follows from our work in that section that a noncompact, connected 1-manifold must be homeomorphic to an open interval and thus to $\mathbf{R}$ by Exercise 5-8. We conclude that S^1 and $\mathbf{R}$ are "essentially" the only connected 1-manifolds.

(e) A number of useful examples can be obtained by forming "product manifolds," a procedure we now described in detail. Suppose X is a manifold of dimension k in $\mathbf{R}^n$ and Y is a manifold of dimension l in $\mathbf{R}^m$. We show that the product $X \times Y$ is a manifold of dimension $k + l$ in $\mathbf{R}^n \times \mathbf{R}^m = \mathbf{R}^{n+m}$. Let (x_0, y_0) be a fixed point in $X \times Y$. There exists an open nbd U_1 of x_0 in X and a diffeomorphism φ of U_1 onto an open subset U_2 of $\mathbf{R}^k$. Similarly, there exists an open nbd V_1 of y_0 in Y and a diffeomorphism ψ of V_1 onto an open subset V_2 of $\mathbf{R}^l$. Then $U_1 \times V_1$ is an open nbd of (x_0, y_0) in $X \times Y$ and $U_2 \times V_2$ is open in $\mathbf{R}^{k+l}$. Define a map $\varphi \times \psi : U_1 \times V_1 \to U_2 \times V_2$ by $(\varphi \times \psi)(x, y) = (\varphi(x), \psi(y))$.

Exercise 5-9. Show that $\varphi \times \psi : U_1 \times V_1 \to U_2 \times V_2$ is a diffeomorphism and conclude that $X \times Y$ is a $(k + l)$-manifold in $\mathbf{R}^{n+m}$.

From this it follows, in particular, that the torus $S^1 \times S^1$ and the cylinder $S^1 \times \mathbf{R}$ are smooth manifolds of dimension two.

Remark: This result extends immediately to products of any finite number of smooth manifolds.

5-6   Tangent spaces and derivatives

Our intuitive conception of a k-dimensional differentiable manifold X in $\mathbf{R}^n$ is of a space that is "locally smooth." Unlike a topological k-manifold that can contain sharp "peaks," X must be nearly "flat" near each of its points, since a diffeomorphism (as opposed to a homeomorphism) cannot smooth out such peaks. It is our intention now to make more precise this notion of "local flatness" that characterizes differentiable manifolds.

Suppose then that X is a smooth k-manifold in $\mathbf{R}^n$ and x is an arbitrary point of X. Choose a local parametrization $\varphi : U \to X$ of a nbd $\varphi(U)$ of x in X, where $U \subseteq \mathbf{R}^k$ is open and $\varphi(u) = x$. Now, φ may be thought of as a map into $\mathbf{R}^n$ and, as such, the derivative $d\varphi_u : \mathbf{R}^k \to \mathbf{R}^n$ is defined. The image $d\varphi_u(\mathbf{R}^k)$ of $\mathbf{R}^k$ under the linear map $d\varphi_u$ is a linear subspace of $\mathbf{R}^n$. We claim that this subspace does not depend on the particular parametrization chosen, that is, that if $\psi : V \to X$, V open in $\mathbf{R}^k$, is another parametrization of a nbd $\psi(V)$ of x in X with $\psi(v) = x$, then $d\psi_v(\mathbf{R}^k) = d\varphi_u(\mathbf{R}^k)$. To see this, note that $\psi^{-1} \circ \varphi$ maps some nbd U_1 of u diffeomorphically onto a nbd V_1 of v. Now, by the chain rule the commutative diagram

$$
\begin{array}{ccc}
 & \mathbf{R}^n & \\
{\scriptstyle \varphi}\nearrow & & \searrow{\scriptstyle \psi^{-1}} \\
U_1 & \xrightarrow[\psi^{-1}\,\circ\,\varphi]{} & V_1
\end{array}
$$

of smooth maps gives rise to the commutative diagram

$$
\begin{array}{ccc}
 & \mathbf{R}^n & \\
{\scriptstyle d\varphi_u}\nearrow & & \searrow{\scriptstyle d(\psi^{-1})_x \,=\, (d\psi_v)^{-1}} \\
\mathbf{R}^k & \xrightarrow[d(\psi^{-1}\,\circ\,\varphi)_u]{} & \mathbf{R}^k
\end{array}
$$

of linear maps. But $d(\psi^{-1} \circ \varphi)_u$ is an isomorphism, so it follows that $d\psi_v(\mathbf{R}^k) = d\varphi_u(\mathbf{R}^k)$ and we have justified the following definition:

Let X be a smooth k-dimensional manifold in $\mathbf{R}^n$ and x an arbitrary point in X. Let U be an open set in $\mathbf{R}^k$ and $\varphi : U \to X$ a parametrization of a nbd $\varphi(U)$ of x in X with $\varphi(u) = x$. The linear subspace $d\varphi_u(\mathbf{R}^k)$ of $\mathbf{R}^n$ is called the *tangent space* to X at x and denoted $T_x(X)$. The elements of $T_x(X)$ are called *tangent vectors* to X at x.

Remark: The geometrical interpretation of $T_x(X)$ is quite simple. Since φ is a diffeomorphism of U onto a nbd $\varphi(U)$ of $x = \varphi(u)$ in X whose best affine approximation near u is $\varphi(u) + d\varphi_u$, the image $\varphi(u) + d\varphi_u(\mathbf{R}^k)$ of this affine map is a hyperplane in $\mathbf{R}^n$, which is the closest "flat" approximation to X near x. But $\varphi(u) + d\varphi_u(\mathbf{R}^k) = x + T_x(X)$, so $T_x(X)$ is the translation to the origin in $\mathbf{R}^n$ of the k-dimensional hyperplane in $\mathbf{R}^n$ that most closely approximates X near x (see Figure 5-8).

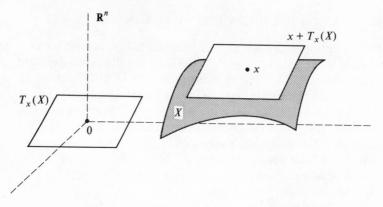

Figure 5-8

Exercise 5-10. Show that the tangent space $T_x(X)$ to a k-dimensional smooth manifold X at any point $x \in X$ has algebraic dimension k. Hint: Use the fact that if $\varphi : U \to X$ is a parametrization of $\varphi(U)$, then $\varphi^{-1} : \varphi(U) \to U$ is smooth.

Now suppose $X \subseteq \mathbf{R}^n$ and $Y \subseteq \mathbf{R}^m$ are smooth manifolds of dimension k and l, respectively, $f : X \to Y$ is a smooth map, and $x \in X$. We intend the "derivative" of f at x to be the linear map whose translation by $f(x)$ best approximates f near x. As such, the appropriate domain and range of this linear map are clearly $T_x(X)$ and $T_{f(x)}(Y)$, respectively. Now, since f is smooth at $x \in X$, there is an open set W in $\mathbf{R}^n$ containing x and a smooth map $F : W \to \mathbf{R}^m$ that extends f. We define the *derivative* $df_x : T_x(X) \to T_{f(x)}(Y)$ of f at x by $df_x = dF_x|T_x(X)$. Naturally, we must justify this definition by showing that dF_x does indeed carry the linear subspace $T_x(X)$ of $\mathbf{R}^n$ into $T_{f(x)}(Y)$ and that, moreover, the linear map $dF_x|T_x(X)$ does not depend on the choice of the extension F. To this end, we choose parametrizations $\varphi : U \to X$ and $\psi : V \to Y$ for nbds $\varphi(U)$ and $\psi(V)$ of x and $f(x)$, respectively, where $U \subseteq \mathbf{R}^k$ and $V \subseteq \mathbf{R}^l$ are open sets. By shrinking U if necessary, we may assume that $\varphi(U) \subseteq W$ and that f maps $\varphi(U)$ into $\psi(V)$. Thus, $\psi^{-1} \circ f \circ \varphi : U \to V$ is a well-defined map. Now, the commutative diagram

$$
\begin{array}{ccc}
W & \xrightarrow{\;\;F\;\;} & \mathbf{R}^m \\
\varphi \uparrow & & \uparrow \psi \\
U & \xrightarrow[\psi^{-1} \circ f \circ \varphi]{} & V
\end{array}
$$

of smooth maps between open subsets of Euclidean spaces gives rise, by the chain rule, to the commutative diagram

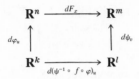

where $u = \varphi^{-1}(x)$ and $v = \psi^{-1}(f(x))$. From this it follows immediately that dF_x carries $T_x(X) = d\varphi_u(\mathbf{R}^k)$ into $T_{f(x)}(Y) = d\psi_v(\mathbf{R}^l)$. Moreover, $dF_x|T_x(X) = d\psi_v \circ d(\psi^{-1} \circ f \circ \varphi)_u \circ (d\varphi_u)^{-1}$, so that our definition of df_x does not depend on the choice of F.

Remark: Note that if U is an open subset of $\mathbf{R}^n$, then, for any $x \in U$, $T_x(U)$ is an n-dimensional linear subspace of $\mathbf{R}^n$ and is therefore all of $\mathbf{R}^n$, that is, $T_x(U) = \mathbf{R}^n$. Thus, if $f : U \to V$ is smooth, where $V \subseteq \mathbf{R}^m$ is open, it follows that the definition of $df_x : T_x(U) \to T_{f(x)}(V)$ just given agrees with the original definition of Section 5-4.

Exercise 5-11. (a) Calculate from the definition the tangent space $T_{(x_0, y_0)}(S^1)$ to the circle S^1 at an arbitrary point $(x_0, y_0) \in S^1$.

(b)   Define $f : \mathbf{R}^2 \to \mathbf{R}$ by $f(x, y) = e^y$ for any $(x, y) \in \mathbf{R}^2$. Calculate $df_{(x_0, y_0)} : \mathbf{R}^2 \to \mathbf{R}$ and show that it is surjective for any $(x_0, y_0) \in \mathbf{R}^2$.

(c)   Let $g = f|S^1$ and compute $dg_{(x_0, y_0)} : T_{(x_0, y_0)}(S^1) \to \mathbf{R}$ for each $(x_0, y_0) \in S^1$. Show that $dg_{(x_0, y_0)}$ may fail to be surjective for certain points $(x_0, y_0) \in S^1$. What is special about these points of S^1?

The chain rule now generalizes immediately to smooth maps between manifolds.

Theorem 5-11 (Chain Rule for Manifolds). Let X, Y, and Z be differentiable manifolds and $f : X \to Y$ and $g : Y \to Z$ smooth maps. Then, for each $x \in X$, $d(g \circ f)_x = dg_{f(x)} \circ df_x$.

Now suppose X is a submanifold of $Y \subseteq \mathbf{R}^n$, $i : X \hookrightarrow Y$ is the inclusion map and $x \in X$. Since $i(x) = x$, di_x carries $T_x(X)$ into $T_x(Y)$. Moreover, since $\mathrm{id}_{\mathbf{R}^n}$ extends i, $di_x = d(\mathrm{id}_{\mathbf{R}^n})|T_x(X) = \mathrm{id}_{\mathbf{R}^n}|T_x(X) = \mathrm{id}_{T_x(X)}$, so $T_x(X) \subseteq T_x(Y)$ and di_x is the inclusion map. Thus, we have proved the following theorem.

Theorem 5-12. If X is a submanifold of Y and $i : X \hookrightarrow Y$ is the inclusion map, then, for each $x \in X$, $T_x(X) \subseteq T_x(Y)$ and $di_x : T_x(X) \to T_x(Y)$ is the inclusion map. In particular, $d(\mathrm{id}_X)_x = \mathrm{id}_{T_x(X)}$.

Exercise 5-12. Describe, in as much detail as possible, the sense in which our construction of $T_x(X)$ and df_x is functorial (see Section 3-4).

Corollary 5-13. Let X and Y be differentiable manifolds and $f : X \to Y$ a diffeomorphism. Then, for each $x \in X$, $df_x : T_x(X) \to T_{f(x)}(Y)$ is an isomorphism. In particular, the dimensions of X and Y are equal.

The proof of Corollary 5-13 is just like that of Theorem 5-8. Corollary 5-10 admits the following extension to manifolds.

Theorem 5-14. Let X and Y be differentiable manifolds, $f : X \to Y$ a smooth map and $x \in X$. Then f is a local diffeomorphism at x iff df_x is an isomorphism of $T_x(X)$ onto $T_{f(x)}(Y)$.

The terms *tangent space* and *tangent vector* that we have introduced in this section are reminiscent of terminology used in advanced calculus, where the tangent space to a smooth surface $\mathscr{S}$ in $\mathbf{R}^3$ at a point $s_0 \in \mathscr{S}$ is the plane in $\mathbf{R}^3$ that contains the velocity vector at s_0 of every smooth curve in $\mathscr{S}$ through s_0. We intend to show next that this interpretation persists even in the case of general manifolds and thereby obtain an alternate description of $T_x(X)$.

First let $U \subseteq \mathbf{R}^k$ be open and consider a smooth curve in U, that is, a smooth map $c : J \to U$, where J is an open interval in $\mathbf{R}$. Let $c = (c_1, \ldots, c_k)$ be the coordinate functions of c, t_0 a fixed point in J, and $x_0 = c(t_0)$ the corresponding point on the curve. The *velocity vector* of c at t_0, denoted $(dc/dt)(t_0)$, is defined by

(12) $$\frac{dc}{dt}(t_0) = (c'_1(t_0), \ldots, c'_k(t_0)).$$

Now observe that, since $c : J \to U$ is smooth, the derivative dc_{t_0} of c at t_0 exists. Moreover, the Jacobian of c at t_0 is

$$c'(t_0) = \begin{bmatrix} c'_1(t_0) \\ \cdot \\ \cdot \\ \cdot \\ c'_k(t_0) \end{bmatrix}$$

so that for any $r \in \mathbf{R} = T_{t_0}(J)$ we have

(13) $$dc_{t_0}(r) = (c'_1(t_0) \cdot r, \ldots, c'_k(t_0) \cdot r).$$

Comparing equations (12) and (13) we find that

(14) $$\frac{dc}{dt}(t_0) = dc_{t_0}(1).$$

Exercise 5-13. Let U be open in $\mathbf{R}^k$ and x_0 a fixed but arbitrary point in U. Show that $T_{x_0}(U)$ consists precisely of the velocity vectors of smooth curves in U through x_0.

Now let X be a smooth k-dimensional manifold in $\mathbf{R}^n$. A *smooth curve* in X is a smooth map $c : J \to X$, where J is an open interval in $\mathbf{R}$. The *velocity vector* of c at $t_0 \in J$ is denoted $(dc/dt)(t_0)$ and defined by

$$(15) \qquad \frac{dc}{dt}(t_0) = dc_{t_0}(1).$$

Thus, $(dc/dt)(t_0) \in T_{x_0}(X)$, where $x_0 = c(t_0)$.

Exercise 5-14. Show that, conversely, every element of $T_{x_0}(X)$ is the velocity vector of some smooth curve in X through x_0. Hint: Parametrize a nbd of x_0 in X and use Exercise 5-13.

Thus, we conclude that $T_{x_0}(X)$ consists precisely of velocity vectors of smooth curves in X through x_0. Now consider a smooth map $f : X \to Y$ between two manifolds and let x_0 be an arbitrary point in X. Then $df_{x_0} : T_{x_0}(X) \to T_{f(x_0)}(Y)$. If these tangent spaces are thought of as sets of velocity vectors, then the action of df_{x_0} is particularly easy to describe. Let $(dc/dt)(t_0)$ be an element of $T_{x_0}(X)$, where c is a smooth curve in X with $c(t_0) = x_0$. Then $df_{x_0}((dc/dt)(t_0)) = df_{x_0}(dc_{t_0}(1)) = df_{x_0} \circ dc_{t_0}(1) = d(f \circ c)_{t_0}(1)$. Consequently,

$$(16) \qquad df_{x_0}\left(\frac{dc}{dt}(t_0)\right) = \frac{d(f \circ c)}{dt}(t_0),$$

and we find that the image under df_{x_0} of the velocity vector $(dc/dt)(t_0)$ is the velocity vector of the smooth curve $f \circ c$ in Y that passes through $f(x_0)$ at $t = t_0$.

5-7   Regular and critical values of smooth maps

Consider the smooth map $f : \mathbf{R}^3 \to \mathbf{R}$ defined by $f(x, y, z) = x^2 + y^2 - z^2$. Although f is clearly surjective, we claim that the image point $0 \in \mathbf{R}$ is of quite a different character than any other image point. To see this, we shall consider the level surfaces $f^{-1}(r)$ of f for all $r \in \mathbf{R}$. First, if $r > 0$, then $f^{-1}(r) = \{(x, y, z) \in \mathbf{R}^3 : x^2 + y^2 = r + z^2\}$, which is the so-called hyperboloid of one sheet shown in Figure 5-9 (a). Notice, in particular, that, for $r > 0$, $f^{-1}(r)$ is a smooth 2-manifold in $\mathbf{R}^3$ since it can be covered by four open subsets of the hyperboloid on which the projection onto some coordinate plane is a diffeomorphism onto its image. Moreover, any two such level surfaces are diffeomorphic: If r_1 and r_2 are positive constants and $h : \mathbf{R}^3 \to \mathbf{R}^3$ is defined by $h(x, y, z) = (r_2/r_1)^{1/2}(x, y, z)$, then h is a diffeomorphism that carries $f^{-1}(r_1)$ onto $f^{-1}(r_2)$. Now suppose $r < 0$. Then $f^{-1}(r) = \{(x, y, z) \in \mathbf{R}^3 : z^2 - (x^2 + y^2) = |r|\}$, the graph of which is the hyperboloid of two sheets shown in Figure 5-9 (b). Again we find that for $r < 0$ the level surfaces $f^{-1}(r)$ are smooth 2-manifolds and that any

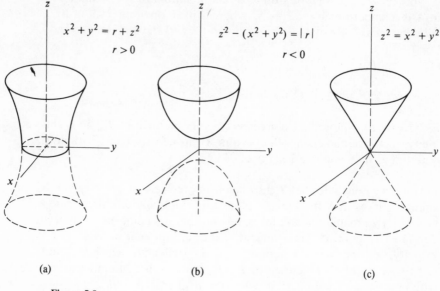

$$x^2 + y^2 = r + z^2$$
$$r > 0$$

$$z^2 - (x^2 + y^2) = |r|$$
$$r < 0$$

$$z^2 = x^2 + y^2$$

(a)   (b)   (c)

Figure 5-9

two such surfaces are diffeomorphic. Finally, note that if $r = 0$, then $f^{-1}(0) = \{(x, y, z) \in \mathbf{R}^3 : z^2 = x^2 + y^2\}$, which is the right circular cone shown in Figure 5-9 (c). Note that $f^{-1}(0)$ is *not* a smooth manifold, since no nbd of the origin in $f^{-1}(0)$ is diffeomorphic to an open subset of $\mathbf{R}^2$.

Recalling that the local behavior of a smooth map is often reflected in properties of its derivative, we shall attempt to account for this phenomenon by examining $df_{(a,b,c)}$ for various points $(a, b, c) \in \mathbf{R}^3$. For any such point $T_{(a,b,c)}(\mathbf{R}^3) = \mathbf{R}^3$ and $T_{f(a,b,c)}(\mathbf{R}) = \mathbf{R}$, so $df_{(a,b,c)}:\mathbf{R}^3 \to \mathbf{R}$. As such, the linear map $df_{(a,b,c)}$ is either surjective or identically zero. Now, the Jacobian of f at the point (a, b, c) is the 1×3 matrix $f'(a, b, c) = [2a, 2b, -2c]$, so that, for any (x, y, z) in $\mathbf{R}^3 = T_{(a,b,c)}(\mathbf{R}^3)$, $df_{(a,b,c)}(x, y, z) = 2ax + 2by - 2cz$. Thus, $df_{(a,b,c)}$ is identically zero iff $(a, b, c) = (0, 0, 0)$. The distinguishing characteristic of the image point $0 \in \mathbf{R}$ is that it is the only value of f whose inverse image contains this "critical point" $(0, 0, 0)$. Any other $r \in \mathbf{R}$ has the property that $f^{-1}(r)$ consists entirely of "regular points" of f, that is, points at which the derivative of f is surjective. Thus motivated, we formulate the following general definitions.

Let X and Y be differentiable manifolds of dimension k and l respectively and let $f : X \to Y$ be a smooth map. A point $x \in X$ is said to be a *regular point* of f if the derivative $df_x : T_x(X) \to T_{f(x)}(Y)$ is surjective; otherwise, x is a *critical point* of f. Thus, the set C of critical points of f is given by $C = \{x \in X : \text{rank } df_x < l\}$. A point $y \in Y$ is a *regular value* of

f if $f^{-1}(y)$ consists entirely of regular points of f. If $f^{-1}(y) \cap C \neq \varnothing$, then y is a *critical value* of f. The set of all critical values of f is therefore $f(C)$. In particular, any $y \in Y - f(X)$ automatically qualifies as a regular value of f despite the fact that such a y is not a value of f at all. Observe that if $k < l$, when df_x cannot be surjective for any $x \in X$ and the only regular values of f are the points in $Y - f(X)$. Obviously, then, we will be interested primarily in the case $k \geqslant l$.

Smooth maps are extremely well behaved at regular points. Indeed, we shall show next that on some nbd of a regular point a smooth map is "essentially" nothing more than a projection.

Theorem 5-15. Let X and Y be differentiable manifolds of dimension k and l, respectively, with $k \geqslant l$, and let $x \in X$ be a regular point of the smooth map $f : X \to Y$. Then there exist open sets $U \subseteq \mathbf{R}^k$ and $V \subseteq \mathbf{R}^l$ and parametrizations $\varphi : U \to X$ and $\psi : V \to Y$ of nbds $\varphi(U)$ and $\psi(V)$ of x and $f(x)$, respectively, such that the map $\psi^{-1} \circ f \circ \varphi : U \to V$ is given by $\psi^{-1} \circ f \circ \varphi(x_1, \ldots, x_l, \ldots, x_k) = (x_1, \ldots, x_l)$.

Remark: The map $\psi^{-1} \circ f \circ \varphi$ is nothing other than f written in terms of the coordinates set up on $\varphi(U)$ and $\psi(V)$ by φ and ψ, respectively.

Proof: Find open sets $U_1 \subseteq \mathbf{R}^k$ and $V \subseteq \mathbf{R}^l$ and parametrizations $\varphi_1 : U_1 \to X$ and $\psi : V \to Y$ of nbds $\varphi_1(U_1)$ and $\psi(V)$ of x and $f(x)$, respectively. Without loss of generality we may assume that U_1 and V are nbds of the origins in $\mathbf{R}^k$ and $\mathbf{R}^l$, respectively, and that $\varphi_1(0) = x$ and $\psi(0) = f(x)$. Define $g : U_1 \to V$ by $g = \psi^{-1} \circ f \circ \varphi_1$. Since df_x is surjective and $dg_0 : \mathbf{R}^k \to \mathbf{R}^l$ is given by $dg_0 = d\psi^{-1}{}_{f(x)} \circ df_x \circ d(\varphi_1)_0$, it follows that dg_0 is also surjective. By performing a change of basis in $\mathbf{R}^k$, we may therefore assume that the $l \times k$ matrix of dg_0 is

$$\begin{bmatrix} 1 & 0 & \cdots & 0 & 0 & \cdots & 0 \\ 0 & 1 & \cdots & 0 & 0 & \cdots & 0 \\ \cdot & \cdot & & \cdot & \cdot & & \cdot \\ \cdot & \cdot & & \cdot & \cdot & & \cdot \\ \cdot & \cdot & & \cdot & \cdot & & \cdot \\ 0 & 0 & \cdots & 1 & 0 & \cdots & 0 \end{bmatrix}$$

Define $G : U_1 \to \mathbf{R}^k$ by $G(a) = (g_1(a), \ldots, g_l(a), a_{l+1}, \ldots, a_k)$, where $a = (a_1, \ldots, a_l, a_{l+1}, \ldots, a_k)$ is in $U_1 \subseteq \mathbf{R}^k$. The matrix of dG_0 relative to the basis just chosen for $\mathbf{R}^k$ is therefore the $k \times k$ identity matrix and is, in particular, nonsingular. By Theorem 5-9 there is an open nbd U of 0 in $\mathbf{R}^k$ on which G^{-1} exists and is a diffeomorphism into U_1. By construction g is the composition of G with the projection $(x_1, \ldots, x_k) \to (x_1, \ldots, x_l)$, so the following diagram commutes:

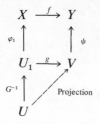

Letting $\varphi = \varphi_1 \circ G^{-1}$, we therefore obtain $\psi^{-1} \circ f \circ \varphi = \psi^{-1} \circ f \circ \varphi_1 \circ G^{-1} = \psi^{-1} \circ \psi \circ g \circ G^{-1} = g \circ G^{-1}$, which is the projection.   Q.E.D.

From Theorem 5-15 we can prove quite easily the result that we anticipated in the example with which we introduced this section.

Corollary 5-16. Let X and Y be differentiable manifolds of dimension k and l, respectively, with $k \geq l$, and let $y \in Y$ be a regular value of the smooth map $f : X \to Y$. Then $f^{-1}(y)$ is either empty or a submanifold of X of dimension $k - l$.

Exercise 5-15. Prove Corollary 5-16. Hint: Suppose $f(x) = y$. Choose parametrizations $\varphi : U \to X$ and and $\psi : V \to Y$ as in the proof of Theorem 5-15, assuming that U and V are nbds of the origins in $\mathbf{R}^k$ and $\mathbf{R}^l$, respectively, and that $\varphi(0) = x$ and $\psi(0) = y$. Now note that $f^{-1}(y) \cap \varphi(U) = \{z \in \varphi(U) : (\varphi^{-1})_1(z) = \cdots = (\varphi^{-1})_l(z) = 0\}$.   Q.E.D.

Observe, however, that the inverse image of a critical value may also be a manifold. Consider, for example, the constant map $f : \mathbf{R}^2 \to \mathbf{R}$ defined by $f(x, y) = 0$ for each $(x, y) \in \mathbf{R}^2$. Since $df_{(a,b)} = 0$ for each $(a, b) \in \mathbf{R}^2$, the sole image point $0 \in \mathbf{R}$ is a critical value of f. Nevertheless, $f^{-1}(0) = \mathbf{R}^2$ is certainly a manifold. The reader should now find a smooth map of $\mathbf{R}^2$ to $\mathbf{R}$ with a critical value whose inverse image is a one-dimensional submanifold of $\mathbf{R}^2$.

Corollary 5-16 is quite a useful device for the production of examples. For instance, if we define $f : \mathbf{R}^{k+1} \to \mathbf{R}$ by $f(x) = \|x\|^2 = x_1^2 + \cdots + x_{k+1}^2$, we obtain a smooth map whose Jacobian at any point $a = (a_1, \ldots, a_{k+1}) \in \mathbf{R}^{k+1}$ is $[2a_1, \ldots, 2a_{k+1}]$. Thus, df_a is surjective unless $a = 0$. Consequently, every nonzero real number is a regular value of f. In particular, we find that $S^k = f^{-1}(1)$ is a smooth manifold of dimension $(k + 1) - 1 = k$ without having to explicitly write down any parametrizations.

The behavior of a smooth map near a critical point can be extraordinarily complicated, and we shall spend the remainder of this chapter investigating such behavior. It is precisely in this analysis of critical points and critical values that our "differential techniques" yield the richest harvest of topological information. Even at this early stage we have some indication of the relationships that exist between the topology of a manifold and the critical point theory for smooth maps defined on it. For instance, we have seen in Exercise 5-11 (b) that a smooth, real-valued function can be entirely without critical points. However, in (c) of that same exercise we saw that such a map acquires critical points when restricted to a compact submanifold because on such a subspace a smooth function must assume maximum and minimum values, and these are necessarily critical values. (Why?) On the other hand, if $f : X \rightarrow Y$ is a smooth map between manifolds of the same dimension with X compact and $y \in Y$ a regular value of f, then $f^{-1}(y)$ is either empty or a submanifold of X of dimension zero (Corollary 5-16), that is, a discrete subspace of X and therefore finite.

At this point an obvious but extremely important question arises. Can a smooth map between two manifolds be so pathological that it has no regular values at all? The crucial result of critical point theory, upon which all of our future investigations rest, assures us that this cannot occur and, indeed, that the set of critical values of any smooth map is "small" in a sense we now intend to make precise.

5-8   Measure zero and Sard's Theorem

Recall that if $a_i < b_i$ for each $i = 1, \ldots, n$, then a set of the form $(a_1, b_1) \times \cdots \times (a_n, b_n)$ is called an *open rectangle* in $\mathbf{R}^n$, while $[a_1, b_1] \times \cdots \times [a_n, b_n]$ is a *closed rectangle* in $\mathbf{R}^n$. If $a_1 = \cdots = a_n$ and $b_1 = \cdots = b_n$, then the rectangle is a *cube*. The *volume* of a rectangle R, denoted $|R|$, is the product of the lengths of its sides, that is, $|(a_1, b_1) \times \cdots \times (a_n, b_n)| = |[a_1, b_1] \times \cdots \times [a_n, b_n]| = (b_1 - a_1) \cdots (b_n - a_n)$. A subset C of $\mathbf{R}^n$ is said to have *measure zero in* $\mathbf{R}^n$ if it can be covered by countably many open rectangles of arbitrarily small total volume, that is, if for every $\epsilon > 0$ there is a sequence R_1, $R_2, \ldots$ of open rectangles in $\mathbf{R}^n$ with

$$C \subseteq \bigcup_{i=1}^{\infty} R_i \quad \text{and} \quad \sum_{i=1}^{\infty} |R_i| < \epsilon.$$

Any subset of a set of measure zero in $\mathbf{R}^n$ also has measure zero in $\mathbf{R}^n$. Moreover, if $C = \cup_{k=1}^{\infty} C_k$, where each C_k has measure zero in $\mathbf{R}^n$, then C also has measure zero in $\mathbf{R}^n$. To see this, let $\epsilon > 0$ be given. For each k choose a sequence R_i^k, $i = 1, 2, \ldots$, of open rectangles in $\mathbf{R}^n$ with

$$C_k \subseteq \bigcup_{i=1}^{\infty} R_i^k \quad \text{and} \quad \sum_{i=1}^{\infty} |R_i^k| < \frac{\epsilon}{2^k}.$$

Then $\quad C \subseteq \bigcup\limits_{i,k=1}^{\infty} R_i^k$

and $\quad \sum\limits_{i,k=1}^{\infty} |R_i^k| = \sum\limits_{k=1}^{\infty} \left(\sum\limits_{i=1}^{\infty} |R_i^k| \right) < \sum\limits_{k=1}^{\infty} \left(\dfrac{\epsilon}{2^k} \right) = \epsilon$

as required.

Exercise 5-16. (a) Show that a subset C of $\mathbf{R}^n$ has measure zero in $\mathbf{R}^n$ iff for every $\epsilon > 0$ there is a sequence $R_1, R_2, \ldots$ of closed rectangles in $\mathbf{R}^n$ with

$$C \subseteq \bigcup_{i=1}^{\infty} R_i \quad \text{and} \quad \sum_{i=1}^{\infty} |R_i| < \epsilon.$$

(b)   Show that a subset C of $\mathbf{R}^n$ has measure zero in $\mathbf{R}^n$ iff for every $\epsilon > 0$ there is a sequence $W_1, W_2, \ldots$ of open (or closed) cubes in $\mathbf{R}^n$ with

$$C \subseteq \bigcup_{i=1}^{\infty} W_i \quad \text{and} \quad \sum_{i=1}^{\infty} |W_i| < \epsilon.$$

Example 5-4. (a) Since a singleton in $\mathbf{R}^n$ has measure zero in $\mathbf{R}^n$ we conclude that any countable set has measure zero in $\mathbf{R}^n$. In particular, the set of rational numbers has measure zero in $\mathbf{R}$.

(b) Suppose $k < n$ and let $\mathbf{R}^k = \{(x_1, \ldots, x_k, x_{k+1}, \ldots, x_n) \in \mathbf{R}^n : x_{k+1} = \cdots = x_n = 0\}$ be the natural copy of $\mathbf{R}^k$ in $\mathbf{R}^n$. We claim that $\mathbf{R}^k$ has measure zero in $\mathbf{R}^n$. Observe that it suffices to show that $\mathbf{R}^{n-1} = \{(x_1, \ldots, x_{n-1}, x_n) \in \mathbf{R}^n : x_n = 0\}$ has measure zero in $\mathbf{R}^n$. Moreover, since $\mathbf{R}^{n-1}$ can be written as a countable union of compact subsets of $\mathbf{R}^{n-1}$, we need only show that any compact set $C \subseteq \mathbf{R}^{n-1}$ has measure zero in $\mathbf{R}^n$. Let $\epsilon > 0$ be given. Since C is compact, there is a rectangle $(a_1, b_1) \times \cdots \times (a_{n-1}, b_{n-1})$ in $\mathbf{R}^{n-1}$ containing C. Let V be the volume of this rectangle in $\mathbf{R}^{n-1}$, that is, $V = (b_{n-1} - a_{n-1}) \cdots (b_1 - a_1)$. Then the rectangle $(a_1, b_1) \times \cdots \times (a_{n-1}, b_{n-1}) \times (-\epsilon/4V, \epsilon/4V)$ in $\mathbf{R}^n$ has volume $(\epsilon/2) < \epsilon$. Thus, C is contained in a single rectangle in $\mathbf{R}^n$ of volume less than ϵ, so it certainly has measure zero in $\mathbf{R}^n$.

(c) *Exercise 5-17.* Show that the Cantor set has measure zero in $\mathbf{R}$. Hint: View the Cantor set as the "middle third set" (see Example 1-1 (f)).

(d) As our final example we exhibit subsets of **R** that are *not* of measure zero in **R**. Let a and b be real numbers with $a < b$. We claim that the interval $[a, b]$ does not have measure zero in **R**. Suppose, to the contrary, that $[a, b]$ does have measure zero in **R**. Then, in particular, there is a sequence $R_1, R_2, \ldots$ of open intervals in **R** with

$$[a, b] \subseteq \bigcup_{i=1}^{\infty} R_i \quad \text{and} \quad \sum_{i=1}^{\infty} |R_i| < b - a.$$

Since $[a, b]$ is compact, some finite number of the R_i cover it, say, $[a, b] \subseteq R_1 \cup \cdots \cup R_n$. The sum of the lengths of $\overline{R}_1, \ldots, \overline{R}_n$ must be less than $b - a$. We obtain a contradiction by showing that this is impossible. For this we may certainly assume that each $\overline{R}_i$ is contained in $[a, b]$. Let $a = t_0 < t_1 < \cdots < t_k = b$ be all of the endpoints of all the $\overline{R}_i$, $i = 1, \ldots, n$. Then each $|\overline{R}_i|$ is the sum of certain $t_j - t_{j-1}$. Moreover, each $[t_{j-1}, t_j]$ lies in at least one $\overline{R}_i$, so

$$\sum_{i=1}^{n} |\overline{R}_i| \leq \sum_{j=1}^{k} (t_j - t_{j-1}) = b - a$$

as required. Thus, $[a, b]$ does not have measure zero in **R**. A trivial consequence of this is that no interval of any kind has measure zero in **R**.

Proposition 5-17. If $C \subseteq \mathbf{R}^n$ has measure zero in $\mathbf{R}^n$ and $f : C \to \mathbf{R}^n$ is smooth, then $f(C)$ has measure zero in $\mathbf{R}^n$.

Proof: Choose an open subset U of $\mathbf{R}^n$ with $C \subseteq U$ and a smooth map $F : U \to \mathbf{R}^n$ with $F|C = f$. Since U is a countable union of closed balls, there is no loss of generality in assuming that C is contained in some compact ball $H \subseteq U$. Let K be a compact ball contained in U and containing H in its interior. Now let $b = \max\{|(\partial F_i/\partial x_j)(x)| : x \in K, i, j = 1, \ldots, n\}$. By applying the Mean Value Theorem for functions of several variables (see Apostol, Theorem 6-17, or Buck, Chapter 5, Theorem 15), we find that, for all $x, y \in K$ and each $i = 1, \ldots, n$,

(17) $\quad |F_i(x) - F_i(y)| \leq b \sum_{j=1}^{n} |x_j - y_j|.$

Now let $\epsilon > 0$ be given. Let $W_1, W_2, \ldots$ be a sequence of cubes in $\mathbf{R}^n$ contained in the interior of K with

$$C \subseteq \bigcup_{k=1}^{\infty} W_k \quad \text{and} \quad \sum_{k=1}^{\infty} |W_k| < \frac{\epsilon}{(bn)^n}.$$

Let W be one of these cubes and a the length of its sides. Since $W \subseteq K$, expression (17) implies that, for all x, $y \in W$ and for each $i = 1, \ldots , n$, $|F_i(x) - F_i(y)| \leq bna$. Thus, $F_i(W)$ is contained in an interval of length bna so $F(W)$ is contained in a cube of volume $(bna)^n = (bn)^n a^n = (bn)^n|W|$. Thus, for each $k = 1, 2, \ldots , F(W_k)$ is contained in a cube of volume $(bn)^n|W_k|$. Now, $f(W_k) \subseteq F(W_k)$ so

$$f(C) \subseteq \bigcup_{k=1}^{\infty} f(W_k) \subseteq \bigcup_{k=1}^{\infty} F(W_k),$$

which is contained in a countable union of cubes of total volume

$$\sum_{k=1}^{\infty} (bn)^n|W_k| = (bn)^n \sum_{k=1}^{\infty} |W_k| < (bn)^n \left(\frac{\epsilon}{(bn)^n} \right) = \epsilon. \qquad \text{Q.E.D.}$$

Exercise 5-18. Show that if $C \subseteq \mathbf{R}^n$ has measure zero in $\mathbf{R}^n$ and $f : C \to \mathbf{R}^n$ is continuous, then $f(C)$ need not have measure zero in $\mathbf{R}^n$.

Lemma 5-18. Any cover of $[0, 1]$ by subintervals that are open in $[0, 1]$ contains a finite subcover

$$[0, 1] \subseteq \bigcup_{j=1}^{k} I_j$$

with $\sum_{j=1}^{k} |I_j| \leq 2.$

Proof: Let $\{I_j\}_{j=1}^{k}$ be a finite subcover of the given open cover that is irreducible in the sense that if any I_{j_0} is removed from $\{I_j\}_{j=1}^{k}$, then the remaining I_j fail to cover $[0, 1]$. Suppose that I_j has endpoints $a_j < b_j$ for $j = 1, \ldots , k$. Renumbering if necessary, we may assume that $a_1 \leq a_2 \leq \cdots \leq a_k$. Now, if $k = 1$, then $I_1 = [0, 1]$ and the result is trivial. Thus, we may assume that $k \geq 2$, and from this it follows that $I_1 = [a_1, b_1) = [0, b_1)$ and $I_k = (a_k, b_k] = (a_k, 1]$. If $k = 2$, then $[0, 1] \subseteq [0, b_1) \cup (a_2, 1]$ and again the result is trivial. We therefore assume that $k \geq 3$. Note that by irreducibility $I_j = (a_j, b_j)$ for $j \neq 1$ and $j \neq k$. We claim that $a_i \neq a_j$ if $i \neq j$. For suppose $a_i = a_j$ with $i \neq j$. Then $b_i \neq b_j$ by irreducibility and the fact that $k \geq 3$. Thus, either $b_i < b_j$, in which case $I_i \subseteq I_j$, or $b_j < b_i$ so that $I_j \subseteq I_i$. In either case we contradict irreducibility. Thus, $a_1 < a_2 < \cdots < a_k$. Note also that $b_1 < b_2 < \cdots < b_k$ since if $b_j \geq b_{j+1}$ for some j, then $(a_{j+1}, b_{j+1}) \subseteq (a_j, b_j)$ and, again, irreducibility is contradicted.

Exercise 5-19. Show that, moreover, $a_j < a_{j+1} < b_j \leq a_{j+2}$ for each j.

It now follows that

$$\sum_{j=1}^{k} |I_j| = \sum_{j=1}^{k} (b_j - a_j) = \sum_{j=1}^{k} (a_{j+1} - a_j + b_j - a_{j+1})$$

$$= \sum_{j=1}^{k} (a_{j+1} - a_j) + \sum_{j=1}^{k} (b_j - a_{j+1})$$

$$\leq \sum_{j=1}^{k} (a_{j+1} - a_j) + \sum_{j=1}^{k} (a_{j+2} - a_{j+1}) \leq 1 + 1 = 2.$$

Q.E.D.

Now suppose $n \geq 2$ is an integer and $n = k + l$, where k and l are positive integers. Then $\mathbf{R}^n = \mathbf{R}^k \times \mathbf{R}^l$. For each $p \in \mathbf{R}^l$ we denote by $\mathbf{R}_p^k$ the "horizontal slice" $\mathbf{R}^k \times \{p\}$ in $\mathbf{R}^n$. Each such slice has measure zero in $\mathbf{R}^n$. The final preliminary result we require is the following much stronger statement.

Theorem 5-19 (Fubini's Theorem). Let $n \geq 2$ be an integer and $n = k + l$, where k and l are positive integers. Let C be a subset of $\mathbf{R}^n = \mathbf{R}^k \times \mathbf{R}^l$ that can be expressed as a countable union of compact subsets of $\mathbf{R}^n$. If $C \subseteq \mathbf{R}_p^k$ has measure zero in the horizontal slice $\mathbf{R}_p^k$ for each $p \in \mathbf{R}^l$, then C has measure zero in $\mathbf{R}^n$.

Proof: We may clearly assume that C is compact. Moreover, by induction on l, it suffices to consider the case $l = 1$, $k = n - 1$. Finally, since C is assumed compact we may, without loss of generality, assume that $C \subseteq \mathbf{R}^{n-1} \times [0, 1] \subseteq \mathbf{R}^n$.

For each $t \in [0, 1]$, let $C_t \subseteq \mathbf{R}^{n-1}$ be such that $C_t \times \{t\} = C \cap \mathbf{R}_t^{n-1}$. Now let $\epsilon > 0$ be given. Cover C_t by open cubes $\{W_t^i\}_{i=1}^{\infty}$ in $\mathbf{R}^{n-1}$ with $\sum_{i=1}^{\infty} |W_t^i| < \epsilon/2$. Let $W_t = \cup_{i=1}^{\infty} W_t^i$. Now, if C happens to be entirely contained in some $W_t \times [0, 1]$, we are finished since

$$C \subseteq \bigcup_{i=1}^{\infty} (W_t^i \times [0, 1])$$

and $\quad \displaystyle\sum_{i=1}^{\infty} |W_t^i \times [0, 1]| = \sum_{i=1}^{\infty} |W_t^i| < \epsilon/2.$

Thus, we may assume that $C - (W_t \times [0, 1]) \neq \emptyset$ for each $t \in [0, 1]$. Let x_n denote the last coordinate function on $\mathbf{R}^n$. For each $t \in [0, 1]$ the function $|x_n - t|$ is continuous on C and vanishes precisely on $C_t \times \{t\}$. Since $C - (W_t \times [0, 1])$ is compact, this function assumes a minimum value α_t with $0 < \alpha_t \leq 1$ on $C - (W_t \times [0, 1])$. Thus,

(18) $\quad C_t \times \{t\} \subseteq C \cap \{x \in \mathbf{R}^n : |x_n - t| < \alpha_t\} \subseteq W_t \times I_t^{\alpha_t};$

where $I_t^{\alpha_t} = (t - \alpha_t, t + \alpha_t) \cap [0, 1]$. The collection of all $I_t^{\alpha_t}$ for $t \in [0, 1]$ covers $[0, 1]$ by subintervals that are open in $[0, 1]$, so by Lemma 5-18 we may select a finite subcover $I_{t_1}^{\alpha_{t_1}}, \ldots, I_{t_k}^{\alpha_{t_k}}$ with $\Sigma_{j=1}^{k}|I_{t_j}^{\alpha_{t_j}}| \leq 2$. The sets $W_{t_j}^{i} \times I_{t_j}^{\alpha_{t_j}}, j = 1, \ldots, k, i = 1, 2, \ldots$, cover C by expression (18) and, moreover, $\Sigma |W_{t_j}^{i} \times I_{t_j}^{\alpha_{t_j}}| = \Sigma |W_{t_j}^{i}||I_{t_j}^{\alpha_{t_j}}| \leq 2\Sigma |W_{t_j}^{i}|$, which is less than ϵ since $\Sigma |W_{t_j}^{i}| < \epsilon/2$ for each $t \in [0, 1]$. Thus, C has measure zero in $\mathbf{R}^n$.   Q.E.D.

Now let X be a smooth manifold of dimension k. Choose countably many parametrizations $\varphi_i : U_i \to X$, where each U_i is open in $\mathbf{R}^k$, such that the coordinate nbds $\varphi_i(U_i), i = 1, 2, \ldots$, cover X. We shall say that a subset C of X has *measure zero in X* if $\varphi_i^{-1}(C \cap \varphi_i(U_i))$ has measure zero in $\mathbf{R}^k$ for each i.

Exercise 5-20. Show that a subset C of a smooth k-dimensional manifold X has measure zero in X iff for each parametrization $\varphi : U \to X$, U open in $\mathbf{R}^k$, $\varphi^{-1}(C \cap \varphi(U))$ has measure zero in $\mathbf{R}^k$. Hint: Use Proposition 5-17.

In particular, our definition does not depend on the choice of the parametrizations ϕ_i.

Finally, we are in a position to state and prove our major result on the "size" of the set of critical values of a smooth map.

Theorem 5-20 (Sard's Theorem). Let X and Y be differentiable manifolds, $f : X \to Y$ a smooth map, and $C = \{x \in X : df_x$ is not surjective$\}$ the set of critical points of f. Then $f(C)$, the set of critical values of f, has measure zero in Y.

Exercise 5-21. Show that Sard's Theorem follows from the special case stated in Theorem 5-21:

Theorem 5-21. Let U be an open subset of $\mathbf{R}^k$, $f : U \to \mathbf{R}^l$ a smooth map, and $C = \{x \in U : \operatorname{rank} df_x < l\}$ the set of critical points of f. Then $f(C)$, the set of critical values of f, has measure zero in $\mathbf{R}^l$.

Proof: The proof is by induction on k. Since the result is obvious for $k = 0$, we will assume the validity of the theorem for $k - 1$ and prove it for k.

For each $i = 1, 2, \ldots$ we let C_i be the set of all x in U such that the partial derivatives of f of order $\leq i$ vanish at x. Each C_i is closed since, if $x \in U - C_i$, then some partial derivative of f of order $\leq i$ is nonzero at x and therefore by continuity is nonzero on some nbd of x. Moreover, $C \supseteq C_1 \supseteq C_2 \supseteq \cdots$. We complete the proof by showing that

(a) $f(C - C_1)$ has measure zero in $\mathbf{R}^l$.

(b) $f(C_j - C_{j+1})$ has measure zero in $\mathbf{R}^l$ for all $j \geq 1$.

(c) $f(C_j)$ has measure zero in $\mathbf{R}^l$ for $j > (k/l) - 1$.

For the proof of (a) we may assume that $l \geq 2$ since, if $l = 1$, $C - C_1 = \varnothing$. For each $x \in C - C_1$ we will find a nbd V of x such that $f(V \cap C)$ has measure zero in $\mathbf{R}^l$. Since $C - C_1$ can be covered by countably many such nbds, this will prove that $f(C - C_1)$ has measure zero in $\mathbf{R}^l$.

Let $f = (f_1, \ldots , f_l)$. Since $x \notin C_1$ some first order partial, say, $\partial f_1/\partial x_1$, is not zero at x. Consider the map $h : U \to \mathbf{R}^k$ defined by $h(x) = (f_1(x), x_2, \ldots , x_k)$ for each $x = (x_1, x_2, \ldots , x_k) \in U$. The Jacobian of h has the form

$$\begin{bmatrix} \partial f_1/\partial x_1 & . & . & . & . & . & . \\ 0 & 1 & 0 & . & . & . & 0 \\ 0 & 0 & 1 & . & . & . & 0 \\ . & & . & . & & & . \\ . & & . & . & & & . \\ . & & . & . & & & . \\ 0 & & 0 & 0 & . & . & . & 1 \end{bmatrix}$$

which is nonsingular at x. Thus, h maps a nbd V of x diffeomorphically onto an open set V' in $\mathbf{R}^l$ (Corollary 5-10 and Exercise 5-7). The composition $g = f \circ h^{-1}$ therefore maps V' into $\mathbf{R}^l$ with the same critical values as $f|V$. We need only show that the set of critical values of g has measure zero in $\mathbf{R}^l$.

Now, any point in V' has the form $h(x) = (f_1(x), x_2, \ldots , x_k)$ for some $x = (x_1, x_2, \ldots , x_k) \in V$. For any such point $g((f_1(x), x_2, \ldots , x_k)) = f \circ h^{-1}((f_1(x), x_2, \ldots , x_k)) = f(x_1, x_2, \ldots , x_k) = (f_1(x), f_2(x), \ldots , f_l(x))$. Thus, g carries points of the form $(t, x_2, \ldots , x_k)$ in V' onto points of the form $(t, y_2, \ldots , y_l)$ in $\mathbf{R}^l$. Consequently, for each t, g induces a map $g^t : (\{t\} \times \mathbf{R}^{k-1}) \cap V' \to \{t\} \times \mathbf{R}^{l-1}$. For each $(t, x_2, \ldots , x_k) \in (\{t\} \times \mathbf{R}^{k-1}) \cap V'$, the Jacobian of g has the form

$$\begin{bmatrix} 1 & | & 0 & \cdots & 0 \\ \hline \vdots & | & & & \\ \vdots & | & \left[\dfrac{\partial g_i^t}{\partial x_j}\right] & \\ \vdots & | & & & \end{bmatrix}$$

so that its determinant is $\det \left[\dfrac{\partial g_i^t}{\partial x_j}\right]$. Thus, a point of $(\{t\} \times \mathbf{R}^{k-1}) \cap V'$ is a critical point of g^t iff it is a critical point of g. By the induction hypothesis the set of critical values of g^t has measure zero in $\{t\} \times \mathbf{R}^{l-1}$ for each t. Thus, the set of critical values of g intersects each $\{t\} \times \mathbf{R}^{l-1}$ in a set of measure zero in $\{t\} \times \mathbf{R}^{l-1}$ and therefore by Fubini's Theorem has measure zero in $\mathbf{R}^l$.

Exercise 5-22. Construct a proof for (b) analogous to that just given for (a). Hint: Fubini's Theorem is not needed.

For the proof of (c) we cover C_j by countably many closed cubes, each of which is contained in U. It will suffice to show that $f(C_j \cap S)$ has measure zero in $\mathbf{R}^l$ for each such cube S. Suppose the sides of S have length δ. Subdivide S into r^k subcubes of side-length δ/r (r is an arbitrary positive integer that we will eventually let approach infinity). Let S_1 be one of the cubes in this subdivision. If S_1 contains no point of C_j then it can be ignored. Suppose now that $S_1 \cap C_j \neq \varnothing$. Choose some point $x \in S_1 \cap C_j$. Any other point of S_1 can be expressed as $x + h$ for some $h \in \mathbf{R}^k$ with $\|h\| < \sqrt{n}\,(\delta/r)$. From Taylor's Theorem (for several variables) and the definition of C_j we have $f(x + h) = f(x) + R(x, h)$, where $\|R(x, h)\| < a\|h\|^{j+1}$, a being a constant that depends only on f and S_1. Thus, $f(S_1)$ is contained in a cube of side length b/r^{j+1}, where $b = 2a(\sqrt{n}\,\delta)^{j+1}$ is a constant. Hence, $f(C_j \cap S)$ is contained in a union of at most r^k cubes having total volume $V(r)$ bounded as follows:

$$V(r) \leqslant r^k \left(\frac{b}{r^{j+1}}\right)^l = b^l r^{k-(j+1)l}.$$

Since we have assumed that $j > (k/l) - 1$, $(j + 1)l > k$ so $k - (j + 1)l < 0$ and $V(r) \to 0$ as $r \to \infty$. The total volume of these cubes containing $f(C_j \cap S)$ can therefore be made arbitrarily small by choosing r sufficiently large. Thus, $f(C_j \cap S)$ has measure zero in $\mathbf{R}^l$ and the proof is complete.   Q.E.D.

Remark: Although the set of critical values of a smooth map $f : X \to Y$ between two manifolds has measure zero in Y, this set can, nevertheless, be dense in Y :

Exercise 5-23. Construct a smooth function $f : \mathbf{R} \to \mathbf{R}$ whose set of critical values is dense in $\mathbf{R}$. Hint: Write the rational numbers in a sequence $r_1, r_2, \ldots$. Now construct a function $f_i : \mathbf{R} \to \mathbf{R}$ that is zero outside $[i, i + 1]$ and has r_i as a critical value.

5-9   Morse functions

Our first major application of Sard's Theorem will be to the analysis of certain real-valued smooth maps on manifolds called Morse functions. We know that if X is a manifold and $f : X \to \mathbf{R}$ is smooth, then, for each $x_0 \in X$, df_{x_0} is either surjective or identically zero. If x_0 happens to be a regular point of f, that is, if df_{x_0} is surjective, then, according to Theorem 5-15, we can choose a coordinate system near x_0 relative to which f is simply the first coordinate function. Consequently, we know essentially all there is to know about the local behavior of f near x_0. Near a critical point, however, the situation is rather more complex (and more interesting). Our object in this section is to isolate a particularly simple class of critical points at which the task of characterizing the local behavior of smooth maps is manageable. For motivation we return once again to elementary calculus.

Consider a smooth map $f : \mathbf{R}^2 \to \mathbf{R}$ and a point (x_0, y_0) in $\mathbf{R}^2$. The Jacobian of f at (x_0, y_0) is given by

$$f'(x_0, y_0) = \left[\frac{\partial f}{\partial x}(x_0, y_0), \frac{\partial f}{\partial y}(x_0, y_0) \right],$$

so a point of $\mathbf{R}^2$ is a critical point of f iff both partial derivatives $\partial f / \partial x$ and $\partial f / \partial y$ vanish there. Recall from the calculus that, at such a point, f has either a relative maximum, a relative minimum, or a saddle point. Also recall that there is a straightforward test, involving the second derivatives of f, that allows us to decide between these three possibilities *provided* the quantity

$$(19) \qquad \left(\frac{\partial^2 f}{\partial x^2} \right) \left(\frac{\partial^2 f}{\partial y^2} \right) - \left(\frac{\partial^2 f}{\partial x \partial y} \right)$$

does not vanish at the critical point (x_0, y_0). For critical points at which this quantity does vanish any of the three possibilities may occur, but a more delicate analysis is required to decide between them. Obviously, then, some critical points are "nicer" than others in the sense that the local behavior of the function near them is more easily discerned. For the purposes of generalization observe that the quantity (19) is nothing more than the determinant of the matrix

$$(20) \qquad \begin{bmatrix} \dfrac{\partial^2 f}{\partial x^2} & \dfrac{\partial^2 f}{\partial y \, \partial x} \\[2ex] \dfrac{\partial^2 f}{\partial x \, \partial y} & \dfrac{\partial^2 f}{\partial y^2} \end{bmatrix}$$

since, by smoothness, the mixed partial derivatives of f are equal. Moreover, if we define the smooth map $\operatorname{grad} f : \mathbf{R}^2 \to \mathbf{R}^2$ by

$$(\operatorname{grad} f)(x, y) = \left(\frac{\partial f}{\partial x}(x, y), \frac{\partial f}{\partial y}(x, y) \right),$$

then the matrix (20) is simply the Jacobian of $\operatorname{grad} f$. We now generalize these ideas as follows: Let $f : U \to \mathbf{R}$ be a smooth function defined on an open subset of $\mathbf{R}^k$. The *gradient* of f is the smooth map grad $f : U \to \mathbf{R}^k$ defined by

$$\operatorname{grad} f = \left(\frac{\partial f}{\partial x_1}, \ldots, \frac{\partial f}{\partial x_k} \right).$$

The Jacobian of grad f, that is, the matrix

$$\begin{bmatrix} \dfrac{\partial^2 f}{\partial x_1^2} & \dfrac{\partial^2 f}{\partial x_2 \, \partial x_1} & \cdots & \dfrac{\partial^2 f}{\partial x_k \, \partial x_1} \\[2mm] \dfrac{\partial^2 f}{\partial x_1 \, \partial x_2} & \dfrac{\partial^2 f}{\partial x_2^2} & \cdots & \dfrac{\partial^2 f}{\partial x_k \, \partial x_2} \\[2mm] \cdot & \cdot & & \cdot \\ \cdot & \cdot & & \cdot \\ \cdot & \cdot & & \cdot \\[2mm] \dfrac{\partial^2 f}{\partial x_1 \, \partial x_k} & \dfrac{\partial^2 f}{\partial x_2 \, \partial x_k} & \cdots & \dfrac{\partial^2 f}{\partial x_k^2} \end{bmatrix}$$

is called the *Hessian* matrix of f and is denoted $H(f)$ or simply H. Now let $x_0 \in U$ be a critical point of f, that is, a point at which $(\partial f/\partial x_1)(x_0) = \cdots = (\partial f/\partial x_k)(x_0) = 0$. We shall say that x_0 is a *nondegenerate critical point* of f if the Hessian matrix H of f is nonsingular at x_0, that is, iff the determinant of H is nonzero at x_0; otherwise, x_0 is a *degenerate critical point* of f.

Observe that a point $x_0 \in U$ is a critical point of f iff grad $f(x_0) = 0$. If x_0 is a nondegenerate critical point of f, then grad $f(x_0) = 0$ and $d(\operatorname{grad} f)_{x_0}$: $\mathbf{R}^k \to \mathbf{R}^k$ is an isomorphism so, in particular, grad f maps a nbd V of x_0 diffeomorphically onto a nbd of 0 in $\mathbf{R}^k$. Now, on V, the map grad f, being one-to-one, can vanish only at x_0, that is, V contains no other critical point of f. We conclude that the nondegenerate critical points of a smooth map $f : U \to \mathbf{R}$, U open in $\mathbf{R}^k$, are isolated in the set of all critical points of f.

Exercise 5-24. Construct examples to show that the degenerate critical points of a smooth function on an open set in Euclidean space may, but need not be isolated in the set of all critical points.

The concept of nondegeneracy extends rather easily to smooth functions defined on manifolds. Consider a smooth function $f : X \to \mathbf{R}$ defined on a k-dimensional smooth manifold X, and let x_0 be a critical point of f. Choose an open set $U \subseteq \mathbf{R}^k$ and a parametrization $\varphi : U \to X$ of a nbd $\varphi(U)$ of x_0 in X with $\varphi(u_0) = x_0$. Observe that u_0 is a critical point for $f \circ \varphi$ since $d(f \circ \varphi)_{u_0} = df_{x_0} \circ d\varphi_{u_0} = 0$. We say that x_0 is a *nondegenerate critical point* of f iff u_0 is a nondegenerate critical point of $f \circ \varphi$; otherwise, x_0

is a *degenerate critical point* of *f*. Show that this definition does not depend on the choice of φ in the next exercise.

Exercise 5-25. Let X be a k-dimensional differentiable manifold, $f : X \to \mathbf{R}$ a smooth function, and $x_0 \in X$ a critical point of f. Let U_1 and U_2 be open sets in $\mathbf{R}^k$ and $\varphi_1 : U_1 \to X$ and $\varphi_2 : U_2 \to X$ parametrizations of nbds $\varphi_1(U_1)$ and $\varphi_2(U_2)$ of x_0 in X with $\varphi_1(u_1) = \varphi_2(u_2) = x_0$. Show that u_1 is a nondegenerate critical point of $f \circ \varphi_1$ iff u_2 is a nondegenerate critical point of $f \circ \varphi_2$. Hint: Assume that $\psi = \varphi_2^{-1} \circ \varphi_1$ is a diffeomorphism of U_1 onto U_2, let $g = f \circ \varphi_2$, and note that $f \circ \varphi_1 = g \circ \psi$. Use the ordinary chain rule for partial derivatives to show that the Hessian H' of $g \circ \psi$ at u_1 and the Hessian H of g at u_2 are related by $H' = (\psi'(u_1))^t H(\psi'(u_1))$, where t means "transpose."

A smooth function $f : X \to \mathbf{R}$ on a manifold X is called a *Morse function* if all of its critical points are nondegenerate.

Example 5-5. We consider the function $h : S^2 \to \mathbf{R}$, called the *height function* and defined by $h(x, y, z) = z$ for each $(x, y, z) \in S^2$. Then h is smooth on S^2 since it extends to a smooth map on all of $\mathbf{R}^3$. We parameterize the hemisphere $z > 0$ by letting $U = \{(x, y) \in \mathbf{R}^2 : x^2 + y^2 < 1\}$ be the open unit disc in $\mathbf{R}^2$ and defining $\varphi : U \to S^2$ by $\varphi(x, y) = (x, y, (1 - x^2 - y^2)^{1/2})$. The critical points of h in $z > 0$ are the images under φ of the critical points of $h \circ \varphi$. But

$$(21) \qquad \operatorname{grad}(h \circ \varphi)(x, y) = \left(\frac{\partial(h \circ \varphi)}{\partial x}, \frac{\partial(h \circ \varphi)}{\partial y} \right)$$

$$= \left(\frac{-x}{(1 - x^2 - y^2)^{1/2}}, \frac{-y}{(1 - x^2 - y^2)^{1/2}} \right)$$

which is zero only at $(x, y) = (0, 0)$. Thus, the only critical point of h in $z > 0$ is $\varphi(0, 0) = (0, 0, 1)$, the "north pole." Now if we parametrize $z < 0$ by defining $\psi : U \to S^2$ by $\psi(x, y) = (x, y, -(1 - x^2 - y^2)^{1/2})$, we find in the same way that the only critical point of h in $z < 0$ is the "south pole" $(0, 0, -1)$. Observe that h has no critical points on any of the other hemispheres $x > 0$, $x < 0$, $y > 0$, or $y < 0$, since on these coordinate nbds z itself is one of the coordinate functions. For example, if we parametrize $y < 0$ by letting $V = \{(x, z) \in \mathbf{R}^2 : x^2 + z^2 < 1\}$ be the open unit disc in the copy $\{(x, y, z) \in \mathbf{R}^3 : y = 0\}$ of $\mathbf{R}^2$ and defining $\chi : V \to S^2$ by $\chi(x, z) = (x, -(1 - x^2 - z^2)^{1/2}, z)$, then $h \circ \chi : V \to \mathbf{R}$ is given by $(h \circ \chi)(x, z) = z$ and $\operatorname{grad}(h \circ \chi)$ is identically equal to $(0, 1)$ so $h \circ \chi$ has no critical points in V and, consequently, h has no critical points in the hemisphere $y < 0$.

We conclude that the height function $h : S^2 \to \mathbf{R}$ has precisely two critical points, that is, $(0, 0, 1)$ and $(0, 0, -1)$, and we claim that both of these critical points are nondegenerate. To show that $(0, 0, 1)$ is a nondegenerate critical point of h, we must show that $(0, 0)$ is a nondegenerate critical point of $h \circ \varphi$. Now, the Hessian of $h \circ \varphi$ is the Jacobian of $\text{grad}(h \circ \varphi)$ which, after computing the derivatives of the components of $\text{grad}(h \circ \varphi)$ shown in equation (21), we find to be

$$
\begin{bmatrix}
\dfrac{y^2 - 1}{(1 - x^2 - y^2)^{3/2}} & \dfrac{-xy}{(1 - x^2 - y^2)^{3/2}} \\[3mm]
\dfrac{-xy}{(1 - x^2 - y^2)^{3/2}} & \dfrac{x^2 - 1}{(1 - x^2 - y^2)^{3/2}}
\end{bmatrix}
$$

The determinant of this matrix is $(1 - x^2 - y^2)^{-2}$, which at $(x, y) = (0, 0)$ is 1, so $(0, 0)$ is indeed a nondegenerate critical point of $h \circ \varphi$ as required. We show that $(0, 0, -1)$ is a nondegenerate critical point of h in the same way. Thus, h is a Morse function with precisely two critical points.

Remark: Entirely analogous arguments show that the height function $h :$ $S^k \to \mathbf{R}$ defined by $h(x_1, \ldots, x_k, x_{k+1}) = x_{k+1}$ for each $(x_1, \ldots, x_k,$ $x_{k+1})$ in S^k is smooth and has precisely two critical points at $(0, \ldots, 0,$ $1)$ and $(0, \ldots, 0, -1)$, both of which are nondegenerate. Thus, any sphere admits a Morse function with exactly two critical points. The culmination of all of our work in differential topology will be a very beautiful theorem of Reeb that states that, among compact k-manifolds, the sphere S^k is characterized topologically by this simple property, that is, that any compact k-manifold admitting a Morse function with precisely two critical points must be homeomorphic (not diffeomorphic) to S^k (see Theorem 5-32).

Next we apply Sard's Theorem to show that Morse functions exist on an arbitrary manifold and, indeed, that they exist in great profusion. First let us consider an arbitrary smooth map $f : U \to \mathbf{R}$, where $U \subseteq \mathbf{R}^k$ is open. Let $x_1, \ldots, x_k$ be the standard coordinate functions on $\mathbf{R}^k$. For each k-tuple $a = (a_1, \ldots, a_k)$ of real numbers we define a new function $f_a : U \to \mathbf{R}$ by $f_a = f + a_1 x_1 + \cdots + a_k x_k$. Now, f itself may or may not be a Morse function on U. Nevertheless, we claim that we can choose the k-tuple $a \in \mathbf{R}^k$ in such a way that f_a is a Morse function on U. In fact, we will show that the set $\{a \in \mathbf{R}^k : f_a$ is not a Morse function on $U\}$ has measure zero in $\mathbf{R}^k$, that is, that f_a is Morse for "almost every" $a \in \mathbf{R}^k$. First note that, for any $p \in U$,

$$
\text{grad } f(p) = \left(\frac{\partial f}{\partial x_1}(p), \ldots, \frac{\partial f}{\partial x_k}(p) \right)
$$

and $\text{grad } f_a(p) = \left(\dfrac{\partial f_a}{\partial x_1}(p), \ldots, \dfrac{\partial f_a}{\partial x_k}(p) \right)$

$= \left(\dfrac{\partial f}{\partial x_1}(p) + a_1, \ldots, \dfrac{\partial f}{\partial x_k}(p) + a_k \right)$

$= \text{grad } f(p) + a.$

Thus, f_a has a critical point at p iff $\text{grad } f(p) = -a$ and f_a has a degenerate critical point at p iff $\text{grad } f(p) = -a$ and the Hessian of f_a at p is singular. Now, f and f_a have the same Hessians since the Jacobians of $\text{grad } f$ and $\text{grad } f_a$ are the same. We find then that f_a has a degenerate critical point at p iff $\text{grad } f(p) = -a$ and $d(\text{grad } f)_p$ is singular, that is, iff $-a$ is a critical value of $\text{grad } f$ with $\text{grad } f(p) = -a$. In particular, f_a can have no degenerate critical points if $-a$ fails to be a critical value of $\text{grad } f$. But, according to Sard's Theorem, the set of critical values of $\text{grad } f$ has measure zero in $\mathbf{R}^k$, so the set of all $a \in \mathbf{R}^k$ for which f_a can have a degenerate critical point has measure zero in $\mathbf{R}^k$. We have therefore proved the following lemma.

Lemma 5-22. Let U be an open subset of $\mathbf{R}^k$ and $f : U \to \mathbf{R}$ an arbitrary smooth function. For each $a = (a_1, \ldots, a_k) \in \mathbf{R}^k$ define a smooth function $f_a : U \to \mathbf{R}$ by $f_a = f + a_1 x_1 + \cdots + a_k x_k$, where $x_1, \ldots, x_k$ are the standard coordinate functions on $\mathbf{R}^k$. Then the set of all $a \in \mathbf{R}^k$ for which f_a fails to be a Morse function on U has measure zero in $\mathbf{R}^k$.

In order to obtain an analogous result for smooth functions defined on arbitrary manifolds, we require the following intuitively obvious result (which you may already have conjectured).

Lemma 5-23. Let X be a k-dimensional, smooth manifold in $\mathbf{R}^n$, $x \in X$ and $x_1, \ldots, x_n$ the standard coordinate functions on $\mathbf{R}^n$. Then there is a nbd of x in X on which the restriction of some k coordinate functions $x_{i_1}, \ldots, x_{i_k}$ forms a local coordinate system at x.

Proof: Viewed as maps on the vector space $\mathbf{R}^n$, the coordinate functions $x_1, \ldots, x_n$ are linear functionals. Indeed, they constitute the basis for the space of linear functionals that is the dual of the standard basis for $\mathbf{R}^n$. In particular, $x_1, \ldots, x_n$ are linearly independent as functionals. Observe that, since each x_i is linear, the derivative at x of x_i is x_i itself. Some k of the linear functionals $x_1, \ldots, x_n$, say, $x_{i_1}, \ldots, x_{i_k}$, are linearly independent when restricted to $T_x(X)$. But the derivative at x of the restriction of x_{i_j} to X is by definition the restriction to $T_x(X)$ of the linear map x_{i_j}, so by linear independence the map $(x_{i_1}, \ldots, x_{i_k}) : X \to \mathbf{R}^k$ is a local diffeomorphism at x.   Q.E.D.

Theorem 5-24. Let X be a k-dimensional differentiable manifold in $\mathbf{R}^n$ and $f : X \to \mathbf{R}$ a smooth function. For each $a = (a_1, \ldots, a_n) \in \mathbf{R}^n$ define a smooth function $f_a : X \to \mathbf{R}$ by $f_a = f + a_1 x_1 + \cdots + a_n x_n$, where $x_1, \ldots, x_n$ are the standard coordinate functions in $\mathbf{R}^n$. Then the set of all $a \in \mathbf{R}^n$ for which f_a fails to be a Morse function on X has measure zero in $\mathbf{R}^n$.

Proof: By Lemma 5-23 we can cover X by countably many open sets U_1, $U_2, \ldots$ such that on each U_i some k of the coordinate functions $x_1, \ldots, x_n$ form a coordinate system. Let U_i be one of these sets. By renumbering if necessary we may assume that $(x_1, \ldots, x_k)$ is a coordinate system on U_i. Let $S_i = \{a \in \mathbf{R}^n : f_a \text{ is not a Morse function on } U_i\}$. It suffices to show that S_i has measure zero in $\mathbf{R}^n$. Consider a function

$$
\begin{aligned}
f_a &= f + a_1 x_1 + \cdots + a_k x_k + a_{k+1} x_{k+1} + \cdots + a_n x_n \\
&= (f + a_{k+1} x_{k+1} + \cdots + a_n x_n) + a_1 x_1 + \cdots + a_k x_k
\end{aligned}
$$

on U_i. Now fix the numbers $a_{k+1}, \ldots, a_n$. Then, since $(x_1, \ldots, x_k)$ is a coordinate system on U_i, each of the functions $f, x_{k+1}, \ldots, x_n$ can be thought of as being defined on an open subset of $\mathbf{R}^k$, namely, the image of U_i under the diffeomorphism $(x_1, \ldots, x_k) : U_i \to \mathbf{R}^k$. The same is therefore true of the function $f + a_{k+1} x_{k+1} + \cdots + a_n x_n$, so Lemma 5-22 implies that the set of all $(a_1, \ldots, a_k) \in \mathbf{R}^k$ for which the function $(f + a_{k+1} x_{k+1} + \cdots + a_n x_n) + a_1 x_1 + \cdots + a_k x_k$ fails to be a Morse function on U_i has measure zero in $\mathbf{R}^k$. Thus, each horizontal slice $S_i \cap \mathbf{R}^k \times \{(a_{k+1}, \ldots, a_n)\}$, $(a_{k+1}, \ldots, a_n) \in \mathbf{R}^{n-k}$, of S_i has measure zero in $\mathbf{R}^k$ so Fubini's Theorem implies that S_i has measure zero in $\mathbf{R}^n$. Q.E.D.

Remarks: In particular, we conclude from Theorem 5-24 that Morse functions exist on any smooth manifold. This simple fact alone has surprisingly deep consequences (see Section 5-11), but the real strength of the theorem lies in the fact that it assures us that nondegeneracy is, in a sense, the common state of affairs for critical points of smooth functions.

The reader will recall that our decision to study nondegenerate critical points was initially prompted by the fact that, at least for functions from $\mathbf{R}^2$ to $\mathbf{R}$, the local behavior of a function near such a critical point is rather easily discerned. We propose now to prove a result analogous to Theorem 5-15 that completely describes the behavior of a smooth function $f : X \to \mathbf{R}$ in a nbd of a nondegenerate critical point in terms of some appropriately chosen local coordinate system. First an elementary lemma from the calculus.

Lemma 5-25. Let V be a convex nbd of the origin in $\mathbf{R}^k$ and $f: V \to \mathbf{R}$ a smooth function with $f(0, \dots, 0) = 0$. Then there exist smooth functions $g_1, \dots, g_k$ on V that satisfy

$$g_j(0, \dots, 0) = \frac{\partial f}{\partial x_j}(0, \dots, 0) \qquad \text{for each } j = 1, \dots, k$$

and such that

$$f(x_1, \dots, x_k) = \sum_{j=1}^{k} x_j \cdot g_j(x_1, \dots, x_k)$$

$$\text{for each } (x_1, \dots, x_k) \in V.$$

Exercise 5-26. Prove Lemma 5-25. Hint: Let $g_j(x_1, \dots, x_k) =$
$$\int_0^1 \frac{\partial f}{\partial x_j}(tx_1, \dots, tx_k)\, dt. \qquad \text{Q.E.D.}$$

Now we show that, in a nbd of a nondegenerate critical point, a smooth function is essentially nothing more than a quadratic polynomial of a particularly simple type.

Theorem 5-26 (Morse Lemma). Let X be a k-dimensional differentiable manifold, $f: X \to \mathbf{R}$ a smooth function, and $x_0 \in X$ a nondegenerate critical point of f. Then there is an open nbd U of $0 \in \mathbf{R}^k$ and a parametrization $\varphi: U \to X$ of a nbd $\varphi(U)$ of x_0 in X with $x_0 = \varphi(0)$ such that the map $f \circ \varphi: U \to \mathbf{R}$ is given by

$$(22) \qquad f \circ \varphi(y_1, \dots, y_k) = f(x_0) - y_1^2 - \cdots - y_\lambda^2 + y_{\lambda+1}^2 + \cdots + y_k^2,$$

where $\lambda \leq k$ is a positive integer that depends only on f and x_0.

Remark: The map $f \circ \varphi$ is simply f written in terms of the coordinates set up on $\varphi(U)$ by the parametrization φ.

Proof: As usual, since the result we are to prove is purely local, there is no loss of generality in assuming that x_0 is the origin in $\mathbf{R}^k$ and that f is defined on a convex nbd of $0 \in \mathbf{R}^k$ and maps 0 onto $0 \in \mathbf{R}$. Applying Lemma 5-25, we write $f(x_1, \dots, x_k) = \sum_{j=1}^{k} x_j \cdot g_j(x_1, \dots, x_k)$, where the g_j are smooth on some nbd of $0 \in \mathbf{R}^k$ and satisfy $g_j(0) = (\partial f/\partial x_j)(0)$. But, since 0 is assumed a critical point of f, $g_j(0) = 0$ for each j. Thus, we may apply Lemma 5-25 to each g_j to obtain $g_j(x_1, \dots, x_k) = \sum_{i=1}^{k} x_i \cdot h_{ij}(x_1, \dots, x_k)$ for functions h_{ij} that are smooth on some nbd of 0 in $\mathbf{R}^k$ and satisfy

$$h_{ij}(0) = \frac{\partial g_j}{\partial x_i}(0) = \frac{\partial^2 f}{\partial x_i \, \partial x_j}(0).$$

Thus, we find that on some nbd of $0 \in \mathbf{R}^k$ we may write

(23) $$f(x_1, \ldots, x_k) = \sum_{i,j=1}^{k} x_i x_j \cdot h_{ij}(x_1, \ldots, x_k),$$

where the matrix $[h_{ij}(0)]$, being the Hessian of f at 0, is nonsingular since 0 is a *nondegenerate* critical point of f. Observe that we can assume that the functions h_{ij} in representation (23) are symmetric in i and j, that is, that $h_{ij} = h_{ji}$, since, if we define $h'_{ij} = \frac{1}{2}(h_{ij} + h_{ji})$, then $h'_{ij} = h'_{ji}$, $f = \Sigma x_i x_j \cdot h'_{ij}$ and the matrix $[h'_{ij}(0)]$ is equal to the matrix

$$\left[\frac{1}{2} \frac{\partial^2 f}{\partial x_i \partial x_j}(0) \right]$$

and is thus also nonsingular.

We now proceed to define the required coordinate system inductively, altering just one of the given coordinate functions at each stage of the induction. Our point of departure is the expansion (23). Since the matrix $[h_{ij}(0)]$ is nonsingular, the quadratic form $\Sigma_{i,j=1}^{k} x_i x_j h_{ij}(0)$ is not identically zero. Thus, we may, after a suitable nonsingular, linear transformation of variables, assume that $h_{11}(0) \neq 0$. (This is an elementary algebraic result for which you may either construct your own proof or consult Birkhoff and MacLane, lemma on p. 252.) Thus, h_{11} is nonzero on some nbd of $0 \in \mathbf{R}^k$, so $(|h_{11}|)^{1/2}$ is a nonzero, smooth function on that nbd. Now define $u_1, \ldots, u_k$ by

$$u_1 = (|h_{11}|)^{1/2} \left(x_1 + \frac{h_{12}}{h_{11}} x_2 + \cdots + \frac{h_{1k}}{h_{11}} x_k \right)$$

$$u_i = x_i \qquad \text{for } i = 2, \ldots, k.$$

(The reason for our choice of u_1 will be clear shortly.) The map $(u_1, \ldots, u_k)$ is smooth on a nbd of 0 in $\mathbf{R}^k$. Moreover, the Jacobian of this map at 0 is

$$\begin{bmatrix} \dfrac{\partial u_1}{\partial x_1}(0) & \dfrac{\partial u_1}{\partial x_2}(0) & \cdots & \dfrac{\partial u_1}{\partial x_k}(0) \\ 0 & 1 & \cdots & 0 \\ \cdot & \cdot & & \cdot \\ \cdot & \cdot & & \cdot \\ \cdot & \cdot & & \cdot \\ 0 & 0 & \cdots & 1 \end{bmatrix}$$

Expanding by the cofactors of the first column, we find that the determinant of this matrix is $(\partial u_1/\partial x_1)(0)$. Computing $\partial u_1/\partial x_1$ by the product rule and evaluating at $(x_1, \ldots, x_k) = (0, \ldots, 0)$, we obtain

$$\frac{\partial u_1}{\partial x_1}(0) = (|h_{11}(0)|)^{1/2} \neq 0.$$

Thus, $(u_1, \ldots, u_k)$ maps some nbd of 0 in $\mathbf{R}^k$ diffeomorphically onto another nbd of 0 in $\mathbf{R}^k$ so that $(u_1, \ldots, u_k)$ is a legitimate local coordinate system at the origin in $\mathbf{R}^k$. Next we compute the form f takes when expressed in terms of these coordinates. By virtue of the symmetry of the h_{ij}, the only terms in expansion (23) that contain the factor x_1 are $h_{11}x_1 + 2\sum_{j=2}^k h_{1j}x_1x_j$. Regard this expression as a quadratic in x_1 and complete the square to obtain

$$
\begin{aligned}
h_{11}x_1^2 + 2\sum_{j=2}^k h_{1j}x_1x_j &= h_{11}\left(x_1^2 + 2\sum_{j=2}^k \left(\frac{h_{1j}}{h_{11}}\right)x_1x_j\right) \\
&= h_{11}\left[\left(x_1 + \sum_{j=2}^k \left(\frac{h_{1j}}{h_{11}}\right)x_j\right)^2 - \left(\sum_{j=2}^k \left(\frac{h_{1j}}{h_{11}}\right)x_j\right)^2\right] \\
&= \pm(|h_{11}|)^{1/2}\left(x_1 + \sum_{j=2}^k \left(\frac{h_{1j}}{h_{11}}\right)x_j\right)^2 \\
&\quad \mp \left((|h_{11}|)^{1/2}\sum_{j=2}^k \left(\frac{h_{1j}}{h_{11}}\right)x_j\right)^2 \\
&= \pm u_1^2 \mp \left((|h_{11}|)^{1/2}\sum_{j=2}^k \left(\frac{h_{1j}}{h_{11}}\right)x_j\right)^2
\end{aligned}
$$

Now, by squaring $(|h_{11}|)^{1/2}\sum_{j=2}^k(h_{1j}/h_{11})x_j$, collecting terms, and substituting $x_j = u_j$ for $j \geq 2$, we obtain

$$
h_{11}x_1^2 + 2\sum_{j=2}^k h_{1j}x_1x_j = \pm u_1^2 + \sum_{j,l=2}^k c_{jl}u_ju_l
$$

for some constants c_{jl}. Now,

$$
\begin{aligned}
f &= h_{11}x_1^2 + 2\sum_{j=2}^k h_{1j}x_1x_j + \sum_{i,j=2}^k h_{ij}x_ix_j \\
&= \pm u_1^2 + \sum_{j,l=2}^k c_{jl}u_ju_l + \sum_{i,j=2}^k h_{ij}x_ix_j
\end{aligned}
$$

and, by changing variables in the last sum and collecting terms, we find that

$$
(24) \qquad f = \pm u_1^2 + \sum_{i,j=2}^k u_iu_jH_{ij}(u_1, \ldots, u_k),
$$

where the H_{ij} are smooth on some nbd of $0 \in \mathbf{R}^k$. Observe that the matrix $[H_{ij}(0)]$ must be nonsingular since otherwise the Hessian of f, computed relative to the coordinate system $u_1, \ldots, u_k$, would be singular at 0. As before we may assume that $H_{ij} = H_{ji}$. Thus, we may repeat the entire procedure for the sum in (24) and thereby isolate one more of the variables. Continuing the process for k-steps, the desired representation for f will emerge.

Exercise 5-27. Explicitly carry out the inductive argument necessary to obtain coordinate functions $y_1, \ldots, y_k$ in a nbd of $0 \in \mathbf{R}^k$ relative to which f has the form $f = \pm y_1^2 \pm y_2^2 \pm \ldots \pm y_k^2$. Hint: Suppose $r < k$ and $v_1, \ldots, v_k$ is a coordinate system in a nbd of $0 \in \mathbf{R}^k$ relative to which f has the form $f = \pm v_1^2 \pm \ldots \pm v_{r-1}^2 + \Sigma_{i,j \geq r} v_i v_j K_{ij}(v_1, \ldots, v_k)$, where the K_{ij} are smooth and satisfy $K_{ij} = K_{ji}$ in a nbd of the origin in $\mathbf{R}^k$.

The map $(y_1, \ldots, y_k)$ carries some nbd V of $0 \in \mathbf{R}^k$ diffeomorphically onto another nbd U of 0. Let $\varphi : U \to V$ be the parametrization that is the inverse of $(y_1, \ldots, y_k)$. Renumbering the y_i if necessary, we may assume that $f \circ \varphi : U \to \mathbf{R}$ is given by $f \circ \varphi(y_1, \ldots, y_k) = -y_1^2 - \cdots - y_\lambda^2 + y_{\lambda+1}^2 + \cdots + y_k^2$ for some $\lambda \leq k$ as required. From this the expansion (22) in the general case is immediate.

All that remains is to show that the integer λ in (22) depends only on f and x_0. Thus, we let $\varphi_1 : U_1 \to X$ and $\varphi_2 : U_2 \to X$ be parametrizations of nbds $\varphi_1(U_1)$ and $\varphi_1(U_2)$ of x_0 such that the maps $f \circ \varphi_1 : U_1 \to \mathbf{R}$ and $f \circ \varphi_2 : U_2 \to \mathbf{R}$ are given by

$$(25) \quad f \circ \varphi_1(y_1, \ldots, y_k) = f(x_0) - y_1^2 - \cdots - y_{\lambda_1}^2 + y_{\lambda_1+1}^2 + \cdots + y_k^2$$

$$(26) \text{ and } \quad f \circ \varphi_2(z_1, \ldots, z_k) = f(x_0) - z_1^2 - \cdots - z_{\lambda_2}^2 + z_{\lambda_2+1}^2 + \cdots + z_k^2.$$

If $\varphi_1(u_1) = \varphi_2(u_2) = x_0$ and $H(f \circ \varphi_1)$ and $H(f \circ \varphi_2)$ are the Hessians of $f \circ \varphi_1$ and $f \circ \varphi_2$ at u_1 and u_2, respectively, then, as we found in Exercise 5-25, there is a nonsingular matrix T such that $H(f \circ \varphi_1) = TH(f \circ \varphi_2) T^t$, that is, the matrices $H(f \circ \varphi_1)$ and $H(f \circ \varphi_2)$ are congruent. According to Sylvester's Law of Inertia (Herstein, p. 310. Theorem 6-z_3), $H(f \circ \varphi_1)$ and $H(f \circ \varphi_2)$ must have the same signature (number of positive diagonal entries minus number of negative diagonal entries). Computing $H(f \circ \varphi_1)$ and $H(f \circ \varphi_2)$ from equations (25) and (26), respectively, we therefore find that $(k - \lambda_1) - \lambda_1 = (k - \lambda_2) - \lambda_2$ so $\lambda_1 = \lambda_2$ as required.   Q.E.D.

Exercise 5-28. The integer λ in (22) is called the *index* of f at x_0. Show that f has a relative maximum at x_0 iff $\lambda = k$ and f has a relative minimum at x_0 iff $\lambda = 0$.

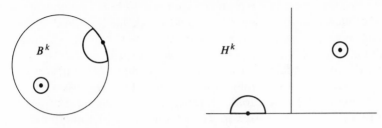

Figure 5-10

Remark: Our major application of the Morse Lemma, to be found in Section 5-12, is a proof of the topological characterization of S^k to which we alluded earlier in this section.

5-10 Manifolds with boundary

In order to carry out the applications we have in mind, it will be necessary to extend the scope of the concepts with which we have been dealing to a class of spaces somewhat more general than that of manifolds. Specifically, we wish to consider such spaces as the closed ball $B^k = \{x \in \mathbf{R}^k: \|x\| \leq 1\}$ and the half-space $H^k = \{(x_1, \ldots, x_k) \in \mathbf{R}^k: x_k \geq 0\}$, which would be manifolds were it not for the presence of boundary points. Although every $x \in B^k$ with $\|x\| < 1$ has a nbd in B^k that is diffeomorphic to an open subset of $\mathbf{R}^k$, this does not appear to be true for those $x \in B^k$ with $\|x\| = 1$. (This second statement will follow from more general results of this section, but you should construct a direct proof at this point.) Observe, however, that every point of B^k has a nbd in B^k that is diffeomorphic to an open set in H^k. (A proof will be given shortly – for now, simply consider Figure 5-10.)

If V is an open set in the space H^k, we define its *boundary* ∂V and *interior* Int V by $\partial V = V \cap \{(x_1, \ldots, x_k) \in H^k : x_k = 0\}$ and Int $V = V - \partial V$. Note that ∂V and Int V thus defined are *not* the topological boundary and interior of V in H^k. A subset X of $\mathbf{R}^n$ is a *k-dimensional manifold with boundary* if each point in X has a nbd in X that is diffeomorphic to an open subset of H^k. Any diffeomorphism φ of an open subset V of H^k onto an open subset U of X is called a *parametrization* of U.

Exercise 5-29. Let X be a k-dimensional manifold with boundary, U an open subset of X, and $\varphi_1 : V_1 \to U$ and $\varphi_2 : V_2 \to U$ two parametrizations of U, where V_1 and V_2 are open in H^k. Show that $\varphi_1(\partial V_1) = \varphi_2(\partial V_2)$.

If X is a manifold with boundary in $\mathbf{R}^n$, then, by virtue of Exercise 5-29, we may define the *boundary* ∂X of X to be the set of all $x \in X$

which belong to $\varphi(\partial V)$ for some parametrization $\varphi : V \to U$ of a nbd U of x in X+ the *interior* of X, denoted Int X, is then defined by Int $X = X - \partial X$. Again note that, in general, ∂X and Int X are not the topological boundary and interior of X in $\mathbf{R}^n$. A manifold as defined in Section 5-5 is also a manifold with boundary, but its boundary is empty. Also observe that if X is a k-dimensional manifold with boundary, then Int X is a k-dimensional (boundaryless) manifold. Moreover, if $x \in \partial X$ and $\varphi : V \to U$ is a parametrization of a nbd U of x in X, then $\varphi|\partial V$ is a diffeomorphism of ∂V, an open set in the copy ∂H^k of $\mathbf{R}^{k-1}$, onto ∂U, an open set in ∂X. Thus, ∂X is a $(k - 1)$-dimensional (boundaryless) manifold. Since Int X is an open dense subset of X, ∂X is closed and nowhere dense. The following is a useful device for the production of examples.

Lemma 5-27. Let X be a k-dimensional boundaryless manifold and $f : X \to \mathbf{R}$ a smooth function that has $0 \in \mathbf{R}$ as a regular value. Then $f^{-1}[0, \infty)$ is a k-dimensional manifold with boundary and $\partial f^{-1}[0, \infty) = f^{-1}(0)$.

Proof: Since $f^{-1}(0, \infty)$ is an open submanifold of X, we need only show that for each $x \in f^{-1}(0)$ there is a parametrization $\varphi : V \to U$, where $V \subseteq H^k$ is open, U is a nbd of x in $f^{-1}[0, \infty)$, and $x \in \varphi(\partial V)$. Since 0 is a regular value of f, Theorem 5-15 implies that there is a nbd V' of 0 in $\mathbf{R}^k$ and a diffeomorphism φ' of V' onto a nbd $\varphi'(V')$ of x in X with $\varphi'(0) = x$ and such that the map $f \circ \varphi' : V' \to \mathbf{R}$ is given by $f \circ \varphi'(x_1, \ldots, x_k) = x_k$. Let $V = (f \circ \varphi')^{-1}([0, \infty)) = \{(x_1, \ldots, x_k) \in U' : x_k \geq 0\}$. Now $\varphi = \varphi'|V$ maps V diffeomorphically onto $U = \varphi'(V') \cap f^{-1}[0, \infty)$ which is a nbd of x in $f^{-1}[0, \infty)$ as required.   Q.E.D.

As an application we consider the map $f : \mathbf{R}^k \to \mathbf{R}$ defined by $f(x) = 1 - \|x\|^2 = 1 - x_1^2 - \cdots - x_k^2$. Then f is smooth and grad $f(x) = (-2x_1, \ldots, -2x_k)$, which vanishes only at the origin. Since $f^{-1}(0)$ does not contain the origin, 0 is a regular value of f. Thus, $f^{-1}[0, \infty) = B^k$ is a k-dimensional manifold with boundary $\partial B^k = f^{-1}(0) = S^{k-1}$.

We now set ourselves the task of defining tangent spaces to manifolds with boundary and derivatives of smooth maps between such manifolds. First consider a smooth map $f : V \to U$, where $V \subseteq H^k$ and $U \subseteq \mathbf{R}^n$ are open sets. Let F denote an extension of f to a smooth map from some open subset of $\mathbf{R}^k$ containing V into $\mathbf{R}^n$. For each $v \in V$ define $df_v : \mathbf{R}^k \to \mathbf{R}^n$ by $df_v = dF_v$.

Exercise 5-30. Show that this definition does not depend on the choice of the extension F. Hint: This is trivial for $v \in$ Int V. For $v \in \partial V$ use the fact that $dF_v(x)$ is continuous in v.

Remark: Note that df_v is a linear map of all of $\mathbf{R}^k$ to $\mathbf{R}^n$ even if v is a boundary point of V. Also observe that the chain rule is still valid for derivatives of smooth maps defined on open subsets of half-spaces.

Now suppose X is a k-dimensional manifold with boundary in $\mathbf{R}^n$ and let x be a point of X. Choose a parametrization $\varphi : V \to X$ of a nbd $\varphi(V)$ of $x = \varphi(v)$ in x, where $V \subseteq H^k$ is open. Then $d\varphi_v$ is a linear map of $\mathbf{R}^k$ to $\mathbf{R}^n$. We define the *tangent space to X at x*, denoted $T_x(X)$, by $T_x(X) = d\varphi_v(\mathbf{R}^k)$. Then $T_x(X)$ is a k-dimensional linear subspace of $\mathbf{R}^n$ (even if $x \in \partial X$) that does not depend on the choice of the parametrization. If $Y \subseteq \mathbf{R}^m$ is another manifold with boundary and $f : X \to Y$ is smooth, we define the *derivative* $df_x : T_x(X) \to T_{f(x)}(Y)$ of f at x as follows: Let F be a smooth map into $\mathbf{R}^m$ that is defined on an open subset of $\mathbf{R}^n$ and extends f and set $df_x = dF_x|T_x(X)$. As usual, we verify that dF_x does indeed send $T_x(X)$ into $T_{f(x)}(Y)$ and that this definition does not depend on the choice of F. (The details are routine and will be left to the reader.) Again, the chain rule remains valid in this more general setting.

We have defined the derivative for smooth maps between manifolds with boundary, and the definitions of regular and critical points and values are precisely the same as in the boundaryless case. Generally, however, the results we have obtained about such points for manifolds without boundary require additional hypotheses when boundaries are nonempty. One particularly important such result is Corollary 5-16. If $f : X \to Y$ is a smooth map between two manifolds with boundary, we shall denote by ∂f the restriction $f|\partial X$. Observe that if $x \in \partial X$, then $d(\partial f)_x = df_x|T_x(\partial X)$. Now the appropriate analog of Corollary 5-16 is the following theorem.

Theorem 5-28. Let X be a k-dimensional manifold with boundary, Y an l-dimensional boundaryless manifold, where $k \geqslant l$, and $f : X \to Y$ a smooth map. Suppose $y \in Y$ is a regular value of both f and ∂f. Then $f^{-1}(y)$ is either empty or a $(k - l)$-dimensional manifold with boundary $\partial(f^{-1}(y)) = f^{-1}(y) \cap \partial X$.

Proof: We need only consider the case in which $f : H^k \to \mathbf{R}^l$ since the result we are to obtain is a local one. Thus, we assume that $y \in \mathbf{R}^l$ is a regular value for both f and $\partial f : \partial H^k \to \mathbf{R}^l$. Suppose $f^{-1}(y) \neq \varnothing$ and let $x \in f^{-1}(y)$. If $x \in \text{Int } H^k$, then it has a nbd that is open in $\mathbf{R}^k$ and contained in H^k. Restricting f to this nbd and applying Corollary 5-16, we find that $f^{-1}(y)$ is a smooth boundaryless manifold of dimension $k - l$ in a nbd of x.

Now suppose $x \in \partial H^k$. Select an open subset U of $\mathbf{R}^k$ containing x and a smooth map $F : U \to \mathbf{R}^l$ that extends f. Then x is a regular point of F so

F is regular on a nbd of x, and by shrinking U if necessary we may assume that F has no critical points on U. Thus, y is a regular value of F, so by Corollary 5-16 $F^{-1}(y)$ is a smooth $(k - l)$-manifold without boundary. Now let π be the restriction to $F^{-1}(y)$ of the projection $(x_1, \ldots, x_k) \to x_k$. We claim that $0 \in \mathbf{R}$ is a regular value of $\pi : F^{-1}(y) \to \mathbf{R}$. To see this, let $z \in \pi^{-1}(0)$. We must show that $d\pi_z$ is not identically zero on $T_z(F^{-1}(y))$. But $d\pi_z$ is by linearity just the restriction to $T_z(F^{-1}(y))$ of the projection $(x_1, \ldots, x_k) \to x_k$, so it will suffice to show that $T_z(F^{-1}(y))$ is not entirely contained in $\mathbf{R}^{k-1} \times \{0\}$.

Exercise 5-31. Let X and Y be boundaryless manifolds, $f : X \to Y$ a smooth map, and $y \in Y$ a regular value of f. Show that, for each $x \in f^{-1}(y)$, $T_x(f^{-1}(y))$ is the null space of $df_x : T_x(X) \to T_{f(x)}(Y)$.

Thus, $T_z(F^{-1}(y))$ is the null space of $dF_z : \mathbf{R}^k \to \mathbf{R}^l$. Suppose the null space of dF_z were entirely contained in $\mathbf{R}^{k-1} \times \{0\}$. Since $\partial f : \mathbf{R}^{k-1} \times \{0\} \to \mathbf{R}^l$ is assumed regular at z, $d(\partial f)_z : \mathbf{R}^{k-1} \to \mathbf{R}^l$ is surjective. Since $d(\partial f)_z = dF_z | \mathbf{R}^{k-1}$, the null spaces of dF_z and $d(\partial f)_z$ coincide. Thus, the dimension of the null space of $d(\partial f)_z$ is $k - l$. But then the rank plus the nullity of $d(\partial f)_z$ is $l + (k - l) = k$, which is impossible. We conclude that 0 is indeed a regular value of $\pi : F^{-1}(y) \to \mathbf{R}$. Thus, $f^{-1}(y) \cap U = f^{-1}(y) \cap H^k = \pi^{-1}[0, \infty)$ is by Theorem 5-28 a smooth $(k - l)$-manifold with boundary $\pi^{-1}(0) = f^{-1}(y) \cap \partial H^k$ as required.   Q.E.D.

From Theorem 5-28 it is clear that for smooth maps $f : X \to Y$, where X is a manifold with boundary, we will be interested primarily in those $y \in Y$ that are regular values for both f and ∂f. Fortunately, Sard's Theorem has a simple generalization that assures us that "most" points in Y have this property.

Theorem 5-29. Let X be a manifold with boundary, Y a boundaryless manifold and $f : X \to Y$ a smooth map. Then the set of all points in Y that are critical values for either f or ∂f has measure zero in Y.

Exercise 5-32. Prove Theorem 5-29. Hint: Note that a critical point of f must be a critical point of either ∂f or $f | \text{Int } X$.   Q.E.D.

We leave it to the reader to convince himself that the concept of nondegeneracy and all the results of Section 5-9 extend without difficulty to smooth, real-valued functions on manifolds with boundary. The only use we make of this fact is to conclude that every compact manifold with

boundary has defined on it a smooth, real-valued function with only finitely many critical points (nondegenerate critical points are isolated).

5-11   One-dimensional manifolds

In this section we prove that there are essentially only two distinct compact, connected one-manifolds with boundary and use this fact to give a
remarkably elegant proof of the No Retraction Theorem for smooth maps
we owe to Hirsch. From this it follows immediately that every smooth
map $f : B^n \to B^n$ has a fixed point. Applying the Stone–Weierstrass
Theorem, we obtain the same result for continuous maps and thus produce yet one more proof of the Brouwer Fixed Point Theorem.

Theorem 5-30. A compact, connected one-dimensional manifold with
boundary is homeomorphic either to the circle S^1 or to a closed, bounded
interval in **R**.

Remark: Our arguments can be refined to show that such a manifold is, in
fact, *diffeomorphic* either to S^1 or a closed, bounded interval in **R,** but
Theorem 5-30 will suffice for our purposes.

   The proof we give, which is essentially that of Guillemin and Pollack, is
quite straightforward, but rather more time consuming than we might expect. As such we shall leave most of the details to the reader and offer
here a broad outline of the ideas involved and a sequence of exercises in
which the gaps are filled.
   Let us then denote by X a compact, connected one-dimensional manifold with boundary. Begin by selecting a smooth function $f : X \to$ **R** with
only finitely many critical points. Let S denote the subset of X consisting
of the critical points of f and the boundary points of X. Observe that,
since ∂X is a zero-dimensional submanifold of X and X is compact, S is
finite. Consequently, $X - S$ is a disjoint union of finitely many connected, boundaryless submanifolds $L_1, \ldots , L_N$ of X (the components of
$X - S$).

Exercise 5-33. Show that f maps each L_i diffeomorphically onto a
bounded open interval in **R.** Hint: $f|L_i$ is a local diffeomorphism; $f(L_i)$ is
a bounded, connected open set in **R.** Show that $f|L_i$ is one-to-one as
follows: Fix $p \in L_i$ and set $c = f(p)$. Let Q be the set of all $q \in L_i$ for
which there is a $d \in$ **R** and a smooth curve $\gamma : [c, d] \to L_i$ (or $\gamma :
[d, c] \to L_i$) with $\gamma(c) = p$, $\gamma(d) = q$ and $f \circ \gamma = $ id. Q is nonempty, open,
and closed in L_i so $Q = L_i$.

At this point we have decomposed X into a finite number of open intervals plus the points of S. The next step is to show that the closure of each L_i in X contains precisely two points not in L_i. For this we let L denote any one of the L_i and $\overline{L}$ its closure in X. Now f maps L diffeomorphyically onto some interval (a, b) in $\mathbf{R}$; let $g : (a, b) \to L$ be the inverse diffeomorphism. Since $\overline{L}$ is compact, $\overline{L} - L \neq \varnothing$ and we may consider a $p \in \overline{L} - L$. Since X is a one-manifold with boundary, we can find a closed subset J of X containing p and a diffeomorphism $\varphi : [0, 1] \to J$ with $\varphi(1) = p$. Then $\varphi(0) = g(t)$ for some $t \in (a, b)$.

Exercise 5-34. Show that J must contain either $g((t, b))$ or $g((a, t))$ and conclude that p must be either $\lim_{s \to b} g(s)$ or $\lim_{s \to a} g(s)$. Hint: Let $S_1 = \{s \in (t, b) : g(s) \in J\}$ and $S_2 = \{s \in (a, t) : g(s) \in J\}$. At least one of these sets must be nonempty. Assume $S_1 \neq \varnothing$ and show that S_1 is open and closed in (t, b).

From Exercise 5-34 we conclude that $\overline{L} - L$ can contain at most two points, namely, $\lim_{s \to b} g(s)$ and $\lim_{s \to a} g(s)$. By using the compactness of $\overline{L}$ and the fact that $f|L$ is a local diffeomorphism, we can see that $\overline{L} - L$ contains precisely two points.

Exercise 5-35. Show that the closure of each L_i in X contains precisely two points not in L_i.

Thus, each $\overline{L}_i - L_i$ consists of precisely two points of S. Moreover, each $p \in S$ is in $\overline{L}_i - L_i$ for at least one L_i. Indeed, we have the following exercise.

Exercise 5-36. Let $p \in S$. Show that if $p \in \partial X$, then $p \in \overline{L}_i - L_i$ for precisely one L_i and if $p \in \text{Int } X$, then $p \in \overline{L}_i - L_i$ for precisely two L_i.

We will call a sequence $L_{i_1}, \ldots, L_{i_k}$ of the L_i a *chain* if each consecutive pair $\overline{L}_{i_j}$ and $\overline{L}_{i_{j+1}}$ have a common boundary point. Since there are only finitely many L_i, we can construct a chain that is maximal in the sense that it cannot be extended by appending any other L_i. By renumbering if necessary, we may let $L_1, \ldots, L_l$ denote this maximal chain. For each $j = 1, \ldots, l - 1$, let p_j denote the common boundary point of $\overline{L}_j$ and $\overline{L}_{j+1}$; let p_0 be the other boundary point of $\overline{L}_1$ and p_l the other boundary point of $\overline{L}_l$.

Exercise 5-37. Show that the maximal chain $L_1, \ldots, L_l$ contains every L_i, $i = 1, \ldots, N$. Hint: Assume $l < N$, show

$$X - \bigcup_{i=1}^{l} \overline{L}_i = \bigcup_{i=l+1}^{N} \overline{L}_i,$$

and contradict connectedness.

By Exercise 5-37 we may again renumber if necessary and assume that $L_1, \ldots, L_N$ is the maximal chain, p_i is the common boundary point of $\overline{L}_i$ and $\overline{L}_{i+1}$ for each $i = 1, \ldots, N-1$, p_0 the other boundary point of $\overline{L}_1$, and p_N the other boundary point of $\overline{L}_N$. Observe that by Exercise 5-36 p_0 and p_N are the only possible boundary points of X. If we now set $f(p_i) = a_i$ for each $i = 0, \ldots, N$, then f maps L_i diffeomorphically onto either (a_{i-1}, a_i) or (a_i, a_{i-1}), whichever interval makes sense. For each $i = 1, \ldots, N$, let $\tau_i : \mathbf{R} \to \mathbf{R}$ be an affine map that carries a_{i-1} to $i-1$ and a_i to i and define $f_i : \overline{L}_i \to [i-1, i]$ by $f_i = \tau_i \circ f$.

Suppose first that $a_0 \neq a_N$. Then $p_0 \neq p_N$ so by Exercise 5-36 ∂X consists of precisely two points. Moreover, the map $F : X \to [0, N]$ defined by $F|\overline{L}_i = f_i$ is well-defined, continuous, one-to-one, and onto $[0, N]$ and is therefore a homeomorphism since X is compact. Finally, suppose $a_0 = a_N$. Then p_0 is a boundary point of $\overline{L}_N$. Since each $\overline{L}_i$ has precisely two boundary points and no p_j is a boundary point of more than two $\overline{L}_i$, it follows that $p_0 = p_N$ (and therefore $\partial X = \varnothing$). Thus, the map $G : X \to S^1$ defined by $G|\overline{L}_i = (\cos(2\pi/N)f_i, \sin(2\pi/N)f_i)$ is a homeomorphism and the proof of Theorem 5-30 is complete.

We have found, in particular, that the boundary of a compact, connected 1-manifold is either empty or contains precisely two points. Since every compact 1-manifold with boundary is a disjoint union of finitely many connected, compact 1-manifolds with boundary, it follows that the boundary of such a manifold is either empty or contains an even number of points. This apparently inconsequential observation now leads us to a very beautiful proof of the following smooth analog of the No Retraction Theorem.

Theorem 5-31. Let X be an arbitrary compact manifold with boundary. Then there exists no smooth map $g : X \to \partial X$ with $\partial g = \mathrm{id}_{\partial X}$.

Proof: Suppose to the contrary that $g : X \to \partial X$ is smooth and $\partial g = \mathrm{id}_{\partial X}$. Since ∂X is boundaryless, Theorem 5-29 implies that there is a $z \in \partial X$ that is a regular value of both g and ∂g. By Theorem 5-28 $g^{-1}(z)$ is a one-dimensional manifold with boundary $\partial(g^{-1}(z)) = g^{-1}(z) \cap \partial X$. Since $g^{-1}(z)$ is compact, $\partial(g^{-1}(z))$ must either be empty or consist of an even number of points. However, since $\partial g = \mathrm{id}_{\partial X}$ by assumption, we have $\partial(g^{-1}(z)) = g^{-1}(z) \cap \partial X = \{z\}$, which is a contradiction. Q.E.D.

In particular, there is no smooth retraction of B^n onto S^{n-1}. Now suppose that $f : B^n \to B^n$ is a smooth map. Following the proof of Theorem 2-28, we find that if f failed to have a fixed point, we could construct a retraction $r : B^n \to S^{n-1}$ that is smooth by virtue of the smoothness of f. Thus, Theorem 5-31 yields the following "Smooth Brouwer Fixed Point Theorem": Every smooth map $f : B^n \to B^n$ has a fixed point. As promised earlier, the Stone–Weierstrass Theorem now serves as a bridge between this and the corresponding topological result. Specifically, we construct our final proof of the Brouwer Fixed Point Theorem as follows: Let $f : B^n \to B^n$ be a continuous map and assume that f has no fixed points, that is, that $\|f(x) - x\| > 0$ for each x in B^n. By the compactness of B^n there is a constant $c > 0$ such that $\|f(x) - x\| > c$ for all $x \in B^n$. By Corollary 5-4 there is a smooth map $g_1 : B^n \to \mathbf{R}^n$ with $\|f(x) - g_1(x)\| < c/4$ for each $x \in B^n$. Thus, $g_1(B^n)$ is contained in the closed ball of radius $1 + (c/4)$ about the origin in $\mathbf{R}^n$. Define a smooth map g on B^n by $g = (1 + (c/4))^{-1}g_1$. Then $g : B^n \to B^n$ and, for each $x \in B^n$,

$$
\begin{aligned}
\|f(x) - g(x)\| &= \|f(x) - g_1(x) + g_1(x) - g(x)\| \\
&\leq \|f(x) - g_1(x)\| + \|g_1(x) - g(x)\| \\
&\leq \frac{c}{4} + \frac{c/4}{1 + c/4}\|g_1(x)\| \\
&\leq \frac{c}{4} + \frac{c}{4} \\
&= c/2.
\end{aligned}
$$

Thus, for each $x \in B^n$,

$$
\begin{aligned}
\|g(x) - x\| &= \|(f(x) - x) - (f(x) - g(x))\| \\
&\geq \|f(x) - x\| - \|f(x) - g(x)\| \\
&> c - (c/2) \\
&> 0
\end{aligned}
$$

so g has no fixed points on B^n, contradicting our result on smooth maps.

5-12 Topological characterization of S^k

Our final objective is to show that the sphere is characterized topologically among all compact boundaryless manifolds by the fact that we can define on it a Morse function with precisely two critical points (see Example 5-5). Specifically, we shall prove the following theorem.

Theorem 5-32. Let X be a k-dimensional compact boundaryless manifold that admits a Morse function with precisely two critical points. Then X is homeomorphic to the k-sphere S^k.

Remark: It is *not* true that such a manifold must be diffeomorphic to S^k. Indeed, Milnor (1956) has constructed 7-manifolds that admit Morse functions with precisely two critical points, but are not diffeomorphic to S^7, although they must be homeomorphic to S^7 by Theorem 5-32.

For the proof of Theorem 5-32 we require the following lemma.

Lemma 5-33. Let f be a smooth, real-valued function on the compact boundaryless manifold X. Let a and b be real numbers with $a < b$, and suppose that $f^{-1}[a, b]$ contains no critical points of f. Then $X^a = \{x \in X : f(x) \leq a\} = f^{-1}(-\infty, a]$ and $X^b = \{x \in X : f(x) \leq b\} = f^{-1}(-\infty, b]$ are diffeomorphic.

Although not particularly difficult, the proof of Lemma 5-33 is of quite a different character than any of our previous results and will be postponed until the next section. Assuming its validity, we proceed with the proof of Theorem 5-32.

Let $f : X \to \mathbf{R}$ be a Morse function on X with precisely two critical points p and q. Since X is compact, p and q must be the two points at which f assumes its minimum and maximum values. Without loss of generality we may assume that $f(p) = 0$ is a minimum and $f(q) = 1$ is a maximum. By Theorem 5-26 and Exercise 5-28 there exist open nbds U and V of 0 in $\mathbf{R}^k$ and parametrizations $\varphi : U \to X$ and $\psi : V \to X$ with $\varphi(0) = p$ and $\psi(0) = q$ and such that

(27) $\qquad f \circ \varphi(u_1, \ldots, u_k) = u_1^2 + \cdots + u_k^2$

(28) $\qquad f \circ \psi(v_1, \ldots, v_k) = 1 - (v_1^2 + \cdots + v_k^2)$

for $(u_1, \ldots, u_k) \in U$ and $(v_1, \ldots, v_k) \in V$. Choose $\epsilon > 0$ such that $X^\epsilon = f^{-1}[0, \epsilon] \subseteq \varphi(U)$ and $f^{-1}[1 - \epsilon, 1] \subseteq \psi(V)$; see Figure 5-11.

By equation (27), $(f \circ \varphi)^{-1}[0, \epsilon] = \varphi^{-1}(f^{-1}[0, \epsilon]) = \{(u_1, \ldots, u_k) \in U : u_1^2 + \cdots + u_k^2 \leq \epsilon\}$, so $X^\epsilon = f^{-1}[0, \epsilon]$ is diffeomorphic to a closed ball of radius ϵ in $\mathbf{R}^k$. Similarly, by equation (28), $(f \circ \psi)^{-1}[1 - \epsilon, 1] = \psi^{-1}(f^{-1}[1 - \epsilon, 1]) = \{(v_1, \ldots, v_k) \in V : 1 - (v_1^2 + \cdots + v_k^2) \geq 1 - \epsilon\} = \{(v_1, \ldots, v_k) \in V : v_1^2 + \cdots + v_k^2 \leq \epsilon\}$, so $f^{-1}[1 - \epsilon, 1]$ is diffeomorphic to a closed ball in $\mathbf{R}^k$. Since any closed ϵ-ball in $\mathbf{R}^k$ is diffeomorphic to B^k, we find that both X^ϵ and $f^{-1}[1 - \epsilon, 1]$ are diffeomorphic to B^k. Now, $f^{-1}[\epsilon, 1 - \epsilon]$ contains no critical points of f, so by Lemma 5-33 X^ϵ is diffeomorphic to $X^{1-\epsilon} = f^{-1}[0, 1 - \epsilon]$ and $X^{1-\epsilon}$ is also diffeomorphic to B^k. Now, $X = X^{1-\epsilon} \cup f^{-1}[1 - \epsilon, 1]$ so we have X written as the union of two subsets, each diffeomorphic to B^k, which intersect only along their common boundary $f^{-1}(1 - \epsilon)$ that is diffeomorphic to S^{k-1}; see Figure 5-12.

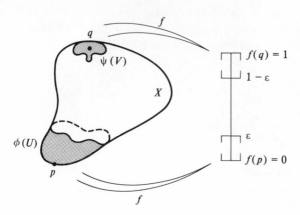

Figure 5-11

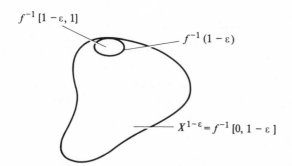

Figure 5-12

Exercise 5-38. Show that any homeomorphism h of S^{k-1} onto S^{k-1} extends to a homeomorphism H of B^k onto B^k.

Observe that each of the closed hemispheres $\{(x_1, \ldots, x_{k+1})) \in S^k : x_{k+1} \geq 0\}$ and $\{(x_1, \ldots, x_{k+1}) \in S^k : x_{k+1} \leq 0\}$ is diffeomorphic to B^k via the projection map. Thus, we may select a diffeomorphism h_1 of $X^{1-\epsilon}$ onto the lower hemisphere in S^k and a diffeomorphism h_2 of $f^{-1}[1 - \epsilon, 1]$ onto the upper hemisphere in S^k such that both h_1 and h_2 carry $f^{-1}(1 - \epsilon)$ diffeomorphically onto the equator $\{(x_1, \ldots, x_{k+1}) \in S^k : x_{k+1} = 0\}$ of S^k. In general, h_1 and h_2 will not agree on $f^{-1}(1 - \epsilon) = X^{1-\epsilon} \cap f^{-1}[1 - \epsilon, 1]$. However, the restriction of $h_2 \circ h_1^{-1}$ to the equator of S^k is a diffeomorphism of S^{k-1} onto itself and, as such, extends to a homeomorphism (*not* a diffeomorphism) H_1 of the lower hemisphere in S^k onto itself. (Extend to B^k by Exercise 5-38 and then compose with projections.) Define H : $X \to S^k$ by

$$H(x) = h_2(x) \qquad \text{if } x \in f^{-1}[1 - \epsilon, 1]$$
$$\quad\;\; = H_1(h_1(x)) \qquad \text{if } x \in X^{1-\epsilon}$$

and observe that, if $x \in f^{-1}(1 - \epsilon)$, then $H_1(h_1(x)) = (h_2 \circ h_1^{-1})(h_1(x)) = h_2(x)$ so H is well-defined. Since H is one-to-one, continuous, and onto S^k, it is a homeomorphism by compactness and the proof of Theorem 5-32 is complete (modulo Lemma 5-33).

5-13   Smooth tangent vector fields

All that remains then is the proof of Lemma 5-33. For this we shall require a basic result from the theory of ordinary differential equations on the relationship between smooth tangent vector fields and one-parameter groups of diffeomorphisms on a compact manifold. (The reader is advised to reexamine Section 5-1, where these concepts are discussed in intuitive terms.) We begin with a definition. Let X be a k-dimensional manifold in $\mathbf{R}^n$. A *smooth tangent vector field* on X is a smooth map $v : X \to \mathbf{R}^n$ such that $v(x) \in T_x(X)$ for each $x \in X$; we shall often refer to such a v simply as a "vector field" on X.

Example 5-6. Consider first an open subset U of $\mathbf{R}^k$ and a smooth map $f : U \to \mathbf{R}$. The map grad $f : U \to \mathbf{R}^k$ defined by grad $f(x) = ((\partial f/\partial x_1)(x), \ldots, (\partial f/\partial x_k)(x))$ for each $x \in U$ is smooth and, since $T_x(U) = \mathbf{R}^k$ for each $x \in U$, is a smooth tangent vector field on U. Now, observe that, by computing the Jacobian of f at x and performing a matrix multiplication, we find that for each $w = (w_1, \ldots, w_k)$ in $\mathbf{R}^k = T_x(U)$,

$$df_x(w) = \frac{\partial f}{\partial x_1}(x) \cdot w_1 + \cdots + \frac{\partial f}{\partial x_k}(x) \cdot w_k = (\text{grad } f(x)) \cdot w,$$

where $(\text{grad } f(x)) \cdot w$ is the usual dot product of the vectors grad $f(x)$ and w in $\mathbf{R}^k$.

Now let X be an arbitrary k-dimensional manifold in $\mathbf{R}^n$ and $f : X \to \mathbf{R}$ a smooth function. For each $x \in X$, $df_x : T_x(X) \to \mathbf{R}$ so df_x is a linear functional on the k-dimensional vector space $T_x(X)$. Since any linear functional on a linear subspace V of $\mathbf{R}^n$ can be represented as the dot product with some unique fixed element of V (see Halmos, theorem, Section 67), we conclude that there exists a unique element v of $T_x(X)$ such that $df_x(w) = v \cdot w$ for each $w \in T_x(X)$. We denote this vector v, which obviously depends on x, by grad $f(x)$ and call it the *gradient* of f at x. Thus, we define a map

$$\text{grad } f : X \to \mathbf{R}^n$$

by the requirement that for each $x \in X$ grad $f(x)$ is the unique element of $T_x(X)$ for which $df_x(w) = (\text{grad } f(x)) \cdot w$ for each $w \in T_x(X)$.

Exercise 5-39. Show that grad $f : X \to \mathbf{R}^n$ is a smooth tangent vector field on X. Hint: Extend f and use the uniqueness of grad $f(x)$.

Now let X be a k-dimensional manifold in $\mathbf{R}^n$, $p \in X$ and v a smooth tangent vector field defined on some nbd of p in X. We are interested in knowing whether or not there exists a smooth curve $\rho : (-\epsilon, \epsilon) \to X$ in X that passes through p at $t = 0$ and has velocity vectors that coincide with the values of v along the curve (see Section 5-1):

$$(29) \qquad \begin{aligned} &\rho(0) = p \\ &\frac{d\rho}{dt}(t) = v(\rho(t)) \qquad \text{for } t \in (-\epsilon, \epsilon). \end{aligned}$$

In order to translate the question into more familiar terms, we choose an open nbd U of 0 in $\mathbf{R}^k$ and a parametrization $\varphi : U \to X$ of a nbd $\varphi(U)$ of p in X that we may assume is contained in the domain of v and satisfies $\varphi(0) = p$. For each $u \in U$, $d\varphi_u$ is an isomorphism of $\mathbf{R}^k$ onto $T_x(X)$. We define a smooth vector field $\varphi_* v$ on U by

$$(30) \qquad (\varphi_* v)(u) = (d\varphi_u)^{-1}(v(\varphi(u))) \qquad \text{for each } u \in U.$$

Let $f_1, \ldots, f_k$ be the coordinate functions for the map $\varphi_* v : U \to \mathbf{R}^k$, that is, $f_i(u) = ((\varphi_* v)(u))_i$ for each $i = 1, \ldots, k$. Now suppose that a smooth curve ρ in X that satisfies (29) does exist. Define a curve $c : (-\epsilon, \epsilon) \to U$ in U by $c = \varphi^{-1} \circ \rho$. Then $c(0) = \varphi^{-1}(\rho(0)) = \varphi^{-1}(p) = 0$. Moreover, for each $t \in (-\epsilon, \epsilon)$, we have

$$\begin{aligned} \frac{dc}{dt}(t) &= dc_t(1) = d(\varphi^{-1} \circ \rho)_t(1) \\ &= d(\varphi^{-1})_{\rho(t)} \circ d\rho_t(1) \\ &= (d\varphi_{\varphi^{-1}(\rho(t))})^{-1}(d\rho_t(1)) \\ &= (d\varphi_{\varphi^{-1}(\rho(t))})^{-1}(v(\rho(t))) \\ &= (\varphi_* v)(\varphi^{-1}(\rho(t))) \\ &= (\varphi_* v)(c(t)). \end{aligned}$$

Thus, c satisfies

$$(31) \qquad \begin{aligned} &c(0) = 0 \\ &\frac{dc}{dt}(t) = (\varphi_* v)(c(t)) \qquad \text{for } t \in (-\epsilon, \epsilon). \end{aligned}$$

Exercise 5-40. Show that, conversely, if $c : (-\epsilon, \epsilon) \to U$ satisfies (31), then $\rho : (-\epsilon, \epsilon) \to X$ defined by $\rho = \varphi \circ c$ satisfies (29).

Thus, we find that the initial value problems (29) and (31) are essentially equivalent in the sense that a solution to either immediately determines a solution to the other. Moreover, if we write $c(t) = (c_1(t), \ldots, c_k(t))$ and recall that $f_1, \ldots, f_k$ denote the coordinate functions of $f = \varphi_* v : U \to \mathbf{R}^k$, then (31) assumes the following more familiar form:

(32)
$$\begin{aligned} c(0) &= 0 \\ c'(t) &= f(c(t)) \qquad \text{for } t \in (-\epsilon, \epsilon) \end{aligned}$$

or, in more detail,

(33)
$$\begin{aligned} c_i(0) &= 0 \\ c_i'(t) &= f_i(c_1(t), \ldots, c_k(t)) \qquad \text{for } t \in (-\epsilon, \epsilon). \end{aligned}$$

Now that we have locally translated problem (29) into an initial value problem for a system of ordinary differential equations, we are in a position to make use of the standard existence and uniqueness theory for such equations. The following result will suffice for our purposes:

Theorem 5-34. Let U be an open subset of $\mathbf{R}^k$ and $f = (f_1, \ldots, f_k) : U \to \mathbf{R}^k$ a smooth map. Let x_0 be a point of U and $a > 0$ a constant such that the closed ball $B_{2a}(x_0)$ of radius $2a$ about x_0 is contained in U. Finally, let L and K be positive constants such that $\|f(x)\| \leq L$ for all $x \in B_{2a}(x_0)$ and $\|f(x) - f(y)\| \leq K\|x - y\|$ for all x and y in $B_{2a}(x_0)$, and select $b > 0$ satisfying $b \leq a/2$ and $b < 1/K$. Then, for each $x \in B_a(x_0)$, there exists a unique smooth map $\alpha_x : (-b, b) \to U$ that satisfies

$$\begin{aligned} \alpha_x(0) &= x \\ \alpha_x'(t) &= f(\alpha_x(t)) \qquad \text{for } t \in (-b, b). \end{aligned}$$

Moreover, the solutions $\alpha_x(t)$ depend smoothly on the initial conditions in the sense that the map $\alpha : (-b, b) \times (\text{Int } B_a(x_0)) \to U$ defined by $\alpha(t, x) = \alpha_x(t)$ is smooth.

Remark: The existence of the constants K and L follows immediately from the smoothness of f (see equation (17) for K).

We shall assume that the reader has had some exposure to results of this genre and will therefore not include a proof. (See, for example, Coddington and Levinson; or Hurewicz.)

Exercise 5-41. Let $U \subseteq \mathbf{R}^k$ be open and $f : U \to \mathbf{R}^k$ smooth. Let $x \in U$ and suppose that $\alpha_1(t)$ and $\alpha_2(t)$ are two smooth maps, both defined on some interval $[a, b]$ about 0 with $\alpha_i([a, b]) \subseteq U$, $i = 1, 2$. Show that if

$$\begin{aligned} \alpha_i(0) &= x \\ \alpha_i'(t) &= f(\alpha_i(t)) \qquad \text{for } t \in [a, b] \end{aligned}$$

for $i = 1$, 2, then α_1 and α_2 agree on all of $[a, b]$. Hint: Let $t_0 = \sup \{t \in [a, b] : \alpha_1(t) = \alpha_2(t)\}$ and suppose $t_0 < b$. Consider the functions $\beta_i(t) = \alpha_i(t + t_0)$ and obtain a contradiction. Similarly, $\inf\{t \in [a, b] : \alpha_1(t) = \alpha_2(t)\} = a$.

In the situation described in Theorem 5-34, we define for each $t \in (-b, b)$ a map $\alpha^t : U_a(x_0) \to U$ by $\alpha^t(x) = \alpha_x(t) = \alpha(t, x)$ for each $x \in U_a(x_0)$. Then each α^t is smooth, but we shall find that a great deal more can be said about these maps. Since the map $\alpha : (-b, b) \times U_a(x_0) \to U$ satisfies $\alpha(0, x) = x$ for each $x \in U_a(x_0)$, α maps the compact set $\{0\} \times B_{a/2}(x_0)$ into $B_{a/2}(x_0) \subseteq U_a(x_0)$. Thus, there is an $\epsilon > 0$ such that α carries $(-\epsilon, \epsilon) \times U_{a/2}(x_0)$ into $U_a(x_0)$. In particular, if $|s| < \epsilon$ and $x \in U_{a/2}(x_0)$, then $\alpha(s, x) \in U_a(x_0)$, and we may define a smooth map γ on $(-\epsilon, \epsilon)$ by $\gamma(t) = \alpha(t, \alpha(s, x)) = \alpha_{\alpha(s,x)}(t)$. Note that by definition γ satisfies

$$\gamma(0) = \alpha(s, x)$$
$$\gamma'(t) = f(\gamma(t)).$$

On the other hand, if $|t + s| < \epsilon$, the map $\beta(t) = \alpha(t + s, x) = \alpha_x(t + s)$ also satisfies

$$\beta(0) = \alpha(s, x)$$
$$\beta'(t) = f(\beta(t)).$$

so, by Exercise 5-41, if $|s|$, $|t|$, and $|t + s|$ are less than ϵ and $x \in U_{a/2}(x_0)$, then $\gamma(t) = \beta(t)$, that is,

(34) $\alpha(t, \alpha(s, x)) = \alpha(t + s, x)$.

Now, $\alpha^{t+s}(x) = \alpha(t + s, x)$ and, assuming that $\alpha^t(x) \in U_{a/2}(x_0)$ also, we have $\alpha^t \circ \alpha^s(x) = \alpha^t(\alpha^s(x)) = \alpha^t(\alpha(s, x)) = \alpha(t, \alpha(s, x))$. We find then that

(35) $\alpha^t \circ \alpha^s(x) = \alpha^{t+s}(x)$.

Thus, on some sufficiently small nbds of 0 in $\mathbf{R}$ and x_0 in U,

(36) $\alpha^t \circ \alpha^s = \alpha^{t+s}$.

Exercise 5-42. Conclude from equation (36) that, for $|t| < \epsilon$, α^t is a diffeomorphism of some nbd of x_0 in U onto another nbd of x_0 in U.

Combining these observations with Theorem 5-34 and transferring the result back to the manifold setting via the "equivalence" of problems (29) and (31) we obtain the following theorem.

Theorem 5-35. Let X be a k-dimensional differentiable manifold, p a point in X, and v a smooth tangent vector field defined on some nbd of p in X. Then there exists a nbd V of p in X, an $\epsilon > 0$ and a unique smooth map $\varphi:(-\epsilon, \epsilon) \times V \to X$ with the following properties:

(a) For each $x \in V$ the map $\varphi_x : (-\epsilon, \epsilon) \to X$ defined by $\varphi_x(t) = \varphi(t, x)$ is a smooth curve in X that satisfies

$$\varphi_x(0) = x$$

$$\frac{d\varphi_x}{dt}(t) = v(\varphi_x(t)) \qquad \text{for } t \in (-\epsilon, \epsilon).$$

(b) If $|s|$, $|t|$, and $|t + s|$ are less than ϵ and x and $\varphi(t, x)$ are in V, then the maps $\varphi^t : V \to X$ defined by $\varphi^t(x) = \varphi(t, x)$ satisfy

$$\varphi^t \circ \varphi^s(x) = \varphi^{t+s}(x).$$

Now, even if the vector field v is defined on all of X we cannot, in general, expect the map φ of Theorem 5-35 to be defined either for all $t \in \mathbf{R}$ or on all of X. However, if X is compact, this can indeed be achieved:

Theorem 5-36. Let X be a compact, k-dimensional differentiable manifold and v a smooth tangent vector field on X. Then there exists a unique smooth map $g : \mathbf{R} \times X \to X$ that satisfies:

(a) For each $x \in X$ the map $g_x : \mathbf{R} \to X$ defined by $g_x(t) = g(t, x)$ is a smooth curve in X that satisfies

$$g_x(0) = x$$

$$\frac{dg_x}{dt}(t) = v(g_x(t)) \qquad \text{for } t \in \mathbf{R}.$$

(b) For each $t \in \mathbf{R}$ the map $g^t : X \to X$ defined by $g^t(x) = g(t, x)$ is a diffeomorphism of X onto itself. Moreover, these maps satisfy

$$g^0 = \mathrm{id}_X$$
$$g^t \circ g^s = g^{t+s} \qquad \text{for all } t, s \in \mathbf{R}.$$

Proof: Since X is compact, we may by Theorem 5-35 select finitely many open sets $V_1, \ldots, V_n$, positive real numbers $\epsilon_1, \ldots, \epsilon_n$, and smooth maps $\varphi_i : (-\epsilon_i, \epsilon_i) \times V_i \to X$ that satisfy (a) and (b) of that theorem and such that $X = \bigcup_{i=1}^{n} V_i$. Let $\epsilon = \min \{\epsilon_1, \ldots, \epsilon_n\}$ and note that if $x \in V_i \cap V_j$, then by uniqueness $\varphi_i(t, x) = \varphi_j(t, x)$ for all t with $|t| < \epsilon$. Thus, we may define $\varphi : (-\epsilon, \epsilon) \times X \to X$ as follows: for each $(t, x) \in (-\epsilon, \epsilon) \times X$, select an i such that $(t, x) \in (-\epsilon, \epsilon) \times V_i$ and set $\varphi(t, x) = \varphi_i(t, x)$. Thus defined, φ is smooth on $(-\epsilon, \epsilon) \times X$ since it is smooth on each element of the cover $(-\epsilon, \epsilon) \times V_1, \ldots, (-\epsilon, \epsilon) \times V_n$. For each t with $|t| < \epsilon$ the maps $\varphi^t : X \to X$ defined by $\varphi^t(x) = \varphi(t, x)$ are diffeomorphisms and satisfy $\varphi^t \circ \varphi^s = \varphi^{t+s}$ as long as $|t|$, $|s|$, and $|t + s|$ are less than ϵ. Defining $\varphi_x : (-\epsilon, \epsilon) \to X$ by $\varphi_x(t) = \varphi(t, x)$, we find that $\frac{d\varphi_x}{dt}(0) = v(x)$. To obtain g, we need only extend φ to $\mathbf{R} \times X$. For this we let $t \in \mathbf{R}$ be arbitrary and write $t = k(\epsilon/2) + r$, where k is an integer and $|r| < \epsilon/2$.

Now, for each $x \in X$, we define

$$g(t, x) = \begin{cases} \varphi^{\epsilon/2} \circ \cdots \circ \varphi^{\epsilon/2} \circ \varphi^r & \text{if } k \geq 0 \\ \varphi^{-\epsilon/2} \circ \cdots \circ \varphi^{-\epsilon/2} \circ \varphi^r & \text{if } k < 0. \end{cases}$$

Since $\varphi^{\epsilon/2}$, $\varphi^{-\epsilon/2}$, and φ^r are diffeomorphisms, each g^t is a diffeomorphism and, in particular, g depends smoothly on x.

Exercise 5-43. Show that g depends smoothly on t.

For $|t| < \epsilon/2$, $g^t = \varphi^t$ so, in particular, $g^0 = \varphi^0 = \text{id}_X$ and $g_x(0) = x$ for each $x \in X$. Moreover, it follows that

$$(37) \qquad \frac{dg_x}{dt}(0) = v(x) \qquad \text{for each } x \in X.$$

Exercise 5-44. Show that $g^t \circ g^s = g^{t+s}$ for all t and s in $\mathbf{R}$.

All that remains is to prove that $(dg_x/dt)(t) = v(g_x(t))$ for each $t \in \mathbf{R}$, and we claim that this follows immediately from equation (37) and Exercise 5-44; see Exercise 5-45.   Q.E.D.

A collection $\{g^t : t \in \mathbf{R}\}$ of diffeomorphisms of a manifold X onto itself is called a *one-parameter group of diffeomorphisms* of X if $g^0 = \text{id}_X$, $g^t \circ g^s = g^{t+s}$ for all $t, s \in \mathbf{R}$ and if the map $g : \mathbf{R} \times X \to X$ defined by $g(t, x) = g^t(x)$ is smooth. A smooth vector field v on X is said to *generate* $\{g^t : t \in \mathbf{R}\}$ if $v(x) = (dg_x/dt)(0)$ for each $x \in X$, where $g_x : \mathbf{R} \to X$ is defined by $g_x(t) = g^t(x) = g(t, x)$ for all $t \in \mathbf{R}$.

Exercise 5-45. Show that if v generates $\{g^t : t \in \mathbf{R}\}$, then $(dg_x/dt)(t) = v(g_x(t))$ for each $t \in \mathbf{R}$.

The content of Theorem 5-36 is that every smooth vector field on a compact manifold X generates a unique one-parameter group of diffeomorphisms of X. With this we are finally in a position to prove Lemma 5-33, restated below for reference.

(Lemma 5-33.) Let f be a smooth, real-valued function on the compact, boundaryless manifold X. Let a and b be real numbers with $a < b$ and suppose that $f^{-1}[a, b]$ contains no critical points of f. Then $X^a = \{x \in X : f(x) \leq a\} = f^{-1}(-\infty, a]$ and $X^b = \{x \in X : f(x) \leq b\} = f^{-1}(-\infty, b]$ are diffeomorphic.

Proof: grad f is a smooth vector field on X defined by the requirement that for each $w \in T_x(X)$, $df_x(w) = (\text{grad } f(x)) \cdot w$ (see Example 5-6).

Observe that if $c : \mathbf{R} \to X$ is any smooth curve in X, then, letting $x = c(t)$, we find by equation (16) that

(38) $\dfrac{d(f \circ c)}{dt} (t) = \operatorname{grad} f(x) \cdot \dfrac{dc}{dt} (t).$

Note that $f(X)$ cannot be contained in $[a, b]$ since f must have critical points. Thus, $f^{-1}[a, b]$ is a proper compact subset of X on which $\operatorname{grad} f$ does not vanish. Therefore, we can construct a smooth function $\rho :$ $X \to \mathbf{R}$ that satisfies

(39) $\rho(x) = \dfrac{1}{\operatorname{grad} f(x) \cdot \operatorname{grad} f(x)}$

on $f^{-1}[a, b]$ and is zero outside a compact nbd of $f^{-1}[a, b]$. Define a smooth vector field v on X by

(40) $v(x) = \rho(x) \operatorname{grad} f(x)$

for each $x \in X$. There exists a unique one-parameter group $\{g^t : t \in \mathbf{R}\}$ of diffeomorphisms of X that generates v. Thus by Exercise 5-45

(41) $\dfrac{dg_x}{dt} (t) = v(g_x(t))$

for each $x \in X$ and $t \in \mathbf{R}$. Fix an $x \in X$. Since $g_x : \mathbf{R} \to X$ is a smooth curve in X, equations (38) and (41) imply

$\dfrac{d(f \circ g_x)}{dt} (t) = \operatorname{grad} f(g_x(t)) \cdot \dfrac{dg_x}{dt} (t)$

$\qquad\qquad = \operatorname{grad} f(g_x(t)) \cdot v(g_x(t))$

so, if $g_x(t)$ is in $f^{-1}[a, b]$, we have by equations (39) and (40),

$\dfrac{d(f \circ g_x)}{dt} (t) = 1.$

We find then that the map $t \to f(g_x(t)) = f(g^t(x)) : \mathbf{R} \to \mathbf{R}$ is linear with derivative 1 as long as $f(g^t(x)) \in [a, b]$.

Exercise 5-46. Complete the proof by showing that the diffeomorphism g^{b-a} carries X^a onto X^b.   Q.E.D.

Supplementary exercises

5-47   Construct a geometrical picture of the vector field determined by the nonlinear system (2) analogous to that shown in Figure 5-1 for the linear system (3). Determine all zeros of this vector field. Sketch the graphs of several of the solution curves g_x and interpret the qualitative behavior of each physically.

5-48 Show that every open ball in $\mathbf{R}^k$ is diffeomorphic to $\mathbf{R}^k$.

5-49 Let X and Y be subsets of $\mathbf{R}^k$ and $\mathbf{R}^l$, respectively, and $f : X \to Y$ a smooth map. The graph of f is the subset of $X \times Y$ defined by graph $(f) = \{(x, f(x)) : x \in X\}$. Show that graph (f) is diffeomorphic to X.

5-50 Let $f : \mathbf{R} \to \mathbf{R}$ be a local diffeomorphism. Show that f maps $\mathbf{R}$ diffeomorphically onto an open interval in $\mathbf{R}$.

5-51 Let (x_0, y_0, z_0) be an arbitrary point on the sphere S^2. Determine a basis for the tangent space $T_{(x_0,y_0,z_0)}(S^2)$.

5-52 Let X and Y be boundaryless manifolds and $(x, y) \in X \times Y$. Show that $T_{(x,y)}(X \times Y) = T_x(X) \times T_y(Y)$.

5-53 Let $f : X_1 \to X_2$ and $g : Y_1 \to Y_2$ be smooth maps between boundaryless manifolds and $(x, y) \in X \times Y$. Show that $d(f \times g)_{(x,y)} = df_x \times dg_y$.

5-54 Let $f : X \to Y$ be a smooth map between boundaryless manifolds with no critical points. Show that f is an open map.

5-55 Let X be a k-dimensional boundaryless manifold and p a point in X. Denote by $C^\infty(p)$ the algebra of all real-valued functions that are defined and smooth on some nbd of p in X. A *derivation* at p is a map $T : C^\infty(p) \to \mathbf{R}$ such that

(i) T is linear, that is, $T(\alpha f + \beta g) = \alpha T(f) + \beta T(g)$ for $\alpha, \beta \in \mathbf{R}$ and $f, g \in C^\infty(p)$.

(ii) T satisfies the product rule, that is, $T(fg) = T(f)g(p) + T(g)f(p)$.

(a) Let $X = \mathbf{R}^k$ and $p \in \mathbf{R}^k$. For each $v \in \mathbf{R}^k$, define $T_v :$ $C^\infty(p) \to \mathbf{R}$ by $T_v(f) = v \cdot \text{grad } f(p)$. Show that T_v is a derivation at p.

(b) Show that if $f \in C^\infty(p)$ is constant on a nbd of p in X, then $T(f) = 0$ for every derivation T at p.

(c) Let $f, g \in C^\infty(p)$ and suppose that f and g agree on some nbd of p in X. Show that $T(f) = T(g)$ for every derivation T at p.

(d) Let c be a smooth curve in X, t_0 a point in its domain and $c(t_0) = p \in X$. Define $T_c : C^\infty(p) \to \mathbf{R}$ by

$$T_c(f) = \frac{d(f \circ c)}{dt}(t_0).$$

Show that T_c is a derivation at p. Conclude that every tangent vector at p gives rise to a derivation at p.

5-56 Let X be a k-dimensional boundaryless manifold in $\mathbf{R}^n$. The *tangent bundle* $T(X)$ of X is the subset of $X \times \mathbf{R}^n$ defined by $T(X) = \{(x, v) \in X \times \mathbf{R}^n : v \in T_x(X)\}$. Let $Y \subseteq \mathbf{R}^m$ be another manifold and $f : X \to Y$ a smooth map. Define $df : T(X) \to T(Y)$ by $df(x, v) = df_x(v)$.

(a) Show that, as a map of $X \times \mathbf{R}^n \subseteq \mathbf{R}^{2n}$ into $Y \times \mathbf{R}^m \subseteq \mathbf{R}^{2m}$, df is smooth.

(b) Show that if f is a diffeomorphism, then so is df.

(c) Show that $T(X)$ is a $2k$-dimensional smooth manifold in $\mathbf{R}^{2n}$.

5-57   Show that the product of two manifolds with boundary need not be a manifold with boundary.

5-58   Let X be a k-dimensional boundaryless manifold and Y an l-dimensional manifold with boundary. Prove that $X \times Y$ is a $(k + l)$-dimensional manifold with boundary $\partial(X \times Y) = X \times \partial Y$.

Guide to further study

Those who are now prepared to embark on a deeper study of topology and its applications will eventually have to assimilate the basic facts of what is known as *general* or *point-set* topology. However, it is possible to postpone entering immediately into this subject. For example, Part I of Hilton and Wylie's classic text on homology theory (see Bibliography) contains a great deal of very fundamental material on algebraic topology that has not been discussed here and most of which is accessible to those who have followed the present text through Chapter 4. Guillemin and Pollack (1974) and Milnor (1956) will take the reader much further into differential topology than Chapter 5 and yet require no additional background.

Alternatively, you might choose to learn just enough point-set theory to advance quickly in some particular area of interest. The superb treatment of classical dimension theory by Hurewicz and Wallman (1948), for instance, requires only a modest familiarity with metric space theory, which might be acquired, for example, from Chapter 2 of Simmons. Each of the texts by Chillingworth (1976), Lefschetz (1949), Simmons (1963), Singer and Thorpe (1967), and Wallace (1968) develops all of the general topology required to cover its topics and each proceeds in a different direction. Chillingworth emphasizes the differential topology required in the global theory of dynamical systems, a subject that has seen some rather beautiful applications recently. The differential topology in Wallace, on the other hand, revolves around the theory of singularities and "surgery" on manifolds and includes a proof of the classical theorem on the classification of two-dimensional manifolds. Lefschetz covers traditional topics in algebraic topology and has a particularly interesting Chapter 6 on manifolds and duality theory. The books by Simmons and Singer and Thorpe have a somewhat different flavor, but are most highly recommended. Simmons' text is beautifully written, applies the general topology developed in its earlier chapters to the study of Banach algebras, and includes proofs of the famous theorems of Gelfand–Neumark and Banach–Stone. Singer and Thorpe proceed rather quickly from elementary point-set topology to a number of rather deep theorems relating the geometry to the algebraic topology of a manifold (the De Rham theory).

220

Should you choose to proceed directly to a detailed study of general topology, there are literally dozens of texts available. The classic work of Kelley (1955) is, of course, an excellent choice, particularly notable for its exceptional Problems. Among the more recent treatments, that of Willard (1970) is perhaps the best and is most strongly recommended. Hocking and Young (1961) cover the basic material, but tend to emphasize those topics of interest in algebraic topology; indeed, the later chapters of the book contain a rather detailed but elementary treatment of homology theory. Dugundji (1970), on the other hand, covers more general topology and includes a very nice treatment of homotopy theory and an impressive collection of classical results on the topology of Euclidean spaces.

A great many doors open to those who have acquired a reasonable knowledge of point-set topology. One can, for example, pursue homotopy theory in the standard work of Hu (1959) or in Massey's (1967) really excellent book. For homology theory, Hilton and Wylie (1960) is favored, but a more concise introduction is available in Mayer (1972). Hirsch (1976) provides a very broad introduction to the methods of differential topology, while the more specialized book by Milnor (1963) gives excellent coverage of the fundamentals of Morse theory. Munkres' (1966) book deals with some of the very important and rather difficult work of Kervaire, Milnor, and Whitney.

Bibliography

Apostol, T., *Mathematical Analysis,* Addison-Wesley, Reading, Mass., 1957.

Birkhoff, G., and S. MacLane, *A Survey of Modern Algebra,* Macmillan, New York, 1971.

Buck, R. C., *Advanced Calculus,* McGraw-Hill, New York, 1965.

Chillingworth, D. R. J., *Differential Topology with a View to Applications,* Pitman, London, 1976.

Coddington, E., and N. Levinson, *Theory of Ordinary Differential Equations,* McGraw-Hill, New York, 1955.

Dugundji, J., *Topology,* Allyn and Bacon, Boston, 1970.

Gans, D., *Transformations and Geometry,* Appleton-Century-Crofts, Meredith, New York, 1969.

Guillemin, V., and A. Pollack, *Differential Topology,* Prentice-Hall, Englewood Cliffs, N.J., 1974.

Halmos, P., *Finite Dimensional Vector Spaces,* Van Nostrand, New York, 1958.

Herstein, I. N., *Topics in Algebra,* Blaisdell, Waltham, Mass., 1964.

Hilton, P. J., and S. Wylie, *Homology Theory,* Cambridge University Press, New York, 1960.

Hirsch, M. W., *Differential Topology,* Springer-Verlag, New York, 1976.

Hocking, J. G., and G. S. Young, *Topology,* Addison-Wesley, Reading, Mass., 1961.

Hu, Sze-tsen, *Homotopy Theory,* Academic Press, New York, 1959.

Hurewicz, W., *Lectures on Ordinary Differential Equations,* MIT Press, Cambridge, Mass., 1958.

and H. Wallman, *Dimension Theory,* Princeton University Press, Princeton, N.J., 1948.

Kelley, J. L., *General Topology,* Van Nostrand, New York, 1955.

Lefschetz, S., *Introduction to Topology,* Princeton University Press, Princeton, N.J., 1949.

Macdonald, I. D., *The Theory of Groups,* Oxford University Press, New York, 1968.

Massey, W. S., *Algebraic Topology: An Introduction,* Harcourt, Brace and World, New York, 1967.

Mayer, J., *Algebraic Topology,* Prentice-Hall, Englewood Cliffs, N.J., 1972.

Milnor, J., "On Manifolds Homeomorphic to the 7-Sphere," *Ann. Math., 64* (1956), 399–405.

Morse Theory, Annals of Mathematics Studies *51,* Princeton University Press, Princeton, N.J., 1963.

Topology from a Differentiable Viewpoint, University of Virginia Press, The University Press of Virginia, Charlottesville, Va., 1965.

Munkres, J. R., *Elementary Differential Topology*, Annals of Mathematics Studies *54*, Princeton University Press, Princeton, N.J., 1961 (revised 1966).

Pontryagin, L. S., *Foundations of Combinatorial Topology*, Graylock Press, Rochester, N.Y., 1952.

Simmons, G. F., *Introduction to Topology and Modern Analysis*, McGraw-Hill, New York, 1963.

Singer, I. M., and J. A. Thorpe, *Lecture Notes on Elementary Topology and Geometry*, Scott, Foresman, Glenview, Ill., 1967.

Spivak, M., *Calculus on Manifolds*, Benjamin, New York, 1965.

Wallace, A. H., *Differential Topology: First Steps*, Benjamin, New York, 1968.

Willard, S., *General Topology*, Addison-Wesley, Reading, Mass., 1970.

Symbols and notation

$\oplus$	Algebraic direct sum
$A_1 \times \cdots \times A_n$	Cartesian product of $A_1, \ldots, A_n$
$\mathbf{R}^n$	Euclidean n-space
$y_0 + V$	$\{y_0 + v : v \in V\}$
$x \cdot y$	Inner product of $x, y \in \mathbf{R}^n$
$\|x\|$	$(x \cdot x)^{1/2}$
$d(x, y)$	$\|y - x\|$
$H(A)$	Convex hull of $A \subseteq \mathbf{R}^n$
I	Closed unit interval $[0, 1]$
$U_r(x_0)$	$\{x \in \mathbf{R}^n : d(x, x_0) < r\}$
$B_r(x_0)$	$\{x \in \mathbf{R}^n : d(x, x_0) \leqslant r\}$
B^n	$B_1(0)$
S^{n-1}	$\{x \in \mathbf{R}^n : \|x\| = 1\}$
$\varnothing$	Empty set
iff	If and only if
nbd	Neighborhood
$\text{int}_X A$	Interior of A in X
$\text{bdy}_X A$	Boundary of A in X
$\text{cl}_X A$	Closure of A in X
$\overline{A}$	Closure of A
$f|A$	Restriction of f to A
π_{X_i}	Projection of $X_1 \times \cdots \times X_n$ onto X_i
$i: A \hookrightarrow X$	Inclusion map of $A \subseteq X$ into X
C_x	Connected component of $\{x\}$
$C(X, Y)$	Set of all continuous maps from X to Y
$C(X)$	$C(X, \mathbf{R})$
$\text{dist}(A, B)$	Distance between $A \subseteq \mathbf{R}^n$ and $B \subseteq \mathbf{R}^n$
$\text{diam } A$	Diameter of $A \subseteq \mathbf{R}^n$
id_X	Identity map on X
$\text{ord } \mathscr{C}$	Order of the cover $\mathscr{C}$
$\dim X$	Topological dimension of the space X
$X \cong Y$	X is homeomorphic to Y
H^p	p-dimensional hyperplane in $\mathbf{R}^n$

$\pi(s)$	Unique p-dimensional hyperplane in $\mathbf{R}^n$ containing $s = \{a_0, a_1, \ldots, a_p\}$
$\lambda_i(x)$	Barycentric coordinates of x
s_p	p-dimensional simplex
$(a_0, a_1, \ldots, a_p)$	p-dimensional simplex spanned by $\{a_0, a_1, \ldots, a_p\}$
$\bar{s}_p$	Closed p-dimensional simplex
$[a_0, a_1, \ldots, a_p]$	Closed p-dimensional simplex spanned by $\{a_0, a_1, \ldots, a_p\}$
$\dot{s}_p$	$\bar{s}_p - s_p$
$s_k \leq s_p$	s_k is a face of s_p
$s_k < s_p$	s_k is a proper face of s_p
s_{p-1}^i	$(p-1)$-face of s_p opposite the vertex a_i
$|K|$	Polyhedron of the geometric complex K
$K(s_p)$	Standard p-dimensional complex
K_r	r-skeleton of the geometric complex K
(K, h)	Triangulation of a topological polyhedron
$v * B$	Cone with vertex v and base B
$\mathrm{st}(s_p)$	Star of the simplex s_p (in a given complex)
$b(s_p)$	Barycenter of s_p
K'	Barycentric subdivision of K
$K^{(n)}$	nth barycentric subdivision of K
$\mu(K)$	Mesh of the geometric complex K
$d(g, h)$	Distance between the maps g and h
ω	Sperner map
$\omega^{(n)}$	Generalized Sperner map
$\dim K$	Algebraic dimension of the geometric complex K
$f_0 \simeq f_1$	f_0 is homotopic to f_1
$F : f_0 \simeq f_1$	F is a homotopy from f_0 to f_1
$X \simeq Y$	X is homotopically equivalent to Y
(X, A)	Topological pair
(X, x_0)	Pointed space
$f_0 \simeq f_1 \text{ rel } A$	f_0 is homotopic to f_1 relative to A
$f_0 \simeq f_1[A]$	f_0 is homotopic to f_1 relative to A
$\alpha \overset{\mathrm{p}}{\simeq} \beta$	α is path homotopic to β
$\alpha\beta$	Product of the paths α and β
α^{-1}	Inverse of the path α
$[\alpha]$	$\overset{\mathrm{p}}{\simeq}$ equivalence class containing α
$\pi_1(X, x_0)$	Fundamental group of X at x_0
$\pi_n(X, x_0)$	nth homotopy group of X at x_0
f_*	Map induced functorially in an algebraic category by a map f
$(\tilde{X}, p)$	Covering space for X

σ_p	Oriented p-dimensional simplex
$a_0 a_1 \ldots a_p$	Oriented p-dimensional simplex spanned by $a_0, a_1, \ldots, a_p$
Z_γ	Group of integers modulo γ
$C_p(K)$	p-chain group of the complex K
∂_p	Boundary operator
$B_p(K)$	All p-boundaries of the complex K
$Z_p(K)$	All p-cycles of the complex K
$H_p(K)$	pth simplicial homology group of the complex K
$c_p \sim c'_p$	c_p is homologous to c'_p
$\pi_p(K)$	pth Betti number of K
vK	Cone complex K over v
vc_p	$(p + 1)$-chain in vK determined by $c_p \in C_p(K)$
$\tilde{K}$	Augmented complex of K
$[\sigma_p, \sigma_{p-1}]$	Incidence number of σ_p and σ_{p-1}
$\chi(K)$	Euler-Poincaré characteristic of K
(C_p, ∂_p)	Chain complex
$\ker \varphi$	Kernel of the map φ
$\operatorname{Im} \varphi$	Image of the map φ
$\varphi \overset{c}{\simeq} \psi$	φ is chain homotopic to ψ
φ_p^0	Map from $C_p(K)$ to $C_p(L)$ induced by the simplicial map $\varphi : \|K\| \to \|L\|$
φ_{*p}	Map from $H_p(K)$ to $H_p(L)$ induced by the simplicial map $\varphi : \|K\| \to \|L\|$
f_{*p}	Map from $H_p(K)$ to $H_p(L)$ induced by the continuous map $f : \|K\| \to \|L\|$
f_*	$\{f_{*p}\}$
$\beta = \{\beta_p\}$	Barycentric subdivision map of $C(K)$ to $C(K')$
$\beta^{(n)}$	Iterated barycentric subdivision map
$H_p(X)$	pth simplicial homology group of the topological polyhedron X
f_{*p}	Map from $H_p(X)$ to $H_p(Y)$ induced by the continuous map $f : X \to Y$
f_*	$\{f_{*p}\}$
$\operatorname{tr} \varphi$	Trace of φ
φ	Homomorphism of G/H to G/H induced by $\varphi : G \to G$, $\varphi(H) \subseteq H$
$L(f)$	Lefschetz number of f
$\deg f$	Topological degree of f
df_{x_0}	Derivative of f at x_0
$f'(x_0)$	Jacobian of f at x_0
$D_h f(x_0)$	Directional derivative of f at x_0 in the direction h

$T_{x_0}(X)$	Tangent space to C at x_0
$\dfrac{dc}{dt}(t_0)$	Velocity vector of the curve c at $t = t_0$
$\|R\|$	Volume of the rectangle R in $\mathbf{R}^n$
$\mathbf{R}_p^k$	k-dimensional slice in $\mathbf{R}^n$ through p
grad f	Gradient of f
$H(f)$	Hessian of f
Int V	Interior of V in the manifold with boundary X
∂V	Boundary of V in the manifold with boundary X
∂f	$f\|\partial X$
$\{g^t : t \in \mathbf{R}\}$	One-parameter group of diffeomorphisms
$T(X)$	Tangent bundle of X

Index